Temperature Biology of Animals

Temperature Biology of Animals

A.R. Cossins
Department of Zoology,
University of Liverpool, UK

and

K. Bowler
Department of Zoology,
University of Durham, UK

London New York
CHAPMAN AND HALL

First published in 1987 by Chapman and Hall
11 New Fetter Lane, London EC4P 4EE
Published in the USA by Chapman and Hall
29 West 35th Street, New York NY 10001

Printed in Great Britain at the
University Press, Cambridge

ISBN 0 412 15900 7

British Library Cataloguing in Publication Data

Cossins, Andrew R.
Temperature biology of animals.
1. Temperature—Physiological effect
I. Title II. Bowler, K.
591.52'22 QP82.2.T4

ISBN 0-412-15900-7

Library of Congress Cataloging in Publication data

Cossins, Andrew R.
Temperature biology of animals.
Bibliography: p.
Includes index.
1. Animal heat. 2. Body temperature—Regulation.
I. Bowler, K. II. Title.
QP135.C8 1987 591.19'12 86-31057
ISBN 0-412-15900-7

Contents

Preface

Temperature is one facet in the mosaic of physical and biotic factors that describes the niche of an animal. Of the physical factors it is ecologically the most important, for it is a factor that is all-pervasive and one that, in most environments, lacks spatial or temporal constancy. Evolution has produced a wide variety of adaptive strategies and tactics to exploit or deal with this variable environmental factor. The ease with which temperature can be measured, and controlled experimentally, together with its widespread influence on the affairs of animals, has understandably led to a large, dispersed literature. In spite of this no recent book provides a comprehensive treatment of the biology of animals in relation to temperature. Our intention in writing this book was to fill that gap. We hope we have provided a modern statement with a critical synthesis of this diverse field, which will be suitable and stimulating for both advanced undergraduate and post-graduate students of biology. This book is emphatically not intended as a monographical review, as thermal biology is such a diverse, developed discipline that it could not be encompassed within the confines of a book of this size.

Initially, we deal with thermal energy and its importance as an environmental factor (Chapter 1). Emphasis is placed upon a description of microclimate, because it is only by appreciating the great complexity and the characteristics of different thermal environments that temperature as a determining or controlling factor in the lifestyles of animals can be understood. The effects of temperature upon the rates of biological processes are described in some detail, together with an overview of the mathematical treatment of this relationship (Chapter 2). Some instances of temperature independence of function are critically assessed.

We then consider in some detail how body temperature is established and regulated in the two broad categories of animals, the bradymetabolic ectotherms and endotherms (Chapter 3) and the tachymetabolic endotherms (i.e. birds and mammals, Chapter 4). These chapters show that very few animals are totally passive with respect to variations in environmental temperature, and that an appreciation of these behavioural and physiological responses to temperature change is critical to an understanding of the survival and evolutionary fitness of animals in a thermally variable and unpredictable environment.

Because their core tissues may experience dramatic fluctuations in temperature, ectotherms often display adaptive physiological responses, both phenotypically and genotypically, which overcome the adverse effects of temperature variations. These adaptive responses are dealt with in two distinct and commonly accepted groupings; the adaptations of capacity for living (so-called 'capacity adaptations' or 'Leitsungadaptation', Chapter 5) and the adaptations of resistance to lethal temperatures (so-called 'resistance adaptations' or 'Resistenzadaptation', Chapter 6). This latter group includes the mechanisms by which animals tolerate or avoid the freezing of their body fluids in winter. Finally, the effect of temperature on such complex processes as reproduction, development and growth in ectotherms are described. These changing reaction systems are shown to be comparatively stenothermal.

Thus, the subject of this book encompasses the effect of temperature over the full range of hierarchical levels, from molecules to the organism, and from physiology to behaviour. We have attempted to provide an overview of most areas of animal thermobiology, but we have inevitably emphasized certain aspects. This stems in part from our own interests and prejudices, but is also because some aspects have received expert coverage elsewhere. We have emphasized, where possible, the ecological consequences of environmental temperature and the adaptive responses animals make to it. Our emphasis and treatment of thermobiology may not suit all teachers and research workers on specific points, but we hope that overall it will provide a balanced and informative view for the view for the reader. The many questions that arise within the book should also present a stimulating challenge to further hypothesis and research, so that we may better understand and appreciate the importance of temperature, not only in the lives of extant species but also as a driving force in animal evolution.

We are grateful to the many authors and publishers who have allowed use of their original material, and beg their understanding of our modification and adaptation of their work. In writing this book we have taken advice from the following authorities: Professors J. Bligh, A.J. Cain, J.N.R. Grainger, J.E. Heath, H. Laudien, G.N. Somero and Dr A. Clarke, who have read and commented on parts or all of the manuscript. The final result owes much to their efforts, though any misconceptions and inaccuracies remain entirely our responsibility.

A.R. Cossins and K. Bowler
July 1986

1 Thermal energy and the thermal environment

1.1 Thermal energy and temperature

Towards the middle of the 19th century, the English botanist Brown noticed that minute particles in the cellular fluids of plant cells were in perpetual motion. Repeated experiments subsequently demonstrated that this so-called Brownian motion of microscopic particles was not due to extraneous factors, such as the vibrations of the building, or to currents caused by microscopic illumination, but to the continuous molecular bombardment which was not quite balanced in all directions. Thus, atoms and molecules are continually in motion, and thermal energy is simply the energy of motion of the atoms and molecules of which matter is composed (to which in some cases must be added the potential energy). The greater the rates of motion of the molecules, the greater is their thermal energy.

Atoms can only exhibit three directions of movement or 'degrees of freedom', that is, movement through space in three dimensions (translation). For diatomic molecules, however, this simple description of kinetic energy is complicated by the rotational motion about two molecular axes. In addition, the atoms of the molecule may vibrate as if they were joined by a spring, so that the molecule possesses both kinetic and potential energy of vibration, resulting in a total of six degrees of freedom (three of translation, two of rotation and one of vibration). In general, any molecule has $3n$ degrees of freedom, where n represents the number of atoms in the molecule, so that for a large and complex biological molecule, such as a protein, a precise description of its kinetic energy from first principles becomes hopelessly complicated.

Molecules and atoms are constantly colliding with one another, and lose or gain energy at each collision, rather like the collision of billiard balls. Within a large population of molecules all velocities from the very high to the very low are represented. The distribution of velocities of straight-line motion within a population can be predicted using the Maxwell–Boltzmann laws. The frequency distribution curve of velocities derived from these laws (Fig. 1.1) shows a definite peak where the majority of molecules possess a

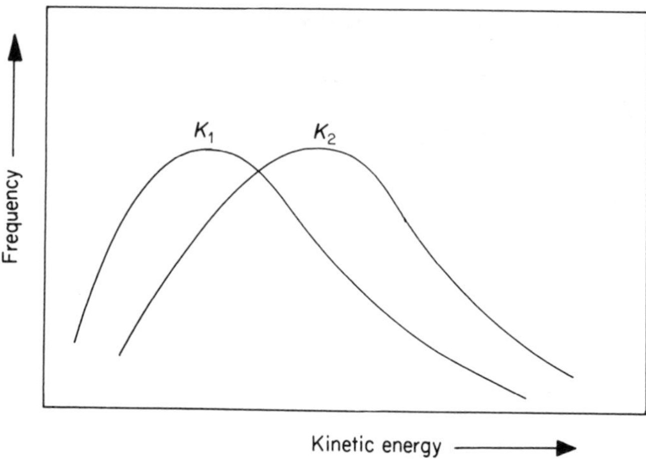

Figure 1.1 The frequency distribution of kinetic energy levels in a population of molecules. K_1 and K_2 represent the distributions observed at a lower and a higher temperature, respectively.

discrete range of velocities, with a much smaller proportion lying outside this range. Thus, although a population of molecules may be extremely heterogeneous with respect to kinetic energy, it may be characterized by an average value – particularly in solids and liquids, where the molecules are so crowded that the motion of each becomes constrained by its neighbours.

Temperature is merely a description of the intensity of kinetic motion of the constituent atoms and molecules of a system, and is an integral measure of their kinetic energy. An object with a high temperature has molecules with a higher average kinetic energy than an object with a lower temperature, though the range of kinetic energies of the individual molecules may overlap (see Fig. 1.1). Temperature is measured by reference to an arbitrary scale whose calibration points are conveniently arranged at the freezing and boiling points of water using a variety of techniques (described in Clark and Edholm, 1985) which depend upon physical (expansion of a liquid) or electrical phenomena (thermocouples and thermistors). The sense of touch is probably the least effective means of measurement. This follows from the common experience that some materials, such as wool, feel warm to touch, whilst others, such as steel, feel cool even though both may be at room temperature. It is the rapid removal of skin heat by the latter that leads to the sensation of coolness.

It is important to recognize that because it contains no volume or mass dimensions, temperature gives no information about the heat content of an object. *Temperature has its greatest value in predicting the direction of heat flow between objects with different kinetic energies*; that is, down a thermal gradient from an object with a higher temperature to one with a lower temperature.

All matter possesses thermal energy and hence exhibits a temperature. In thermodynamics heat is defined as energy transfer between communicating systems, arising solely from a temperature difference. This transfer only lasts until the two systems come into thermal equilibrium, at which stage their temperatures are, by definition, equal. Thus, heat is a transient phenomenon which is manifest as a change in temperature.

Historically, the unit of thermal energy is the calorie (cal) which is defined as the amount of heat required to raise the temperature of 1 g of water from 14.5°C to 15.5°C at an atmospheric pressure of 760 mmHg. The specified temperature interval is necessary because the amount of heat required varies slightly with temperature. The joule (J) is the equivalent SI unit of energy (1 cal approximately equals 4.18 J), but because biology largely deals with aqueous systems the calorie has a more direct and obvious meaning and is a popular unit. The quantity of heat required to raise the temperature of a standard quantity of water (1 g) by 1°C is termed its specific heat capacity. This is an important coefficient because it provides a means of calculating the heat required to change the temperature of a known quantity of water, thus to increase the temperature of 100 g water by 30°C requires $100 \times 30 \times 4.18 = 12\,540$ J. The specific heat of other substances varies widely (Table 1.1) which in practical terms

Table 1.1 The thermal properties of various materials.

Material	Thermal conductivity ($Wm^{-1}K^{-1}$)	Specific heat ($Jkg^{-1}K^{-1}$)	Emissivity
Copper	385	383	0.03
Aluminium	204	396	0.2
Stainless steel	16	460	0.07
Granite	3	816	—
Ice	2.2	1926	0.92
Concrete	1.4	880	0.8
Wood (pine)	1.4	2805	0.9
Glass	0.8	837	0.94
Brick	0.7	840	0.7–0.8
Cork (expanded)	0.036	1884	0.95
Polyurethane foam	0.026	—	—
Cotton	0.06	1298	—
Fur	0.04	—	—
Air	0.026	1006	—
Freon-12	0.071	—	—
Water	0.6	4182	0.95
Human tissue	0.42	—	—

Data from Simonson (1975), Hammell (1955) and Cornwell (1977).

means that the application of a given quantity of heat to equal masses of different substances will produce quite different increases in temperature. Knowing the mass of a particular body and its specific heat allows the calculation of the heat required to change temperature by a given amount. Because of its anomalous solvent properties, the specific heat of water is very high, even compared with metals. The specific heat of air is 1 J g^{-1} °C^{-1}, but because the density of air is only 1.2 g l^{-1}, the heat capacity of 1 litre is only 1.2 J °C^{-1}. The corresponding value for water is 4184 J °C^{-1}. The 4000-fold difference is of great importance to the temperature biology of animals. The rate of heat transfer is usually expressed as J s^{-1} or watts (W).

1.2 Life, the low-temperature phenomenon

The concept of what is hot and what is cold is entirely subjective and based on thermal experience. In the universe the range of temperatures varies from close to absolute zero (– 273°C or 0 K) in the depths of space to several thousand degrees at the surface of stars. The core temperatures of stars are several orders of magnitude hotter still. On the Earth the lowest temperatures have been recorded in the polar regions at about – 170°C whilst the highest recorded temperatures are found at the surface of deserts (80°C), geothermal springs (90–100°C) or deep-sea hydrothermal vents (up to 350°C).

Whilst some bacteria and blue-green algae live and reproduce well at temperatures up to 110°C (Brock, 1985), eukaryotic organisms are generally restricted to those areas where their body temperatures do not fall much below the freezing point of water or rise much above 45°C. As we shall see, the upper limit of life is probably determined by the stability of non-covalent bonds, which are important in maintaining the complex structure of macromolecules (see Chapter 2). The lower limits are probably set by the need for most animals to avoid ice-crystal damage to their cellular structures (see Chapter 6). The full biokinetic range of body temperatures can only be withstood by a few animals, even over a seasonal timescale, and most individuals live within a much narrower range of temperatures.

1.3 Mechanisms of heat transfer

The transfer of heat between living organisms and their environment conforms to normal physical laws; that is, there is a net flow from an area of higher kinetic energy (or temperature) to an area of lower energy (or temperature). The rate of heat transfer depends upon the routes of heat transfer and the physical characteristics of the objects in question. There are two important and fundamentally different mechanisms of heat exchange – radiation and conduction. An understanding of these basic processes, and of convection and evaporation, is crucial to an appreciation of how animals interact with their thermal environment. A more-detailed description of these physical processes is provided by Cornwell (1977) and by Simonson (1975).

1.3.1 *Radiation*

Electromagnetic radiation in the infrared wavelengths is able to interact with matter in such a way as to alter its kinetic energy and its temperature. Like all electromagnetic radiations it travels at the speed of light and requires no medium of propagation, so that two bodies may exchange heat without physical contact. Radiation therefore provides a means of heat exchange when two bodies are separated by a non-absorbing medium such as air. Thus, the surface of the Earth directly receives radiation from the Sun even though the upper atmosphere is extremely cold. Similarly, the glass of a greenhouse remains colder than its contents because radiant energy from the Sun is transmitted to the absorbing surface below.

All physical objects with a temperature above 0 K emit radiation in the visible and infrared wavebands. As an object, say a metal bar, is heated its visible colour changes from a dull red to white. This familiar phenomenon has its basis in the fact that as surface temperature increases the spectrum of the radiation not only increases in height, but also shifts to progressively shorter wavelengths, a relationship known as *Wein's displacement law*. Emissions from objects at normal biological temperatures (0–40°C) lie between 3 and 60 μm which is in the non-visible region. Objects with temperatures above 1500°C, most notably the Sun, emit radiations which extend well into the yellow-red region of the visible spectrum (Fig. 1.2(a)). Thus, in the present context it is convenient to distinguish two bands of radiation, short-wavelength radiations from the Sun and long-wavelength radiations from objects of low surface temperature on the Earth's surface (Fig. 1.2(b)).

The rate of heat transfer by radiation depends upon a number of surface characteristics. At this point it is helpful to define a surface which is a perfect emitter (the so-called 'black body') as one that emits the maximum power at a specified temperature. The Stefan–Boltzmann law states that the heat emitted from such a surface by radiation is proportional to the fourth power of the absolute temperature. The proportionality constant (the Stefan–Boltzmann constant, σ) thus provides the emissive power of unit surface area of a black body, and has a value of $5.67 \times 10^{-8}\,\mathrm{W\,m^{-2}\,K^{-4}}$. The emissivity of any other (non-black) surface is the amount of radiation emitted relative to a black body (see Table 1.1 for examples). The rate of radiative heat loss from unit area of a non-black surface (Q_r) then depends upon its surface, absolute temperature (T) and its emissivity (ε)

$$Q_r = \varepsilon\sigma T^4 \qquad (1.1)$$

The emissive power of familiar objects is surprisingly large. For example, this equation predicts an emissive power for the human body of approximately 1 KW (300 K, surface area $2\,m^2$, emissivity = 1). Of course, this loss of heat is partially compensated for by absorption of radiation from the surroundings, so the net loss is much smaller.

There are three possibilities for radiation striking a body – absorption, reflection and transmission. The proportion of incident radiation which follows each path is termed the absorptivity, the reflectivity and transmissivity, respectively. Solids generally transmit no radiation, so that only the first two processes are of biological importance. When an object has surface properties such that no incident radiation is reflected then it has an absorptivity of unity. This type of object is also termed a 'black body'. It can be shown that emissivity and absorptivity are numerically equal (*Kirchoff's law*), which for our purposes means that most animals have high absorptivities. By contrast, a surface which absorbs no incident radiation is a perfect reflector and has an absorptivity close to zero. Surfaces which are black for radiation purposes are not necessarily black to the human eye, since we have already distinguished between the visible waveband provided

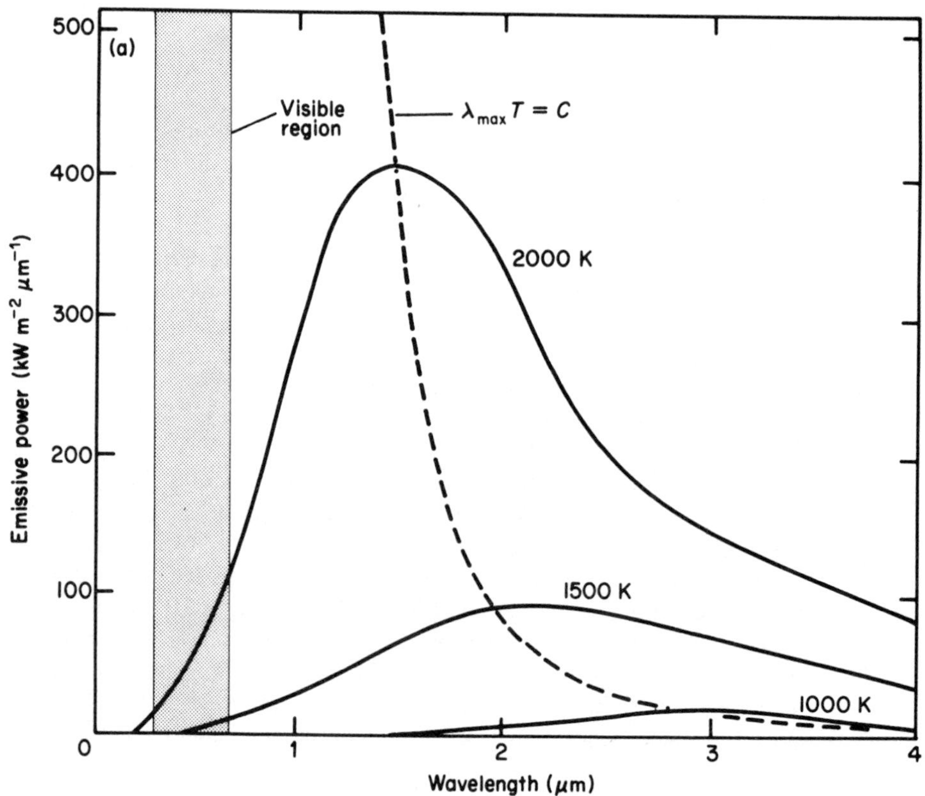

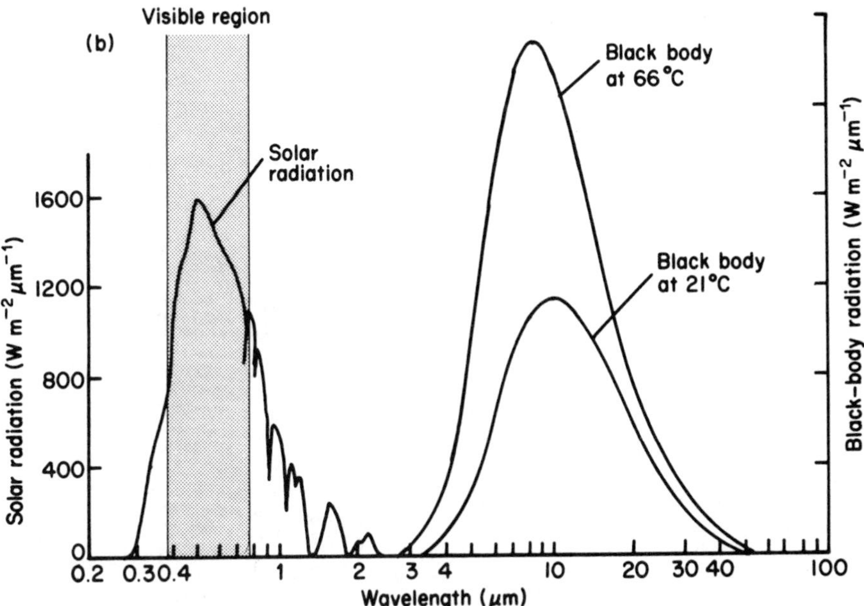

Figure 1.2(a) The effect of surface temperature upon the intensity and the spectrum of emitted radiation. Note that as surface temperature increases the spectrum shifts towards the visible waveband. The broken line was calculated from Wein's displacement law, which states that the wavelength for maximal emission = $2897/T$ (in K). (After Cornwell, 1977.) (b) The distinctive radiation spectra of radiation from the Sun and from bodies at biological temperatures, which are due to their different surface temperatures. Note that the scale for solar radiation is approximately 40 times greater than for the black bodies. (After Bond *et al.*, 1967.)

by the Sun and infrared radiation emitted by cool objects. White paper, for example, is nearly radiator black, with an absorptivity of 0.97. Human skin has an absorptivity of 0.9–0.95 and is virtually independent of skin colour, and a similar situation exists for fur, feathers and clothing. However, visible colour has a major influence on the absorptivity of sunlight; darker colours absorb a higher fraction of incident solar radiation than lighter colours. The term 'albedo' refers specifically to the reflectivity of a material for sunlight, and is roughly equivalent to the glare of a body.

The biological importance of all this is that surface visible colour of animals has a negligible effect upon their radiative heat exchange with their immediate environment. However, surface colour does directly effect the absorption of solar radiation where a high proportion of the spectrum is in the visible region. Dark-coloured surfaces absorb a greater proportion of the incident solar radiation than light-coloured surfaces, a fact that is

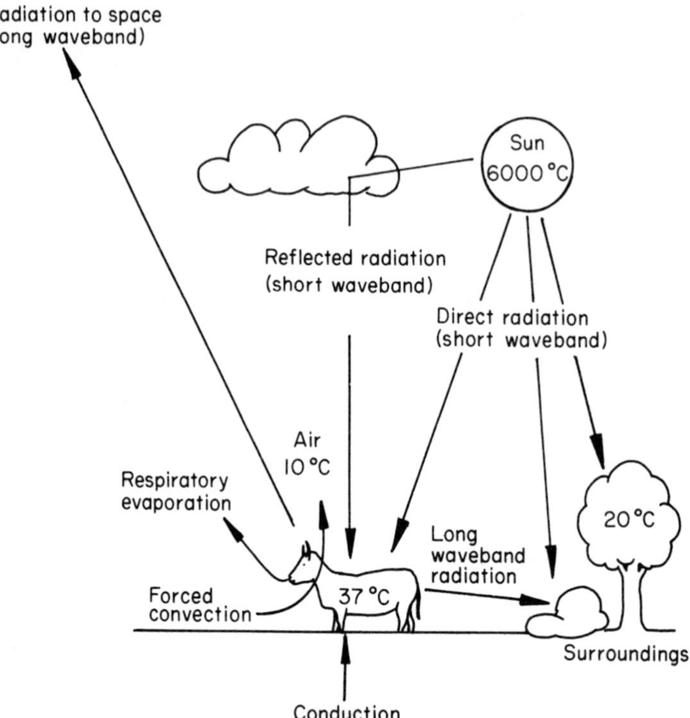

Figure 1.3 The exchange of heat between an animal and its environment. The arrows indicate the direction of net transfer for the temperatures shown. If the surroundings became warmer than the animal, then the arrows between the animal and its immediate environment would be reversed.

frequently used to advantage by animals in situations where radiative heat flux is important.

For the biologist the important question is the net radiative exchange between the animal and its immediate environment. The great complexity of this exchange for a terrestrial animal is schematically illustrated in Fig. 1.3. The interaction between the animal and any specific object is obviously complicated by the fact that, because of their geometrical arrangement, only a small proportion of total radiation from one object reaches the other object. The complexity of interactions with many such objects can be reduced by treating the total radiative environment as a single emittor with an average surface temperature. It is necessary to include a term to account for the effective area of radiative exchange, since part of that environment will comprise the sky, which does not provide emissions. Thus

$$H_r = \sigma \varepsilon_1 \varepsilon_2 (T_1^4 - T_2^4) A \tag{1.2}$$

where H_r is the net radiation, σ is the Stefan–Boltzmann constant, ε_1 and ε_2 are the emissivities, T_1 and T_2 are the surface temperatures in kelvin of the interacting bodies and A is an expression taking into account the effective radiating areas of the surfaces. Because of the variable and complex nature of temperature, surface area and emissivity both of animals and of their environments, it is not easy to use this relationship in any practical way. The important point to remember is that net exchange between an animal and its environment is proportional to the difference in the fourth power of their respective absolute temperatures. A modest increase in the surface temperature of an animal increases the radiative heat loss considerably.

1.3.2 *Conduction*

The second basic mechanism of heat flow occurs by the interaction or collision of adjacent molecules and the transfer of kinetic energy. This type of heat transfer requires a direct physical contact between the interacting objects. The rate of conductive heat transfer is proportional to the temperature difference between them $(T_1 - T_2)$ and is inversely proportional to the distance over which conduction takes place. This leads to the basic equation of heat conduction in one dimension, the so-called Fourier equation

$$H_c = -kA(T_1 - T_2)/l \tag{1.3}$$

where H_c is the rate of conductive heat transfer (W), A is the cross-sectional area across which conduction occurs (m^2), T_1 and T_2 are the temperatures of the two bodies and l is the distance over which conduction takes place (m). The term k, the thermal conductivity, reflects the ease of heat flow in any particular material and has units of $W\ m^{-1}\ °C^{-1}$. The values of k for different materials vary widely (Table 1.1). The conductivity of air is extremely low compared with that of water, and this fact is of great importance in reducing conductive heat loss from terrestrial animals, particularly from those which maintain their body temperature above that of their surroundings. The conductivity of insulative materials used by these animals is very similar to that of air, mainly because they act by trapping a layer of air within their matrix. It is significant that as wool is compressed to remove the trapped air its conductivity is dramatically increased. Hammell (1955) has shown that the principal avenues for heat transfer across fur is air conduction and natural convection of the entrapped air. Heat transfer along the hairs and by radiation is negligible. Fat has a conductivity which is similar to that of air, and is also an important insulative material.

The main problem in applying Equation 1.3 to animals with any rigour is the difficulty of estimating their effective surface area. In addition, the conductive surface area often varies with time as the animal moves, and the thermal conductivity of the integument may vary with the physiological status of the animal. The key factor from a biological point of view is that conductive heat transfer depends upon the magnitude of the thermal gradient; the greater the gradient the greater the heat flux.

1.3.3 *Convection*

Conductive heat transfer in fluids generally forms a very small fraction of the total heat transfer, convection being the predominant mechanism. This is a process in which a fluid is warmed by conductive heat flow from a warmer body. This reduces the density of the fluid and causes it to rise, its place being taken by cooler fluid. The replacement of warmer fluid by cooler fluid maintains the temperature gradient at a high value and facilitates conductive heat transfer. In some cases, fluid may be forced over the surface of an object, either as a result of bulk fluid movements, or because of motion through the fluid. The net effect of this 'forced' convection is similar to free or 'natural' convection, though it produces a roughly tenfold (air) to 100-fold (water) increase in the rate of heat transfer.

Since convection is limited by the movement of a fluid, the process is governed by complex physical laws of fluid dynamics. Fluid viscosity, surface curvature, thermal conductivity, density and specific heat are all important in this respect, but of particular significance in a biological context is the structure and thickness of the heat-transfer surface. Any interruption of the free movement of fluid over the surface of an object will impede convection and lead to the formation of 'boundary' layers of semi-immobilized fluid. The most important boundary layer rests on the surface of objects and is only a few millimetres thick. Motion of fluid in this layer is very restricted, so that heat transfer occurs entirely by conduction. More extensive but less effective boundary layers extend away from the surface, depending upon the morphology of the surface.

The thickness of the boundary layers is greatly affected by the type of fluid flow. Turbulence greatly disturbs the structure of the boundary layers and thereby increases convective heat transfer. The boundary layer becomes thinner as the curvature of the surface increases and when the rate of fluid movement increases. In animals the formation of extended boundary layers of air is the principle behind the creation of insulation, and this has been a potent evolutionary strategy, especially in birds and mammals. Fur, feathers and clothing are all means of preventing the free movement of air next to the skin of birds and mammals so that the distance over which conductive heat transfer occurs is increased and the temperature gradient term in Equation 1.3 is reduced.

1.3.4 *Evaporation*

Radiation, conduction and convection all depend upon a temperature gradient between the interacting bodies, and are collectively termed 'sensible' heat transfer mechanisms. Evaporative heat transfer, however, does not. Instead it relies upon heat taken up by a liquid (water) when it changes to the vapour state with a corresponding cooling of the evaporative surface. Consequently, heat loss can occur even when the surroundings are at a higher temperature.

To change the physical state of water into water vapour requires an enormous quantity of thermal energy. The relevant quantity so far as animals are concerned is the latent heat of evaporation of water, which for 1 g of water at room temperature is approximately 2.4 kJ. This is more than five times as great as that required to heat 1 g of water from room temperature to its boiling point, which illustrates the enormous amounts of heat that may be lost through this process at the general body surface of animals and, more particularly, at moist respiratory surfaces. Evaporation

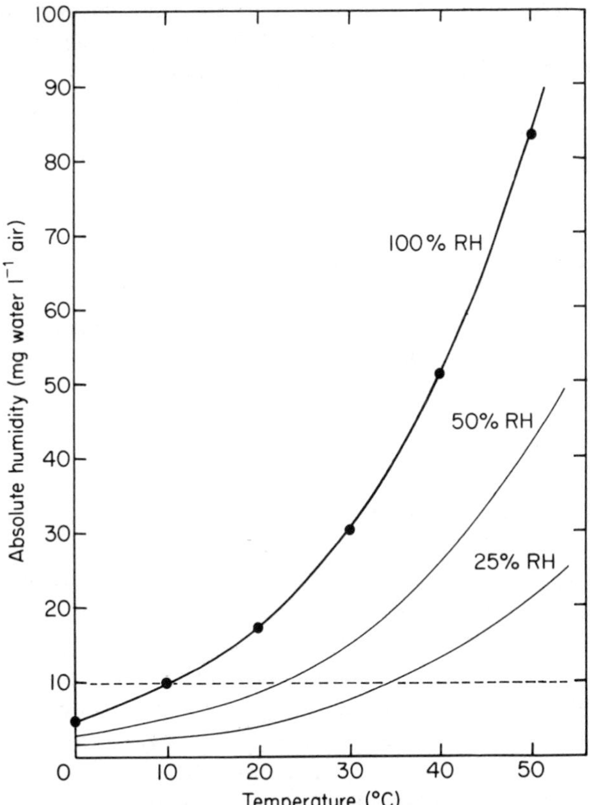

Figure 1.4 The humidity of air saturated with water vapour at different temperatures. Water-saturated air at 10°C becomes only 25% saturated when warmed up to 40°C. The driving force for evaporation is proportional to the amount by which water vapour is less than saturated. Thus, at a given absolute humidity the rate of evaporation is greatly increased by warming. RH – relative humidity.

is a very important means of thermoregulation in terrestrial animals (see Chapter 4), though it does require the ready availability of water, a situation that does not always apply in those circumstances in which heat must be lost.

Knowing the latent heat of evaporation, it is relatively easy to estimate the magnitude of this route of heat transfer by simply measuring the quantity of water that has been vaporized over a given period. The rate of evaporation depends on the extent to which the atmosphere is already saturated with water vapour – the drier the air, the greater the rate of evaporation is. This is often expressed as the relative humidity (RH), which is the *ratio* of the quantity of water vapour present to that present in air that is fully saturated at the same pressure and temperature. It is conventionally expressed as a percentage. A relative humidity of 50% means that the air contains only half the amount of water vapour it could contain. Because the amount of water vapour in fully-saturated air varies with temperature (Fig. 1.4), RH automatically varies with temperature even though the quantity of water vapour per unit volume remains constant. Thus, when cold saturated air is warmed it contains the same quantity of water vapour but its relative humidity becomes rather lower (see example in Fig. 1.4). Such circumstances lead to enhanced evaporative water loss from animals, particularly from their exposed mucous membranes.

A more useful concept is the saturation deficit, which is the *amount* by which water vapour concentration or partial pressure is below saturation. This provides a more satisfactory measure of the drying power of the atmosphere and of the rate of evaporative heat exchange, because it indicates how much more water vapour can be formed.

1.4 The thermal environment

The thermal characteristics of different environments are determined by the thermal capacities and conductivities of the surrounding medium, be it water or air. The surface temperature of the Earth is principally determined by a balance between the gain of radiant energy from the Sun and the loss of longer-wavelength radiation to space. The energy of sunlight received at the Earth's surface at right angles to the Sun's rays is known as the solar constant, and has a value of $1360 \mathrm{W\,m^{-2}}$. However, of the incoming radiation only about 19% penetrates to the Earth's surface directly and most (37%) is scattered in the atmosphere either to space or to the ground (Fig. 1.5). The balance is reflected by clouds, which in some climates greatly reduces the annual radiative flux at the ground. Over southern England on a cloudless summer day the solar input may reach $900 \mathrm{W\,m^{-2}}$ whilst on a cloudy day this is reduced by 30–60%. The mean annual radiation received over 24 h at the Earth's surface is of the order $100 \mathrm{W\,m^{-2}}$ in temperate areas and two-to-three times this in subtropical areas.

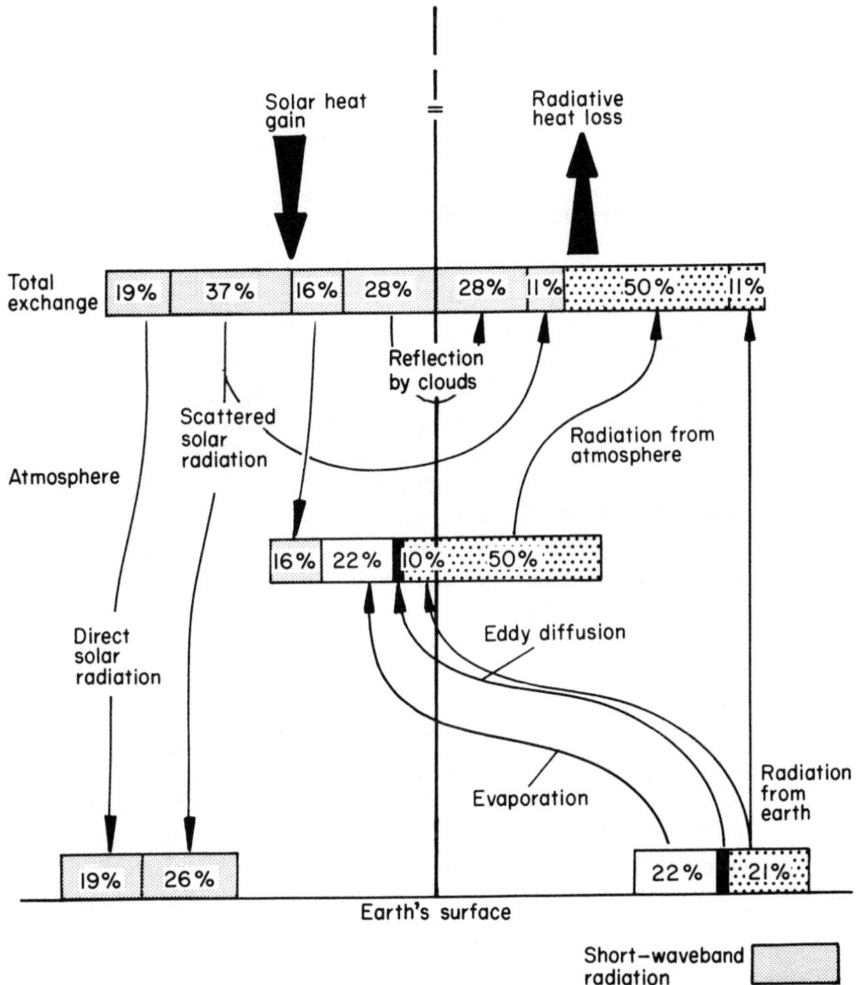

Figure 1.5 The annual balance of heat exchange for the Earth. The gain of energy is shown to the left of the figure and the losses to the right. (After Geiger, 1965.)

The absorptivity and the area of surface that is irradiated determine the amount of radiative uptake. Ice and snow are very highly reflective, and polar regions consequently absorb only a small proportion of incident sunlight. In addition they receive only a small fraction of the radiative input per unit area received in equatorial regions, simply because of the respective angles at which the sunlight strikes the Earth's surface.

The Earth has a surface temperature of $-10°C$ to $+45°C$, and thus acts as a radiator in the long-wavelength infrared waveband. Much of this radiation is lost directly to space but some is absorbed in the atmosphere and subsequently lost to space (see Figs 1.3 and 1.5). Because heat exchange between space and the Earth is entirely by radiation the greatest diurnal and seasonal variations in climate occur at the effective radiative surface, and are attenuated above and below this surface as determined by the thermal characteristics of the surrounding media.

1.4.1 *The terrestrial environment*

The low specific heat of air means that its temperature may be altered by comparatively small amounts of heat. The low thermal conductance of air means that in the absence of convection, heat will not be easily conducted away from the irradiated surface, giving rise to steep thermal gradients. The most dramatic diurnal changes in temperature are experienced at the radiative surface, since during the day solar radiation warms the ground but during the night the loss of infrared radiation causes cooling.

These factors, together with the presence of vegetation, make terrestrial habitats the most thermally varied and complex. In consequence they offer a particularly rich variety of climatic niches for animals to exploit, which may not be at all obvious from climatological measurements. The true air temperature used by meteorologists is that measured in a standardized shelter at a set height above the ground. Not surprisingly, these conditions have been carefully chosen to minimize the influences of microclimate, so the observations are characteristic of a wider geographical area. Climatological temperatures therefore bear little relationship to the temperatures that animals experience in their natural habitats, and in any particular locality there are usually certain microenvironments which show fluctuations in humidity, temperature and wind speed that are considerably more dramatic than climatological information would indicate. This concept of microclimate is of great importance in all aspects of biology and agriculture, and is described in great detail in the classic work by Geiger (1965) and more recently by Monteith (1973).

The extreme complexity of terrestrial microclimates is illustrated in Figs 1.6 and 1.7 by a specific example from a cool temperate area in southwestern England where the temperature at different points above and below a grass lawn in southwestern England was continuously measured over a 3-day period during autumn. Days 1 and 2 were clear and sunny and showed circadian variations of almost 40°C at the surface of the grass, which is not too different from the daily variations found in much more extreme environments. These variations were progressively attenuated with increasing height above the surface. The temperature 10 cm beneath the surface of the soil was relatively constant during this time, mainly because of the low thermal conductance and large heat capacity of soil. At depths of 1 m and more even seasonal variations are almost undetectable.

On day 3 the sky became very overcast and cloudy. This largely blocked

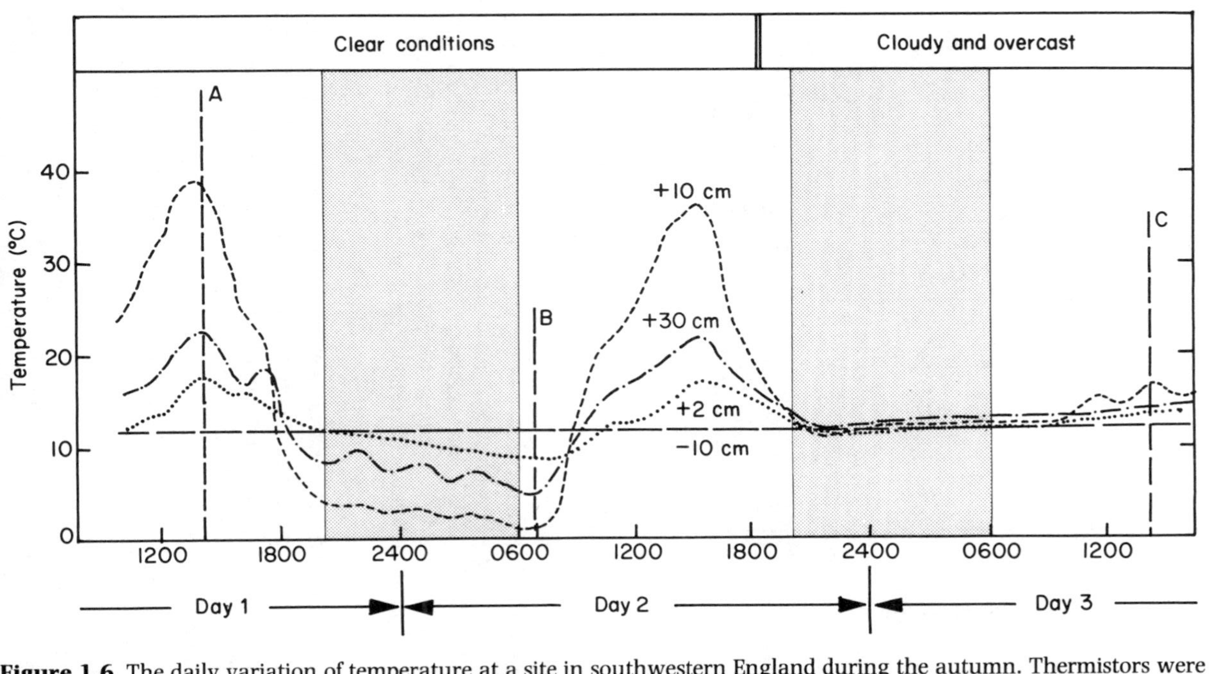

Figure 1.6 The daily variation of temperature at a site in southwestern England during the autumn. Thermistors were placed at different positions above and below the surface of a grass lawn, and ambient temperatures were automatically recorded over a 48-h period. The vertical lines marked A, B and C indicate the times for which vertical temperature gradients have been constructed in Fig. 1.7. (University of Liverpool, Class of 1979.)

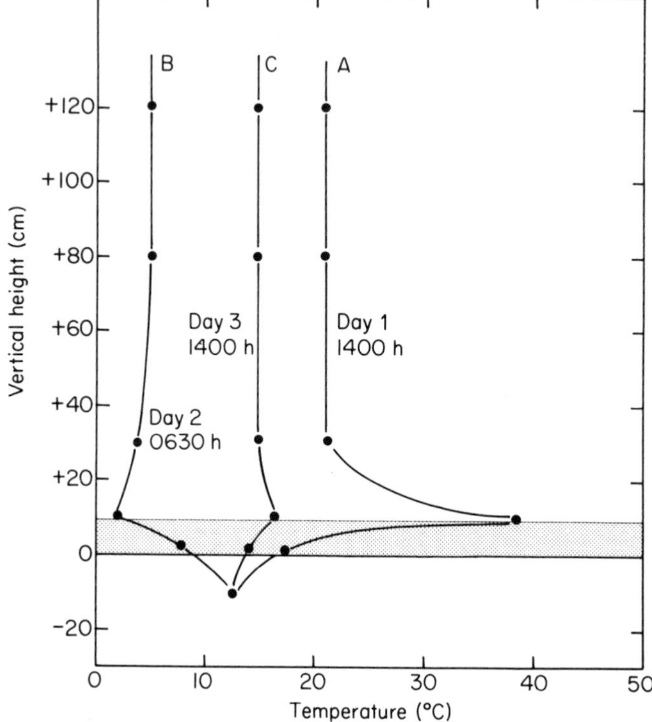

Figure 1.7 The 'tautochrones' or vertical temperature gradients observed during the night and day with clear conditions, and during the day on day 3 when conditions were cloudy and overcast. The shading indicates the height of the grass above the soil surface. The presence of both solar input and long-wave emissions have a profound influence on the thermal microclimate. See Fig. 1.6 for details.

solar input during the day as well as radiative heat loss to space during the night. As a result, the diurnal oscillations of temperature were greatly attenuated and the vertical temperature gradients disappeared.

The magnitude of the thermal gradients that existed at any one time may be appreciated from Fig. 1.7. These curves of temperature against vertical distance are termed 'tautochrones'. The gradients were greatest just above and just below the radiative surface; on day 1 it was approximately 2.5°C cm^{-1} during the day and 0.8°C cm^{-1} during the night. The additional complexity of a terrestrial environment that arises from the presence of different substrates, rocks, stones and small patches of uncovered earth can only be appreciated with the use of a sensitive thermistor or thermocouple. Moving the probe a few centimetres from close to a stone to the underside of a nearby leaf can mean a temperature change of over 10°C. The temperature difference between the insolated and shaded surface of a stone

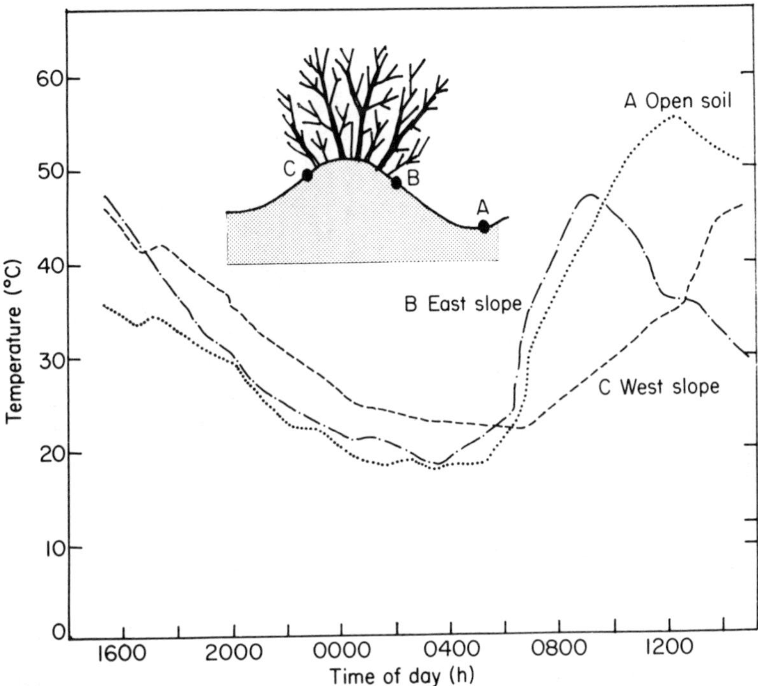

Figure 1.8 The daily changes in soil surface temperature at three positions next to a shrub in the pre-Saharan steppes of Tunisia. Note that the peak temperature was greatest on the open eastern, ground (position A) and that the side of the shrub (B) warms up more rapidly than the opposite side (C). (After Heatwole and Muir, 1979.)

may be over 20°C. Figure 1.8 shows the diurnal temperature variations on the eastern and western side of a small desert shrub. The morning rise in temperature is more rapid on the eastern side; at 1000 h the temperature was almost 20°C higher than on the western edge. The presence of broken cloud has a profound effect upon the time-course of solar input. Figure 1.9 shows this effect with a series of rapid, high-frequency and large-amplitude variations in solar radiation. Not only does this lead to a lower time-averaged solar radiation, but it also creates a major problem for small basking animals whose body temperatures rapidly drop in the shade.

The factors that determine the thermal characteristics of soil are twofold; first, those that relate to soil type such as water content, texture and surface absorptivity and, secondly, those that are related to the position of the site with respect to the Sun (latitude, elevation and aspect) and to the wind (exposure). In general, soils that have high thermal conductivity have smaller diurnal and seasonal fluctuations in temperature and smaller vertical temperature gradients, since the heat is conducted away more rapidly. The second factor can have great significance. For example, a

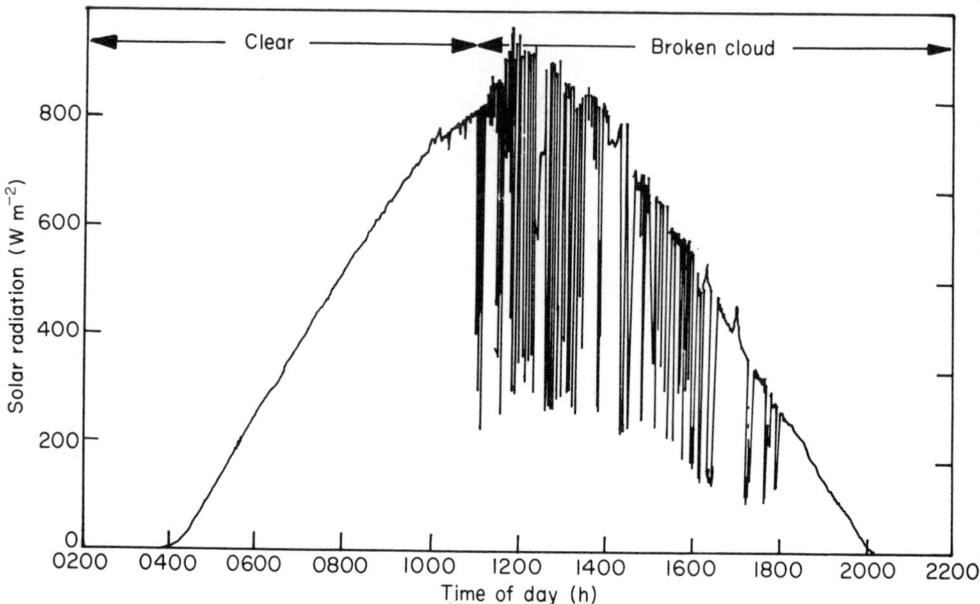

Figure 1.9 Solar radiation on a day of broken cloud in England. The presence of cloud at 1100 h onwards leads to high-frequency variations in solar input, resulting in a bimodal distribution of solar input. Solar radiation can be a very unreliable resource for basking animals in temperate regions. (After Monteith, 1973.)

surface inclined at 20° facing south may receive up to twice as much radiation as a horizontal surface.

Vegetation has, perhaps, the most profound influence on microclimate by the creation of boundary layers of air and by the absorption of sunlight. In such circumstances, which of course, are the rule rather than the exception, the upper surface of the vegetation assumes the role as the effective radiative surface. Thus, it is necessary to define an 'outer effective surface' (Geiger, 1965) which is the vertical position where most radiative exchange occurs and where the most extreme variations in temperature occur. Beneath this surface heat exchange occurs by convection, the structure of the vegetative column dictating the magnitude of forced (i.e. wind-driven) and natural air movements.

These principles are well illustrated in the classic work of Geiger (1965) who measured the tautochrones of a bed of antirrhinum plants (Fig. 1.10). In July, when the plants were small and formed an incomplete cover, the midday profiles were similar to that of bare earth. As the plants became fully grown, as in August, the dense leaf structure shifted the region of highest midday temperatures to higher levels. At night, as the leaves radiate, the cool air sinks to the ground and the night-time minimum lies at the ground

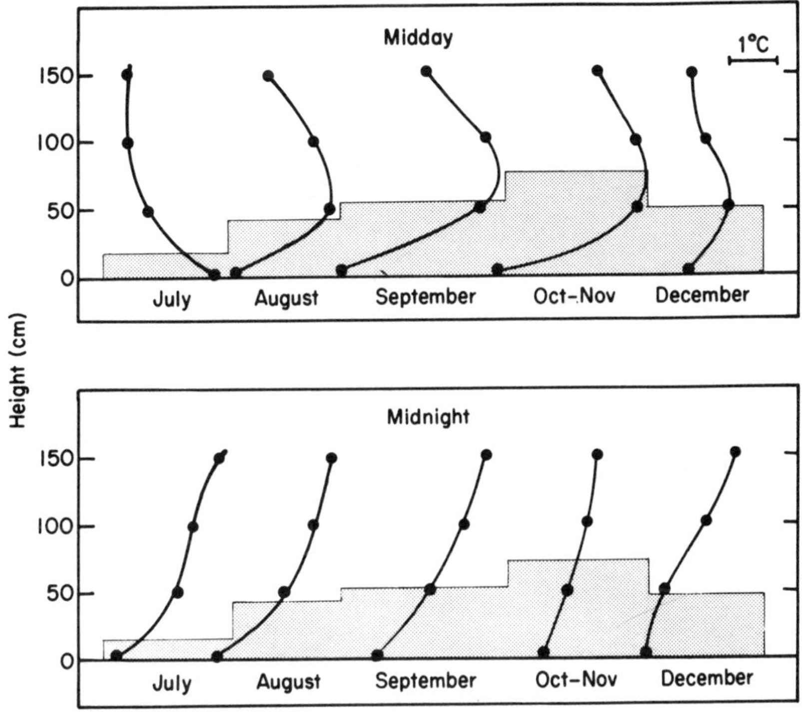

Figure 1.10 Tautochrones in an antirrhinum flower bed near Munich. The profiles were recorded monthly and superimposed over histograms which show the height of the vegetation. The midday maximum temperatures were recorded at the surface of the vegetation, whilst midnight minimum temperatures were recorded at ground surface. (After Geiger, 1965.)

surface. The integrity of the outer surface and the precise form of the tautochrones depends largely upon the physical form of the vegetative loft. The more-open, vertical arrangement of rye (Fig. 1.11) leads to an outer effective surface which is well below the surface of the grain, since both sunlight and wind can penetrate well into the cover.

It is clear from this description that small terrestrial invertebrates can easily take advantage of this complex thermal mosaic to avoid stressful temperatures. This can be achieved by

(1) Adjusting their activity periods either diurnally or seasonally;
(2) Moving through the vertical temperature gradient (burrowing, climbing up objects or vegetation);
(3) By moving from one side of an opaque object to the other as the position of the Sun changes during the day.

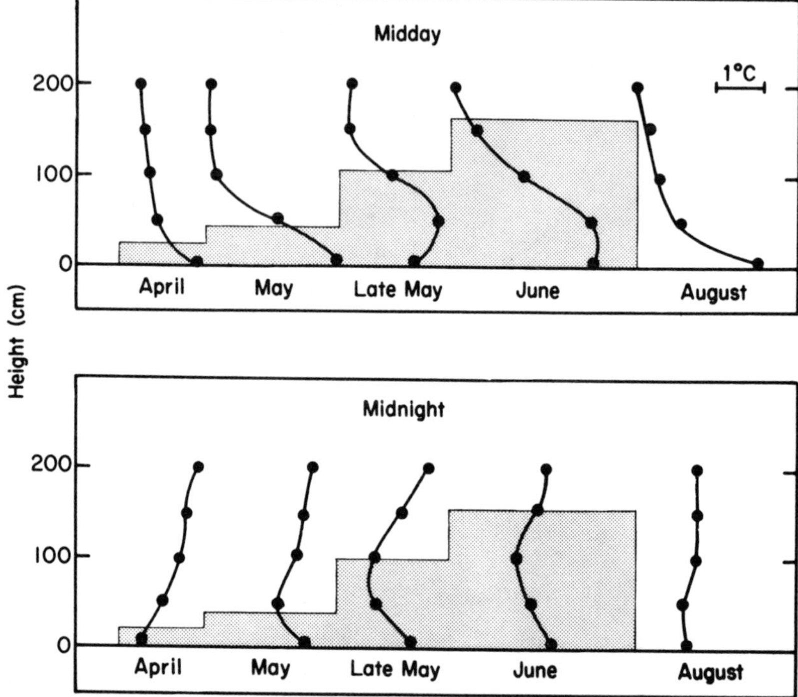

Figure 1.11 Tautochrones in a field of winter rye. Again the temperature profiles have been superimposed over histograms of vegetation height. Because of the loose and vertical structure of the plants, the maximum midday temperature was recorded well down in the loft. (After Geiger, 1965.)

1.4.2 *The aquatic environment*

Water has several properties of profound importance to animals, not least because they are largely composed of water. It has a relatively high specific heat capacity, so water can absorb and lose relatively large amounts of heat with only small changes in temperature. This, together with the low thermal conductivity of water, makes aquatic environments very thermostable places relative to terrestrial environments; the larger the body of water, the more stable is its temperature. Because the thermal conductivity of water is so much higher than that of air, aquatic environments provide a much less rich variety of microclimates for exploitation.

The thermal structure of large bodies of water is complicated by its anomalous thermal properties. For example, water has a maximal density at 4°C, so water of a different temperature, whether warmer or cooler, will rise to overlie water at 4°C. Water expands near its freezing point. This causes ice to float in water, so water cools from the top downwards. Natural waters seldom have temperatures extreme enough to be outside the range tolerated by living organisms. In summer, they warm only slowly and in winter animals can avoid being frozen by living in the water beneath the ice cover.

The oceans, because of their size, are important in the Earth's heat balance. They form a capacious heat store, and their relative constancy has a profound moderating influence on the climate of maritime regions. In contrast with air, water is heated only from its surface since water strongly absorbs solar radiation, particularly in the longer wavelengths. Most solar energy is absorbed in the top 100 m of sea water (Fig. 1.12). Cooling of sea water occurs by evaporation, radiation and the melting of polar ice. Surface waters are, therefore, subject to the greatest temperature fluctuations. Polar waters are close to 0°C throughout the year, but always above − 1.89°C, for at that temperature sea water becomes frozen. Tropical seas have surface temperatures of about 30°C, with only small diurnal and seasonal fluctuations. Temperate seas are more variable, but even so the diurnal fluctuation at the surface is rarely greater than 4°C. Seasonal fluctuations may be as great as 20°C.

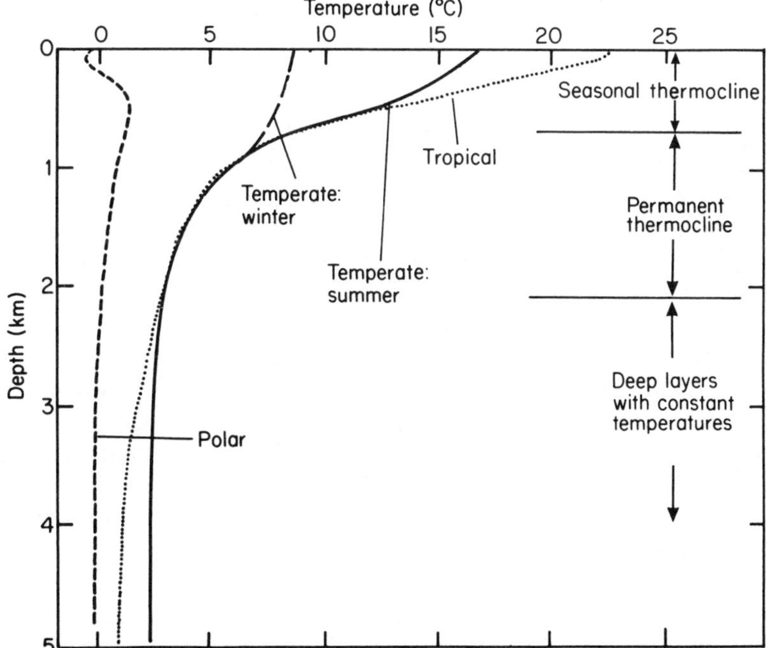

Figure 1.12 The vertical temperature gradient in the upper ocean. It is only the upper 200 m that experiences seasonal variations in temperate regions, deeper zones being extremely thermostable. In polar seas the thermocline is almost abolished. (After Monin, Kamenovitch and Kort, 1974.)

Figure 1.12 shows how water temperature declines with increasing depth between 300 m and 1200 m, the so-called permanent thermocline. At greater depths water temperature varies only slightly between about -2.5°C and 4°C. A temporary seasonal thermocline is also established in the surface waters of tropical and temperate seas. This is similar to the thermocline that develops seasonally in many lakes, and occurs because the less dense warmer water overlies colder water and, in the absence of mixing, results in a marked thermal stratification.

1.5 Conclusions

A great diversity of thermal conditions exists in both terrestrial and aquatic habitats. From the standpoint of the organism two thermal factors seem to be of great significance in affecting its lifestyle. The most obvious is the absolute range of temperatures it will experience, but equally important is the rate of change of temperature in these thermally-fluctuating habitats.

It is the lack of constancy, both in time and in space, that makes for this complexity and which creates difficult but very important problems for the biologist. First, what temperature(s) should be used to characterize the thermal conditions in a particular habitat. The complexity is such that frequently the biologist needs to use mean daily, monthly or annual temperatures as an index of the integrated thermal experience of an organism. It may well be that the daily maximum or even the daily temperature range has the greatest impact on the life of an organism. The fact that mean temperatures are generally unsatisfactory is clear from a comparison of two areas each with a daily mean of, say, 15°C. In one the temperature range may be small and in the other it may be large. These two environments pose completely different problems for animals living there, problems that are not obvious from climatological data.

The second problem is posed by the potentially large number of distinct thermal environments in any particular climatic zone. Microenvironmental conditions may bear very little relationship to the conditions that characterize a generalized habitat and that may be obtained from the meteorologist. Very detailed studies are necessary to understand fully the thermal experience of terrestrial animals.

2 The direct effects of temperature changes

2.1 Introduction

Intuitively, we all know that an increase in temperature generally results in increased animal activity; the increase in kinetic energy of their constituent molecules being translated into increased rates of metabolism and other processes. We should also be aware of the fact that there is an upper thermal limit to this 'rate' effect of temperature, above which biological processes become progressively more affected by the debilitating effects of temperature. Therefore, a graph relating the rate of a particular biological process or activity to temperature, the so-called 'rate–temperature curve' or 'R–T curve', typically consists of two parts (Fig. 2.1); over the normal range of

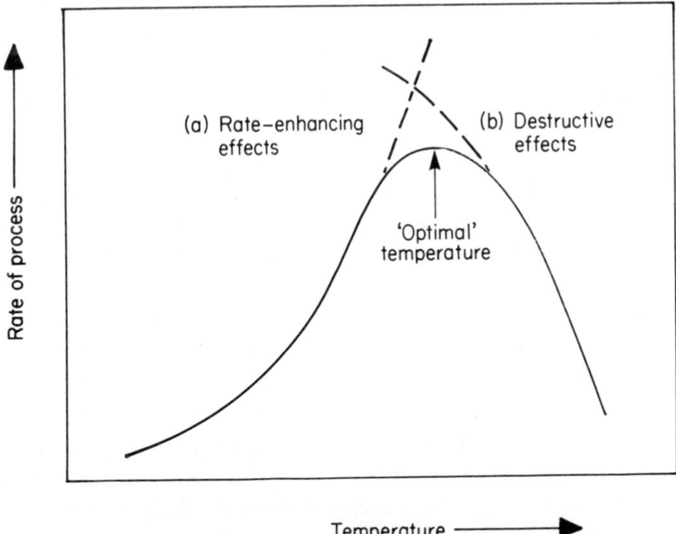

Figure 2.1 Hypothetical rate–temperature curve illustrating the biphasic effects of temperature upon biological processes.

temperatures an increased temperature results in increased rate, whilst above the normal range an increased temperature results in a decreased rate (see Fig. 2.1). The temperature at which the rate is greatest is known as the optimal temperature.

In this chapter we shall consider in some detail how rate–temperature curves are experimentally determined, the causes of the rate effects and the debilitating effects of temperature, and finally some notable exceptions to the general observation of increased rate with increased temperature.

2.2 Rate–temperature curves – experimental considerations

For enzymatic reactions it is a relatively simple matter to construct rate–temperature curves merely by measuring the rate of reaction at a series of incubation temperatures. A convenient apparatus for the incubation of reaction mixtures at a large number of temperatures is a metal bar in which a series of holes have been drilled to accommodate test tubes. The bar is heated at one end and cooled at the other, thereby creating a thermal gradient along its length. The stability of the gradient depends mainly upon the effectiveness of its surrounding insulation and the thermal capacity of the bar, so that iron bars are considerably more stable than aluminium bars.

Experiments on whole animals are considerably more difficult to perform. The irritability that is a characteristic of animals endows them with an ability to respond to an unfamiliar environment or to changed circumstances, usually with much-increased activity. *Thus, almost all processes are disturbed to varying extents both by the manipulations involved in the actual rate measurement* (i.e. mounting in the respirometer) *and by sudden shifts in temperature.* Grainger (1956) has monitored the rate of oxygen consumption of various crustaceans following a sudden shift in temperature, and found that a sudden rise or fall in temperature caused a marked overshoot or an undershoot, respectively, in the rate of oxygen consumption, which was followed by small oscillations until a new stabilized rate was achieved (Fig. 2.2). In crustaceans Grainger found that the period of overshoot and oscillations was approximately 1–2 h. In fish the disturbance after manipulation and thermal shock is more severe and may last for several days or perhaps weeks (Holeton, 1974). Therefore, in studying temperature effects it must be demonstrated that handling stress has been eliminated and that the rate has been measured during the new stabilized state following a temperature shift. The severity of shock reaction depends to a great extent upon the rate and magnitude of the temperature shift; slow adjustments will produce much less disturbance than sudden temperature changes.

A second important consideration is the occurrence of adaptive changes

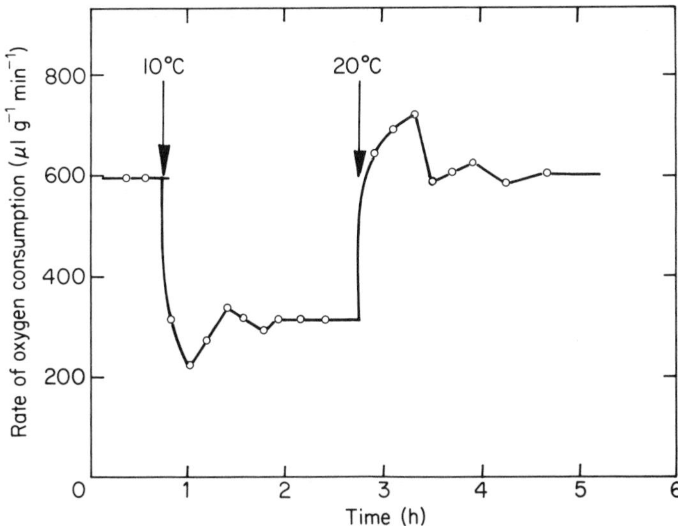

Figure 2.2 The effects of sudden temperature changes upon the rate of oxygen consumption of the crustacean *Hemimysis lamnornae*. The animals were initially held at 20°C. Note the transient oscillations in rate immediately after the temperature shift. (After Grainger, 1956.)

in the rate of metabolic processes as a result of the temperature change. Animals which are exposed to increased temperatures for long periods tend to reduce their rate of metabolism to compensate for the rate-increasing effects of the temperature change (see Chapter 5). These physiological adjustments are part of an organism's normal response to seasonal changes in temperature, and generally take several days or perhaps weeks to become apparent. Rate measurements taken during or after this period of physiological adjustment will not estimate the true effects of altered temperature, and the resulting rate–temperature curve will typically have a much reduced slope. The construction of acute (or sudden) rate–temperature curves is therefore often a compromise between avoiding the effects of the initial shock reaction and avoiding the physiological adjustments that occur with chronic or longer-term exposure to a given temperature.

Finally, the considerable physiological plasticity of animals in a seasonally-varying environment means that an animal will not have identical physiological properties throughout the seasons. It is necessary, therefore, to control the environmental experience of organisms before the experiment by maintaining them under specified conditions of temperature, photoperiod, humidity (in terrestrial animals) and diet, a process usually termed acclimation (see Chapter 5 for a detailed discussion).

The rate and temperature dependence of many biological processes will depend to some extent upon the level of activity of an animal. For example, the rate of oxygen consumption of active fish may be two- to four-times greater than that of quiescent fish (Fry, 1957) whilst during flight insects may increase their oxygen consumption by more than 100-fold. In respect of activity, three levels of metabolism are commonly recognized (Fry, 1957):

(1) Standard metabolism, which is that required for vital functions in the absence of spontaneous activity;
(2) Routine metabolism, which also includes that required for normal, unrestrained activity;
(3) Active metabolism which includes that required for specified, forced levels of locomotory activity.

Any measure of the metabolic rate which is not performed at a constant and

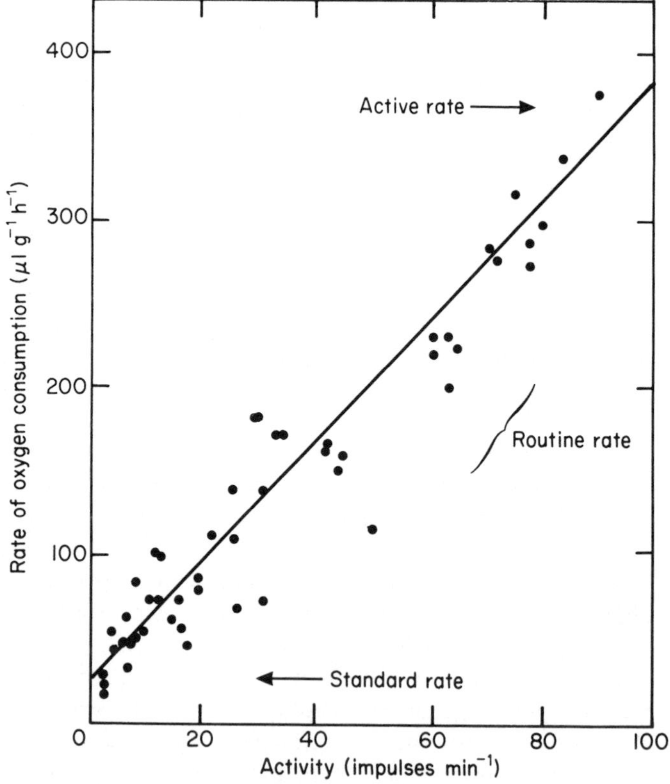

Figure 2.3 The effects of activity level upon the rate of oxygen consumption of the goldfish *Carassius auratus*. Activity was measured using an aluminium paddle which was deflected by the water currents set up by the motion of the fish. Standard and active metabolism may be estimated by extrapolation, as indicated. (After Spoor, 1946.)

specified level of activity will be subject to some variability, and thus considerable uncertainty.

To estimate standard metabolism, it is usually not enough to ensure the quiescence of an animal by placing it in a darkened chamber under constant conditions, because much spontaneous activity can still occur (Beamish and Mookerjii, 1964). Spoor (1946) has developed an 'activity meter' which records the deflections of an aluminium paddle caused by motion of a fish. He plotted a graph of rate of oxygen consumption against the activity level (Fig. 2.3) and estimated the respiratory rate at zero activity (i.e. standard metabolism) by extrapolation. Corti and Weber (1948) designed a more elaborate device which measured activity by recording the movements, caused by fish activity, of a delicately suspended aquarium. Other approaches to this problem include immobilization by the use of anaesthetics and by transection of the spinal cord. A further complication is the effect of body size since, as Rao and Bullock (1954) and Bullock (1955) point out, the rate of many processes such as ventilation by the mussel *Mytilus* and standard metabolism of *Talorchestia* (Edwards and Irving, 1943) and their temperature dependence vary with body weight. Unless rates are determined for animals of identical size, the effect of temperature will be obscured. This size dependence presumably has a cellular basis which might also obscure the rate–temperature curves for cellular and subcellular processes.

Experiments on non-active animals such as eggs and diapausing insects are not totally exempt from the disturbing effects of rapid shifts in temperature. Grainger (1956) found that the eggs of the freshwater crayfish *Astacus* and the adult brine shrimp *Artemia* that were anaesthetized with ether showed overshoots in oxygen consumption even though there was no detectable locomotory activity. This suggests that the shock effect may be a cellular property brought about by a disturbance of some metabolic steady state.

2.3 Empirical descriptions of rate effects

Scientists have long sought a quantitative description of the effects of temperature upon biological processes, either for descriptive purposes or to achieve a deeper understanding of the underlying mechanism. Most early attempts by biologists were purely empirical, fitting particular equations to the data and where necessary introducing additional constants and correction factors to improve the agreement.

2.3.1 Linear relationships

Perhaps the earliest quantitative approach was the rule of thermal summation used by De Reaumur in 1735 and by Boussingault in 1851.

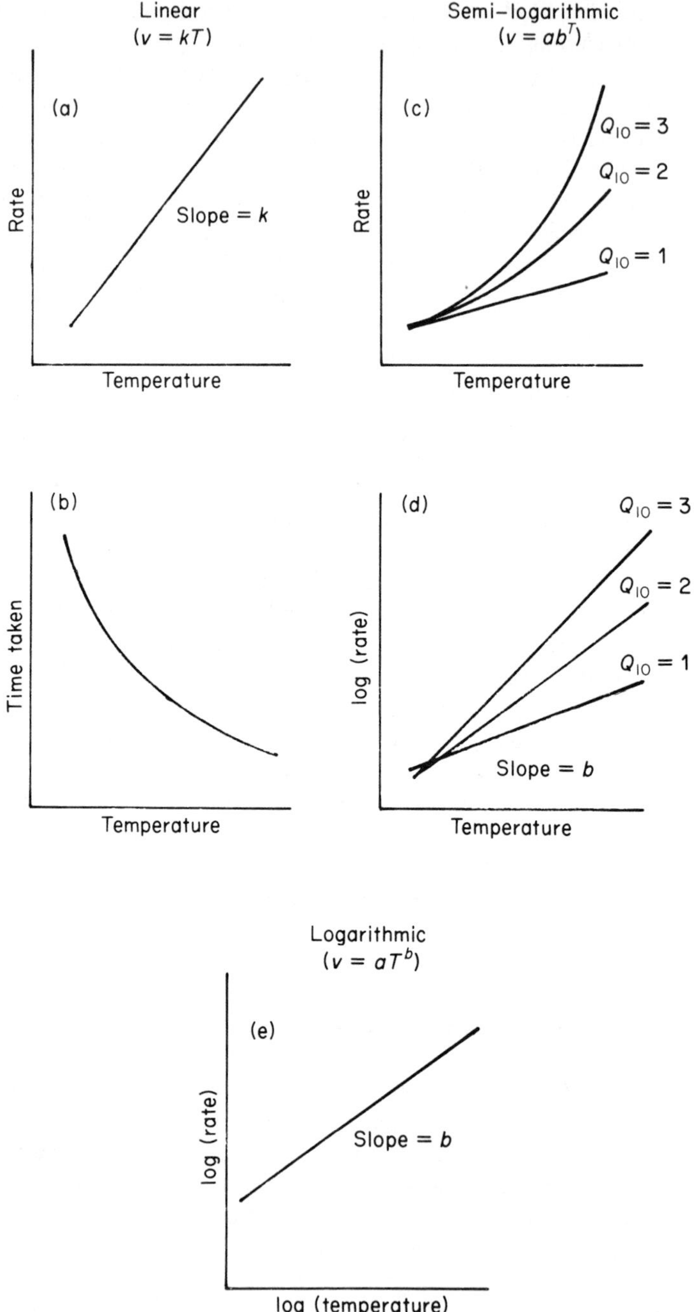

Figure 2.4 Graphs to show the various linear and exponential formulae used to describe the rate effect of temperature.

This stated that the time taken for a biological process to occur (y) multiplied by the temperature (T °C) gives a constant (K). i.e.

$$yT = K = \text{thermal constant} \tag{2.1}$$

Many authors have suggested that the rule fits the observations more closely if the temperature is reckoned from a biological zero where the process is stopped by cold rather than from 0°C, so that

$$y(T - T_0) = K \tag{2.2}$$

where T_0 is the theoretical temperature at which the process is stopped by the cold (i.e. the 'null point'). This so-called rule is commonly used in agriculture and entomology to predict the completion of developmental phases of insect pests in a varying thermal environment, since it is easy to apply and gives reasonably accurate results. However, it has been superseded by more-satisfactory empirical descriptions (see Chapter 7).

A number of workers have claimed that the rate of biological processes is linearly related to temperature. For developmental processes, Krogh (1914a) laid special emphasis on a linear relationship of the form

$$v = kT \tag{2.3}$$

where v is the rate or velocity and k is a constant. Since the rate of a process may be calculated as the reciprocal of the time taken, Krogh's linear relation may be written as

$$1/y = kT \tag{2.4}$$

and if K is defined $1/k$ this becomes

$$yT = K \tag{2.5}$$

which is identical to the rule of thermal summation. Graphically, the rule of thermal sums results in a rectangular hyperbola (Fig. 2.4(b)) which is the reciprocal curve of the straight line of Krogh's linear equation (Fig. 2.4(a)).

2.3.2 *Semi-logarithmic relationships*

These linear approximations were shown to be reasonably good in some cases, but usually only over the medium range of temperatures. The rates at lower temperatures were usually greater, and the rates at higher temperatures were less than those predicted by extrapolation of a linear rate–temperature curve from the middle range of temperatures. This problem was partially overcome when Berthelot (1862), studying the rate of fermentation by yeast, found an exponential effect of temperature:

$$v = ab^T \tag{2.6}$$

where a and b are constants. If we plot v against T we obtain a rapidly rising curve, the value of b describing the rate of increase (Fig. 2.4(c)). Exponential terms are best treated by taking logarithms, thus

$$\log v = \log a + T \log b \tag{2.7}$$

so that log v is now a linear function of temperature with a slope of log b and an intercept of log a on the ordinate (Fig. 2.4(d)). For this equation, the ratio of rates over a given temperature interval will be constant, a fact which forms the basis of the well-known Q_{10} relationship for a 10°C interval or the Van't Hoff rule:

$$Q_{10} = v_{t+10}/v_t \tag{2.8}$$

where v_{t+10} and v_t are the rates at temperature $T + 10$°C and T°C, respectively. Obviously, it is not necessary to determine the rates over an exact 10°C interval in order to calculate Q_{10}. Any two temperatures can be used, provided the interval is sufficiently far apart that the difference in rates can be reliably measured. In the case Q_{10} can be determined from

$$Q_{10} = \frac{v_1^{10/(T_1 - T_2)}}{v_2} \tag{2.9}$$

$$\log Q_{10} = 10 \frac{(\log v_1 - \log v_2)}{(T_1 - T_2)} \tag{2.10}$$

Q_{10} has a fixed relationship with the value of the constant b of Berthelot's equation, which is the increase in log rate for a 1°C interval, so that

$$\log Q_{10} = 10b \tag{2.11}$$

It is obviously cumbersome to calculate Q_{10} for each pair of observations over the entire temperature range, and a more convenient means of calculating Q_{10} and assessing the agreement of data with the Q_{10} relationship is to plot the data according to Equation 2.7. Figure 2.4(c) shows that a Q_{10} of 2 leads to a doubling of rate over a 10°C interval and a quadrupling over a 20°C interval. A Q_{10} of 3 has even more dramatic effects, so that it is clear that small differences in Q_{10} have very large effects upon temperature dependence. A Q_{10} of between 2 and 3 is the rule rather than the exception in biology, and these values provide a useful rule-of-thumb in predicting the effects of temperature variation.

The Q_{10} has been widely used in biology as a convenient and easily understood measure of temperature effects, and is certainly useful as an empirical description. However, Q_{10} is often given a precision and meaning which is quite unjustified, since for the vast majority of biological processes and even for some chemical processes it varies with temperature. This discrepancy was first attributed to poor experimental technique, and some workers averaged the Q_{10} values observed over a wide range of temperature. Subsequent studies, initially by Krogh (1914b) and Ege and Krogh (1914),

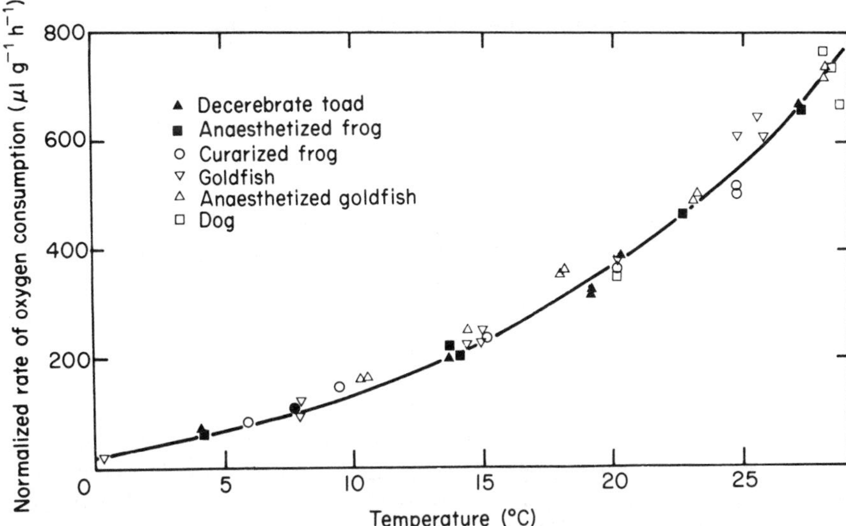

Figure 2.5 Krogh's normal curve comparing the effects of temperature upon the rate of oxygen consumption of fish, frogs, a toad and a dog. The curves for each animal have been normalized to facilitate the comparison of their shapes. (After Krogh, 1914b.)

indicated that the inconstancy of Q_{10} was not artifactual but systematic, so that as temperature increased Q_{10} tended to decrease in a progressive manner. Krogh found that although the rates of oxygen consumption of a variety of vertebrates were different, the shapes of the rate–temperature curves were nearly identical (Fig. 2.5). This general curve where Q_{10} varies from 9 at low temperatures to 2 at higher temperatures, is called Krogh's 'normal curve' and represents a relationship of general significance.

The variation of Q_{10} with temperature severely limits its usefulness for descriptive purposes, since comparisons can only be made over two similar temperature intervals, so that it is now conventional to quote the temperature range over which Q_{10} was determined (i.e. $Q_{10(5-15°C)}$). Even so, comparisons of Q_{10} between different species are suspect because a given temperature interval may fall in quite different portions of their respective rate–temperature curves. Clearly, 25°C would be quite favourable for many temperate fish species, but is injurious or even lethal for Antarctic fish or for mammals. As Bělehrádek (1930) put it, only the well-known mathematical timidity of biologists can explain the widespread popularity of Q_{10}.

2.3.3 *Logarithmic relationships*

In view of the inadequacy of Berthelot's formula and Q_{10}, Bělehrádek

(1930) proposed an alternative exponential formula of the form

$$v = aT^b \tag{2.12}$$

where a and b are constants. Again, by taking logarithms this becomes

$$\log v = \log a + b \log T \tag{2.13}$$

so that a graph of log v against log T results in a straight line with a slope b and an intercept on the log y axis of log a (Fig. 2.4(e)). In some instances the fit can be improved if the temperature is corrected so that it becomes zero when the process effectively ceases rather than at zero Celsius. The corrected equation becomes

$$v = a(T - T_0)^b \tag{2.14}$$

where T_0 is the 'biological' zero temperature. For processes with a 'biological' zero temperature of approximately 0°C it is obviously unnecessary to introduce the correction. The constant b is a temperature coefficient and usually has a value between 0.6 and 2.0, but is usually close to unity, in which case the equation becomes identical with the linear equations described previously. Clearly, Krogh's linear equation and the rule of thermal sums are but special cases of the more general relationship expressed by Bělehrádek's equation. As Bělehrádek (1930, 1935) points out, his formula, though empirical, often results in a much improved description of experimental data, especially in cases where the Q_{10} decreases with rising temperature. In other instances the Bělehrádek formula is considerably less good and, therefore, is no more general in its application than other equations. Progressively better empirical descriptions for developmental processes have been developed, such as the 'catenary curve' of Janisch (1925) and the logistic curve of Davidson (1944), and these are considered in Chapter 7.

Of the principal relationships described here the semi-logarithmic relationship is by far the most widely used, partly for reasons which will become apparent in the next section. Linear relationships have been occasionally noted in studies of heart beat rate in vertebrates (e.g. Harri and Talo, 1975) and more frequently in developmental studies. The logarithmic relationship of Bělehrádek has also been successfully applied to developmental rates of amphibian embryos, but is comparatively underused in other areas of study.

2.4 Theoretical equations

Calculating the rates of processes from first principles has obvious attractions with respect both to predictive ability and to a fundamental understanding of the nature of the processes involved. As we shall see, the sheer complexity of biological processes largely prevents the application of such a reductionist approach to more-complex biological processes, although much progress has been made in describing how temperature affects simple chemical and enzymatic processes. This approach can be traced back to the Dutch physical chemist Van't Hoff who, in 1884, developed a theoretical equation for the temperature-dependence of equilibrium reactions (Van't Hoff, 1896). In a reversible reaction where k_1 and k_2 represent the forward and reverse rate constants, we have

$$A + B \underset{k_2}{\overset{k_1}{\rightleftharpoons}} C$$

At equilibrium

$$k_1[A][B] = k_2[C] \tag{2.15}$$

where the brackets denote the concentration of the reactants and products. The equilibrium constant K_{eq} becomes

$$K_{eq} = k_1/k_2 = [C]/[A][B] \tag{2.16}$$

Now the temperature dependence of this reaction is given by

$$\frac{d(\ln k_1)}{dT} - \frac{d(\ln k_2)}{dT} = \frac{q}{RT^2} \tag{2.17}$$

where R is the universal gas constant (8.3 J K^{-1} mol^{-1}), T is now the absolute temperature and q is the quantity of heat (J) which is liberated when 1 mole of A and B is converted into C. Van't Hoff pointed out that although Equation 2.17 does not give the relationship between a single rate constant k and temperature, it has the form

$$\frac{d(\ln k)}{dT} = \frac{a}{T^2} + b \tag{2.18}$$

where a and b are constants. Now if $a = 0$ the equation becomes identical with that proposed by Berthelot, which then becomes a special case of the Van't Hoff equation. However, Arrhenius (1889, 1915) showed empirically that this equation described the observed data particularly well when $b = 0$, so that

$$\frac{d(\ln k)}{dT} = \frac{\mu}{RT^2} \tag{2.19}$$

where μ is a constant. By integration between the rate k_1 at absolute temperature T_1 and rate k_2 at absolute temperature T_2, Arrhenius found that

$$k_1 = k_2 \exp\left[\frac{\mu}{R}\left(\frac{T_1 - T_2}{T_1 T_2}\right)\right]$$

or

$$\ln\left(\frac{k_1}{k_2}\right)=\frac{\mu}{R}\left(\frac{1}{T_1}-\frac{1}{T_2}\right) \tag{2.20}$$

A graph of ln k against $1/T$ yields a straight line with a slope of μ/R (if the rates are plotted as a $\log_{10}$ term the slope becomes $\mu/2.3R$). The constant μ was variously termed the 'critical thermal increment', the 'temperature characteristic' or more commonly the 'Arrhenius activation energy', and is conventionally determined from the so-called Arrhenius plot of $\log_{10} k$ against $1/T$ (Fig. 2.6). We should note at this point that the meaning of μ was rather ill-defined and was certainly not equivalent to the constant q derived for equilibrium reactions by Van't Hoff.

It will be remembered that even for chemical reactions the Q_{10} progressively reduces as temperature increases. From the Arrhenius formula, as transformed in Equation 2.20, it may be deduced that for higher temperatures the increase in rate constants for a 10°C rise in temperature will be smaller than at lower temperatures, which reflects more closely the natural situation. However, this advantage of the Arrhenius equation over the Q_{10} formula is only apparent when

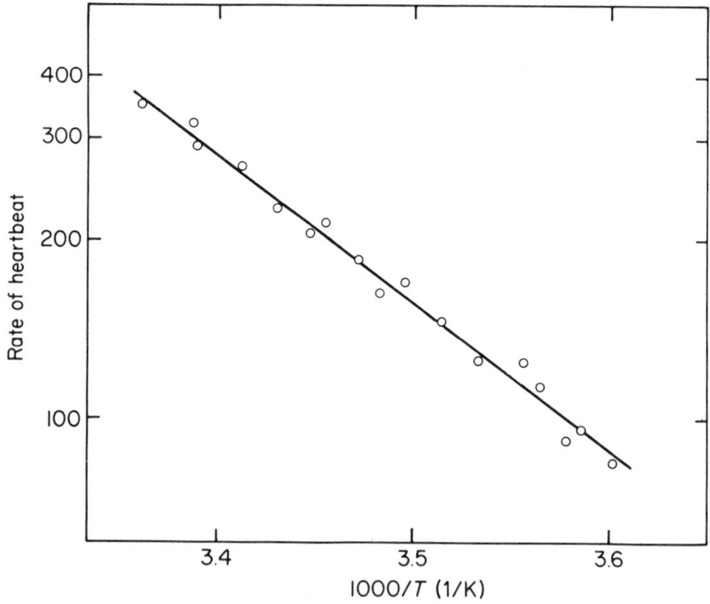

Figure 2.6 The Arrhenius plot for the effects of temperature upon the rate of heartbeat of the water-flea *Daphnia pulex*. The data points conform to a straight line, the slope of which provides an estimate of the 'thermal increment' or 'activation energy'. (University of Liverpool, Class of 1983.)

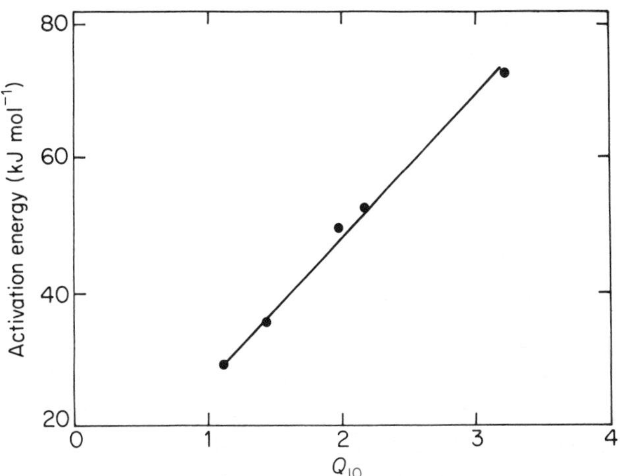

Figure 2.7 The relationship between Q_{10} and the Arrhenius activation energy for a biological process. Note that the relationship is nearly linear over the range of temperatures examined, as might be expected from the nearly linear relationship between temperature and the reciprocal temperature. Thus, for processes with the normal level of variability it is usually not possible to distinguish between the two approaches on statistical grounds. (After Bělehrádek, 1930.)

experimenting over a very wide range in temperatures such as 0–200°C, and when experimental variability is low. Over the biological range of temperatures the reciprocals of absolute temperatures (273–313 K) are almost a linear function of the celsius temperature (0–40°C) so that Q_{10} and μ vary to practically the same extent (Fig. 2.7). The differences in the agreement of the Arrhenius equation and the Q_{10} formula with the observed data are generally well within the limits of error for biological observations, and it is often not possible to justify the use of one formula in preference to the other on a statistical basis.

Kavanau (1950) has empirically altered the Arrhenius equation by introducing a new term, T_0, which is the effective zero temperature for biological processes rather than 0 K, so that

$$k = D\exp - \frac{E_0}{R(T - T_0)} \tag{2.21}$$

where D is the pre-exponential constant. E_0 may be derived from a modified Arrhenius plot of $\log k$ against $1/(T - T_0)$, but it no longer has the same meaning as the Arrhenius activation energy. This equation increases the curvilinearity of the modified Arrhenius plot and more accurately describes the effect of temperature upon a number of processes ranging from enzymatic reactions to the rates of development of insect pupae and the rate of CO_2 production by seedlings (Kavanau, 1950). The zero temperature for most biological processes was found to be between $-10°$ and 0°C, but for *in vitro* enzymatic reactions it lies between -30 and -70°C. This formula has not been widely used in biology and the reasons for this are not entirely clear, since it holds the promise of describing rate processes which deviate from the Berthelot and the Arrhenius equations. It also resolves a major criticism levelled at the Arrhenius formula (Bělehrádek, 1930, 1935) that the zero temperature of 0 K (-273°C) bears no relation to the temperature at which most biological processes effectively cease.

Perhaps the major reason for the preferred use of the Arrhenius formula in biochemistry and biophysics is the quasi-theoretical framework developed by Arrhenius for the temperature coefficient, μ, in simple chemical reactions. Obviously, biological processes are made up of innumerable individual reactions and the observed μ can have no simple significance, but the success of the Arrhenius equation in describing biochemical and biological processes suggests that they often behave kinetically as simple reactions, and can be profitably interpreted as such. Arrhenius noted that the effect of temperature upon chemical reactions was much too great (12% $°C^{-1}$) to be accounted for by the increase in kinetic energy of the constituent molecules (1/6% $°C^{-1}$). He introduced the new concept of the 'activated state' in which the reactants combined to form an intermediary complex, but only those complexes with sufficient energy, the energy of

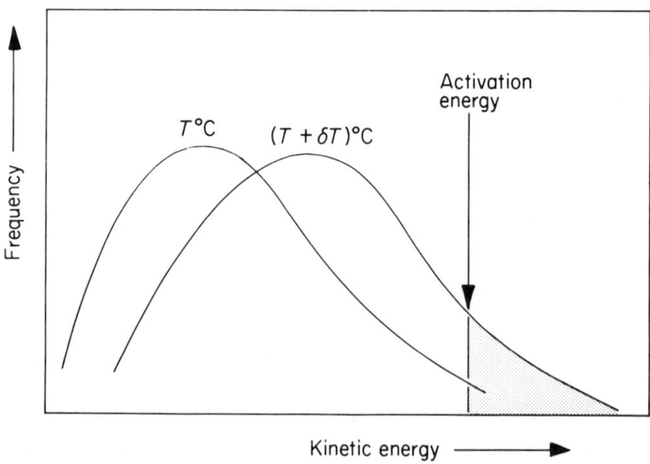

Figure 2.8 The frequency distribution of energy levels for a population of molecules calculated by the Maxwell–Boltzmann distribution law at temperatures T°C and $(T + \delta T)$°C. As temperature increases the curve moves to the right, so the proportion of molecules with energy levels greater than the energy of activation increases exponentially.

activation, were able to complete the reaction and form the products. Thus

$$\begin{array}{ccccc} A+B & \rightleftharpoons & AB^{\neq} & \rightleftharpoons & C \\ \text{Reactants} & & \text{Activated complex} & & \text{Product} \end{array} \tag{2.22}$$

where the superscript ($^{\neq}$) conventionally indicates a property of the activated state. If this is the case, the exponential effect of temperature upon rate can easily be explained by reference to the frequency distribution of energy levels for a population of reacting molecules that is predicted by the Maxwell–Boltzmann distribution law (Fig. 2.8). At temperature T only a small proportion of molecules with energy levels greater than the activation energy may react. However, at the increased temperature ($T + \delta T$) the frequency distribution curve moves to higher energy levels, so the proportion of molecules with an energy greater than the activation energy increases exponentially.

2.5 The classification of biological processes by temperature coefficients

2.5.1 *Q_{10} values*

An obvious goal of biologists is the use of temperature coefficients to assist in the identification of the underlying chemical and physical mechanisms of biological processes. The earliest attempts to draw such conclusions followed the demonstration that certain physical processes, such as diffusion, had very low temperature coefficients, whilst chemical reactions typically had higher coefficients. Thus, Snyder (1908, 1911) argued that because heart beat had a high Q_{10} its underlying basis was chemical, whilst the conduction of nerve impulses relied primarily on physical processes because it exhibited a Q_{10} approaching unity. In some instances processes which one might have expected to have a physical basis yielded high Q_{10} values, and it was concluded that in reality they were chemical.

The hope of any specific identification of mechanism has been partially realized in only two cases. First, most photochemical reactions have a Q_{10} of 1–1.2, and on this basis some biological processes such as photosynthetic light reactions were shown to have photochemical components. However, many other chemical processes including thermoneutral reactions may, in principle, be independent of temperature and recently a number of studies have shown that certain whole-animal processes such as respiration may have Q_{10} values approaching unity, as described later in this chapter. Secondly, the death of organisms at high temperatures typically yields a very high Q_{10} of 30–100 which is similar to the values for heat denaturation of proteins, and it seems reasonable to expect some causal relationship between the two.

The identification of mechanism by measuring Q_{10}, then, is only possible in the very broadest sense and even then is not unequivocal. This is not only because of the inconstancy of Q_{10} over a wide range of temperatures but also because there are numerous exceptions to the general Q_{10} values quoted above; some chemical reactions have Q_{10} lower than 2 and a Q_{10} above 2 does not exclude a physical mechanism with any certainty. Clearly, the division of biological processes into 'physical' and 'chemical' has no real basis in fact, the more so because numerous examples of both types of reaction will occur sequentially or simultaneously in a biological situation.

2.5.2 *Master reactions and Arrhenius activation energy*

The classification of biological processes by use of their temperature coefficients was a problem which attracted much attention in the early part of this century, Crozier and his colleagues (Crozier, 1924, 1925a, b; Crozier and Federighi 1925a, b; Crozier and Stier 1925a–d) were particularly taken with the Arrhenius activation energy, which these workers believed to be more accurate reflection of the underlying molecular events than the Q_{10}. The idea that complex processes ultimately depend upon a series of consecutive reactions each with its particular μ value had previously been developed by Blackmann (1905) and Pütter (1914). The rate of the overall process was thought to be limited by that of the slowest step, the so-called 'rate-limiting step', so that the activation energy of the overall process would be that of this 'master' reaction. Crozier postulated, first, that certain processes of the same general type have similar μ values, and claimed that this was evidence that each process was rate-limited by a common step (Crozier, 1924). He felt that it might be possible to identify master reactions in different instances by agreement of their μ values. For example, the locomotion of ants and the creeping of diplopods as well as other neuromuscular processes were all shown to have a specific value (Crozier, 1925a) and this, Crozier maintained, was the temperature characteristic of central neural discharges. Secondly, Crozier suggested that curvilinear Arrhenius plots were best interpreted as a series of intersecting straight lines with gradually increasing slopes at lower temperatures. These breaks in the graphs were interpreted as sudden transitions from one rate-limiting step to another, each with quite different μ values, so that over a lower range of temperatures a reaction with a high μ was rate-limiting whilst at higher temperatures another reaction with a lower μ value became rate-limiting.

We have seen that in situations where there is some biological variability the Arrhenius μ is no better a description of rate–temperature curves than Q_{10}, so that it is not entirely surprising that Crozier was no more successful than earlier workers. However, it is instructive to discuss the criticisms that were levelled against Crozier's work, since they have important implications not only to present practices of constructing rate–temperature curves, but also to the mechanistic deductions that may be drawn from them. The main objections were as follows.

(a) Although Crozier maintained that μ values for similar processes were identical, it is commonplace to find, in fact, that μ varies with a variety of factors. The temperature dependence of whole-animal processes depends to some extent upon the season, their reproductive status, their general condition, and even age. For enzymatic reactions μ is sensitive to pH, ionic strength, relative concentrations of enzymes and substrates, and the presence of inhibitors. The gradual accumulation of information by many workers has shown that the uniformity of μ for common processes described by Crozier was largely illusory, and the μ values for similar processes in different organisms were often sufficiently different to reject the assumption of common master reactions. Some of these differences, to be sure, were due to different experimental techniques and to a lesser extent to scatter of the data. However, one is forced to the conclusion that the concept of master reactions, despite its fundamental correctness, is much too simple to apply with rigour to anything other than elementary rate processes.

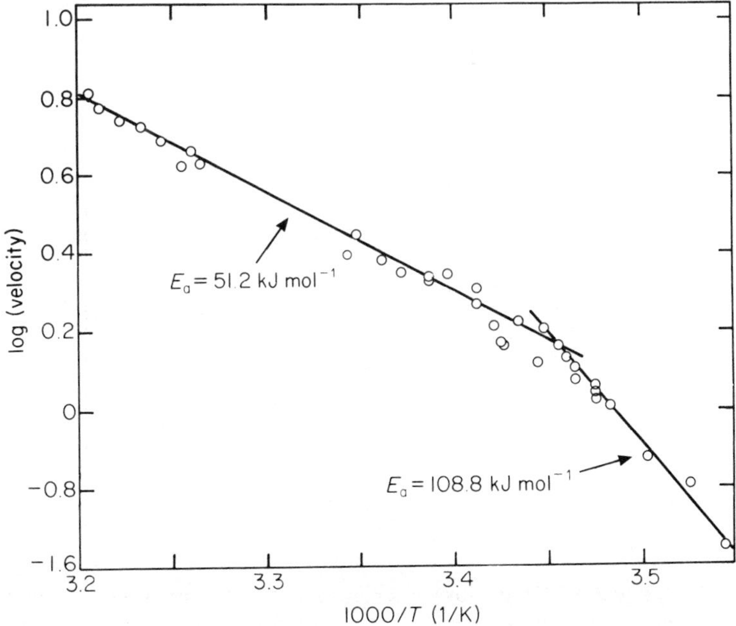

Figure 2.9 The Arrhenius plot for the rate of creeping of the ant *Liometopum apiculatum*, as interpreted by Crozier. (After Crozier, 1925a.)

(b) The graphical procedures adopted by Crozier for the interpretation of non-linear Arrhenius plots were highly questionable. Figure 2.9 shows the Arrhenius plot for the rate of creeping of the ant *Liometopum*, over the temperature range 8–40°C, which was interpreted by Crozier (1925a) as two straight lines which intersected at 16°C. Above this temperature the rate was thought to be controlled by a 'master' reaction with $\mu = 51\,\text{kJ}\,\text{mol}^{-1}$, whilst below 16°C a different reaction became rate-limiting with $\mu = 109\,\text{kJ}\,\text{mol}^{-1}$; the transition between the two reactions was sudden. However, bearing in mind the scatter of the data, it seems that a smooth curve is equally appropriate. This problem applies particularly when the difference in slope of the intersecting lines is small and Crozier's interpretation is even more difficult to justify, particularly when the graph is presented with a contracted ordinate scale (Fig. 2.10). However, the same dubious assignment of biphasic or multiphasic Arrhenius plots is common practice even today.

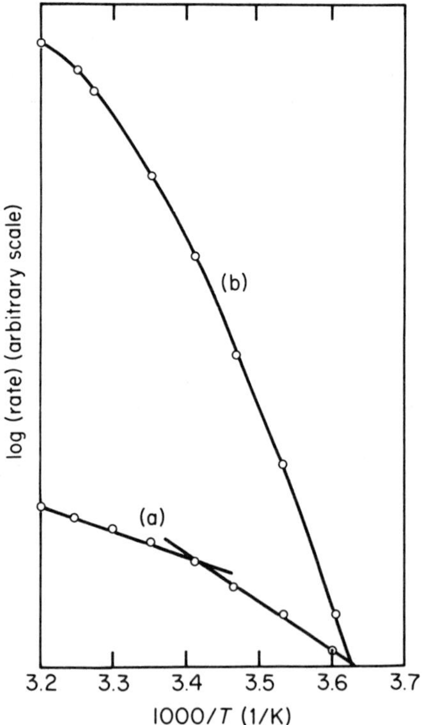

Figure 2.10 Arrhenius plots, plotted on two different scales, for the rate of oxygen consumption by yeast. The two intersecting straight lines (a) are revealed as a curvilinear plot as the ordinate is expanded (b). (After Buchanan and Fulmer, 1980.)

(c) In principle, non-linear Arrhenius plots can result from the simultaneous activity of two or more parallel processes, or perhaps from two successive reactions with different activation energies (Dixon and Webb, 1979). Kistiakowsky and Lumry (1949) have shown on theoretical grounds that sharp inflections in these cases can only be observed if the activation energy above and below the break temperature differs by hundreds of thousands of joules per mole instead of the 20–60 kJ mol^{-1} that is usually observed. Indeed, Kavanau (1950) calculated that for an enzyme reaction with two sequential reactions of $\mu = 44$ and 240 kJ mol^{-1}, respectively, the changeover in the rate-limiting reaction occurs over a 20°C range. Theoretically, therefore, it is highly unlikely that Crozier's inflections are sharp.

2.5.3 *Membrane phase transitions and Arrhenius discontinuities*

A number of biochemical and biophysical processes do show significant changes in slope over a restricted range of temperatures and a few convincingly show two intersecting straight lines with a more or less sharp

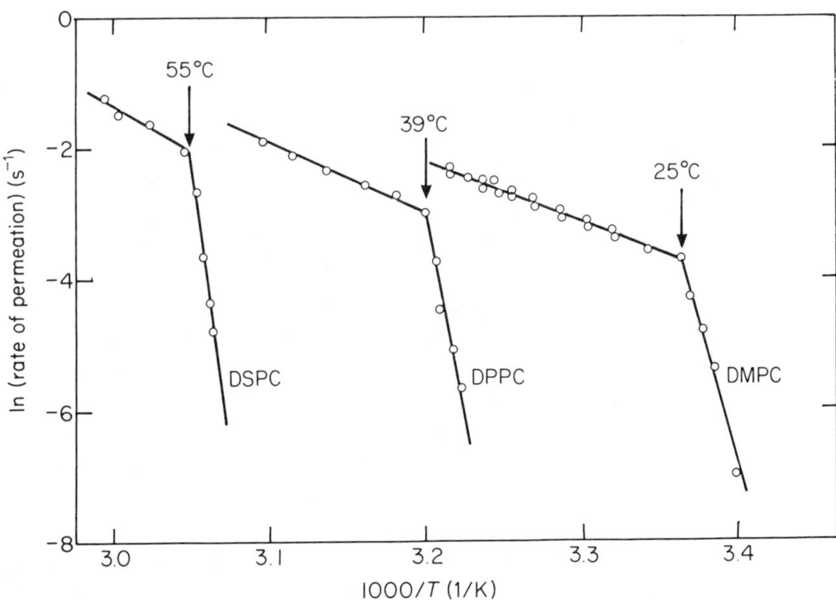

Figure 2.11 Discontinuous Arrhenius plots for permeation of a sugar across artificial bilayers prepared from pure phospholipids containing fatty acids with 14, 16 or 18 carbon atoms (DMPC, DPPC and DSPC, respectively). There was a sharp break which occurred close to the respective phase transition temperatures of the vesicles. (After Jähnig and Bramhall, 1982.)

discontinuity (see Fig. 2.11). Recently, much attention has focused upon the properties of cellular membranes, such as permeation or the activity of membrane-bound enzymes such as Na^+, K^+ – ATPase, since it is possible that the membrane lipids, which form an important solvent compartment for these processes, may undergo changes in physical state (i.e. from gel to liquid-crystalline) over the physiological temperature range. Thus, the existence of sharp breaks in Arrhenius plots was thought to provide important evidence of a regulatory role of membrane phase-state in enzymatic properties. Above the transition temperature the lipid environment of the enzyme was thought to be fluid, whilst below it the lipid assumes a more ordered, crystalline arrangement which restricts the conformational flexibility of the enzyme and thereby increases the activation energy for the reaction (Kumamoto *et al.*, 1971). A similar argument has been proposed to account for complex Arrhenius plots of a variety of biological processes in which changes in the structure of 'bound' water in cellular systems occur at specific temperatures of 15, 30 and 45°C (Etzler and Drost-Hansen, 1979).

Evidence in favour of this interpretation in respect of membranes is of three types. First, spectroscopic probes of membrane physical structure exhibit sudden changes in properties at temperatures which approximate the break temperature of the enzyme. Secondly, the break temperature can be reduced by agents which are known to cause an increase in membrane fluidity (such as ethanol) or by dietary manipulation of membrane lipid composition. Thirdly, breaks which have been observed in membranes of non-hibernating species are absent in the equivalent membranes of hibernators (Lyons and Raison, 1970; Raison and Lyons, 1971; Raison 1973). The suggestion is that the latter have adaptively shifted the phase-transition temperature to below the biologically important range in order to avoid the adverse effects of a solidified membrane.

More recent determinations of the phase transitions by calorimetric techniques have generally found that the bulk transition is generally well below the temperature of the Arrhenius break. In some cases these transitions only include about 2% of the membrane lipid, so it seems more likely that Arrhenius breaks are due to changes in the properties of microdomains rather than to the whole bilayer. Moreover, the theoretical basis for sharp Arrhenius discontinuities is not well understood. Silvius and McElhaney (1981) have recently developed mathematical descriptions of membrane rate processes which predict the shape of the Arrhenius plot from first principles. They noted that biphasic linear plots could be produced in systems which do not undergo a phase transition and discontinuities of this sort have been observed both in soluble and membrane-bound enzymes which have been detergent-solubilized to strip off the lipid shell (Dean and Tanford, 1978). These transitions are, at least in some cases, probably due to structural transitions of the protein rather than of its solvent environment. A biphasic plot does not therefore in itself, provide unequivocal evidence of the nature of the underlying process.

There is no doubt that the discontinuity shown for artificial membranes in Fig. 2.11 represents the effects of a bulk phase transition. However, because phase transitions in biological membranes tend to occur over a broad temperature range (10–20°C), and because the change in slope is usually much smaller than that shown in Fig. 2.11, it is often easy to produce convincing fits of data with a moderate degree of experimental variation with a linear biphasic plot (see, for example, Fig. 2.12). On the other hand, a curvilinear plot often fits just as closely, and statistical techniques are then required to decide which provides the most appropriate plot. Silvius and McElhaney (1980) found that Arrhenius plots of a membrane-bound ATPase in *Acholeplasma*, which were previously reported

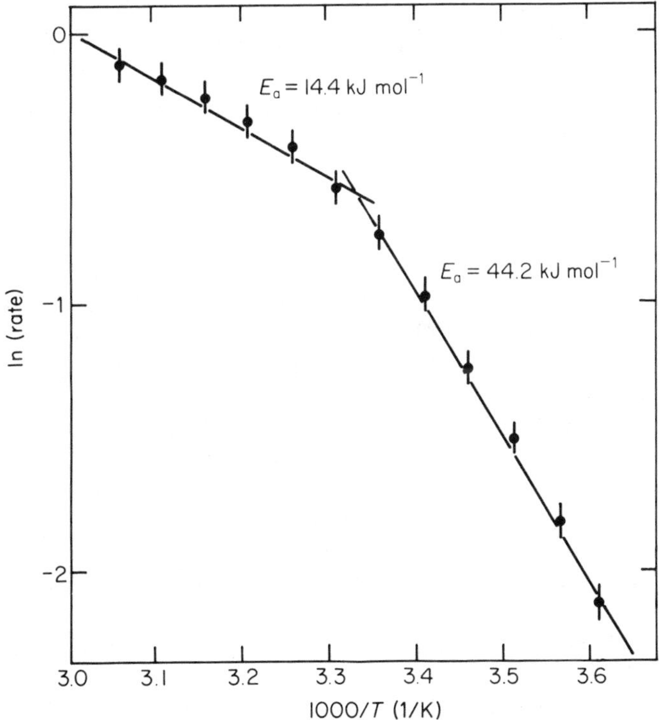

Figure 2.12 The application of two intersecting straight lines to a curved Arrhenius plot. The data points were calculated from a curvilinear function. This curved data set can be fitted by the two lines within the 5% error bars as shown, even though no break actually exists. Great care should be exercised in justifying the interpretation of non-linear Arrhenius plots as two intersecting straight lines. (After Silvius and MacElhaney, 1981.)

as biphasic linear, proved to be curvilinear when experimental error was reduced. This important example shows that the common assumption of intersecting straight lines as the best way of representing non-linear Arrhenius plots is unwarranted. Because several models predict roughly similar forms of Arrhenius plot it is not possible to differentiate between them on statistical grounds, particularly when there is appreciable experimental error.

Arrhenius plots are undoubtedly more widely used in biochemical and biophysical studies than at higher levels of organization. The reason for this is not difficult to see, since the interpretation of Arrhenius plots is not without difficulties, even at the biochemical level of organization. With more complex processes of tissues and whole animals the assumptions on which the Arrhenius equation was originally formulated have progressively less relevance.

2.6 Theory of absolute reaction rates

The Arrhenius equation (Equation 2.19) can be rewritten

$$v = a \exp(-E_a/RT) \tag{2.23}$$

where v represents the observed velocity, R the universal gas constant, T is the absolute temperature and the constant E_a is identical with μ, the Arrhenius activation energy. The 'collision theory' of reactions has provided some insight into the meaning of the pre-exponential constant a as being composed of two factors: the frequency of collisions of the reactants and a 'steric' factor which determines the probability of a collision resulting in the creation of the activated state.

The concept of the activated state has also been considerably refined and modified from the original empirical form described by Arrhenius, to provide the basis of the modern theory of absolute reaction rates (for a complete discussion see Johnson, Eyring and Stover (1974)). This theory postulates that all elementary rate processes such as diffusion, hydrolysis and oxidation can be thought of as an equilibrium between the reactants in their normal state and in their activated state, as shown in Equation 2.22. This idea of an activated state, or transition complex as it is usually called by physical chemists, may be precisely described only in terms of quantum mechanics but can be more easily comprehended by reference to Fig. 2.13, which schematically describes the changes in overall free energy that occur during the reaction. The transformation of the reactants into products results in a substantial decrease in the overall free energy of the system, ΔG°, the magnitude of which describes the extent to which the reaction will be completed but gives no information on how rapidly the transformation

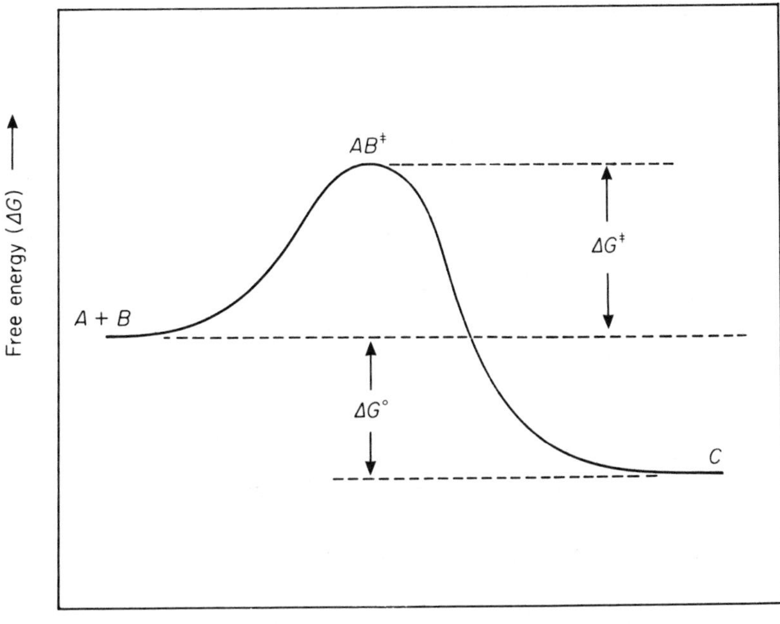

Figure 2.13 Schematic diagram to show the changes in free energy of the reactants, the activated complex ($AB^{\neq}$) and products during the course of the reaction. $\Delta G^{\neq}$, represents the change in free energy necessary to form the activated complex and ΔG° represents the overall change in free energy for the reaction, the heat of the reaction.

will occur. ΔG° may be calculated from the equilibrium constant K_{eq}

$$\Delta G^{\circ} = -RT \ln K_{eq} \tag{2.24}$$

ΔG° is the gain or loss of free energy of the reaction per mole of reactants under certain specified conditions of pH, ion concentration, etc. ΔG° is composed of two parts, the heat of reaction ΔH° and the change in entropy ΔS°. The former is the heat released during the conversion of 1 mol reactants to products, and the latter is a measure of the increase in randomness of the system. These quantities are related by

$$\Delta G^{\circ} = \Delta H^{\circ} - T\Delta S^{\circ} \tag{2.25}$$

We have seen that in order to react, the reactants must achieve sufficient energy through the numerous collisions with their neighbours to form a transition complex. This is a state in which the reactants are contorted and strained in such a manner that their constituent bonds become more susceptible to modification through reaction. This activation step is usually rate-limiting to the overall reaction, so the height of the free energy barrier, $\Delta G^{\neq}$, provides a measure of the rate of the reaction. The theory of absolute reaction rates postulates that the transition complex ($AB^{\neq}$) exists for only a very brief moment (10^{-13} s) and rapidly breaks down with a rate $k_B T/h$ (where k_B is Boltzmann's constant, h is Planck's constant and T is the absolute temperature) that is constant for all reactions and is independent of the nature of the complex. The probability that the transition state will break down to form products rather than reactants is usually taken as equal to unity. The theory postulates that the overall forward rate constant, k, may be accounted for on the basis of a pseudo-equilibrium (with a constant $K^{\neq}$) between the rate of formation of the transition state and its universal rate of breakdown, so that

$$k = k_B T K^{\neq}/h \tag{2.26}$$

Note that $K^{\neq}$ is not a true equilibrium constant since the reactants and activated complex are not in equilibrium in the sense that the reaction occurs in both directions to equal extents. Rather, $K^{\neq}$ is the ratio of the rate of formation of the activated complex from the reactants, k, to the universal rate of breakdown of the activated complex to products ($k_B T/h$). It is assumed that the breakdown of the activated complex to form the products has no direct influence on the rate of forward reaction. However, since $K^{\neq}$ behaves as an equilibrium constant for the overall reaction, it is possible to apply the laws of thermodynamics, so that

$$\Delta G^{\neq} = -RT \ln K^{\neq} \tag{2.27}$$

$$K^{\neq} = \exp\left(-\frac{\Delta G^{\neq}}{RT}\right) = \exp\left(-\frac{\Delta H^{\neq}}{RT} + \frac{\Delta S^{\neq}}{R}\right) \tag{2.28}$$

where $\Delta G^{\neq}$, $\Delta H^{\neq}$ and $\Delta S^{\neq}$ are the changes in free energy, heat and entropy, respectively, for the activation step. The forward rate constant now becomes.

$$k = \frac{k_B T}{h} \exp\left(-\frac{\Delta G^{\neq}}{RT}\right) = \frac{k_B T}{h} \exp\left(-\frac{\Delta H^{\neq}}{RT} + \frac{\Delta S^{\neq}}{R}\right) \tag{2.29}$$

Now if $\Delta S^{\neq}$ does not vary with temperature, differentiation of this equation with $\Delta H^{\neq} - T\Delta S^{\neq}$ substituted for $G^{\neq}$ yields

$$(\Delta H^{\neq} + RT) = 2.303\, RT \frac{\mathrm{d}(\log k)}{\mathrm{d}T} \tag{2.30}$$

which is identical with the Arrhenius equation except that

$$\mu = E_a = (\Delta H^{\neq} + RT) \tag{2.31}$$

So, $\Delta H^{\neq}$ can be estimated directly from an Arrhenius plot of the rate constant, k. Over biokinetic temperatures the difference between E_a and $H^{\neq}$

is approximately 2.4 kJ mol^{-1}, which is usually well within the limits of error of most biological observations. $\Delta S^{\neq}$ can then be easily calculated from

$$\Delta G^{\neq} + \Delta H^{\neq} = T\Delta S^{\neq} \tag{2.32}$$

Thus, and this is the important point, for a simple elementary rate process it is possible to relate the empirical temperature coefficient derived by Arrhenius to a specific thermodynamic quantity, $\Delta H^{\neq}$. This relationship endows the Arrhenius formula with a powerful theoretical basis which is often extended to include the very complicated processes that are commonly observed in biology. However, the calculations required to predict the rates of even the simplest chemical reactions are extensive and complicated, and the enormous complexity of biological processes precludes the estimation of their rates from first principles. In principle the theory applies to complicated processes as well as to simple reactions, and Johnson, Eyring and Stover (1974) suggest that it can be profitably used to gain a clearer insight into the mechanism of biological processes. The thermodynamic quantities of such complex processes as growth, development, heat death and locomotion have been calculated, since they often behave kinetically as simple equilibrium reactions, though the precise interpretation of their thermodynamic quantities is not clear. They certainly do not clarify the mechanism of the process except in the most general way.

There is no doubt of the utility of the absolute rate theory in describing the simpler biological processes. For biochemical studies the theory of absolute reaction rates has proved useful, first, in providing a satisfactory quantitative description of the complete rate–temperature curve and, secondly, in providing a rational account of the fundamental aspects of temperature effects including, as we shall see, the reduction in rate at temperatures above the optimum.

The first application of the theory was to the luminescence of bacterial cells, a convenient process in which the intensity of luminescence is an instantaneous measure of the rate of activity of the enzyme luciferase. These studies were carried out before the identity of the reactants and the reaction sequence had been firmly established. Nevertheless, the theory quantitatively accounted for variations in luminescence caused not only by temperature but also by a number of other factors, such as hydrostatic pressure, inhibitors and pH. Furthermore, it did this without a precise knowledge of the reaction sequence or the components of a reaction sequence.

The theory has been of great benefit in research into the molecular aspects of evolutionary temperature adaptation (see Chapter 5). Much recent work has shown that the thermodynamic parameters of the activation process for several cellular enzymes and for myofibrillar ATPase of different species are adjusted over evolutionary time to achieve similar enzymatic activities in animals from such diverse environments as polar seas and hot springs.

2.7 Destructive effects of temperature

It is obvious that no single reaction scheme can display the biphasic rate–temperature curve that is characteristic of biological processes. A minimum of two reactions is necessary; one which accelerates the rate process in question and the other which decelerates the same process. However, because of the constraints of kinetic theory both reactions must increase in effect with increasing temperature. This idea was recognized by the earliest workers, who correctly assumed that the decrease in rate was caused by the destruction of the catalyst (enzyme). In general, enzymes are very thermolabile structures. This is because most, if not all, functional properties of biological macromolecules depend upon their tertiary and quaternary structures; features that are determined by the existence of large numbers of low-energy or non-covalent or 'weak' bonds between their constituent parts.

As their name suggests, weak bonds are characterized by the relatively small amounts of thermal energy that are necessary for their destruction. The energy of the strongest weak bond is only about an order of magnitude greater than that of the average thermal energy of molecules at 25°C (approximately 2.4 kJ mol^{-1}). The important role of non-covalent bonds in biology may be accounted for by the fact that special (i.e. catalytic?) mechanisms are not required for their formation and destruction (for a clear treatment of the role of non-covalent bonds in biology see Watson (1975). For example, enzyme activity requires a degree of flexibility of the three-dimensional configuration of the polypeptide, which would be impossible if its tertiary structure were determined by covalent bonds. Since the frequency curve of kinetic energy levels described previously (see Fig. 1.1) has an extended tail in the higher-energy values, many molecules with sufficient kinetic energy to break even the strongest weak bonds will exist even at physiological temperatures. Thus, individual weak bonds are continually broken and reformed, and have an average lifetime of fractions of a second. As temperature is increased the proportion of weak bonds that are broken at any one time increases exponentially to a point where the normal configurations of proteins, nucleic acids and membranes are disrupted and their 'life-supporting' functions break down, resulting in cellular damage and, ultimately, the death of the organism.

Enzymes are particularly labile structures and exposure to high temperatures usually causes a complete loss of activity. This inactivation process may be treated as an equilibrium between the active and inactive forms of the enzyme, and by applying the theory of absolute reaction rates it is possible to calculate the enthalpy and entropy of activation for the inactivation process. Both values are exceptionally high which, together with the large increases in molar volume, indicates the breaking of a large number of non-covalent bonds, such as hydrogen bonds, to produce an unfolded and disordered enzyme.

In principle, protein inactivation is reversible so long as the correct conditions of pH, temperature, ionic strength, etc., are provided. A particularly striking example is the nucleotide pyrophosphatase of *Proteus vulgaris* which may be completely inactivated by exposure at 70°C for 10 min but is completely reactivated on cooling to 37°C. In practice, however, the rate of renaturation under physiological conditions is usually negligible. The most familiar example of denaturation, that of egg albumin, has not yet been successfully reversed, and this has supported the general but mistaken view that the denaturation of proteins is irreversible.

The role of reversible enzyme inactivation in the form of rate–temperature curves was initially and most convincingly demonstrated for luminescence of the bacterium *Photobacterium phosphoreum* (Brown, Johnson and Marsland, 1942). This shows a typical intensity versus temperature graph (see Fig. 2.14) with an increase in luminescence with increasing temperature followed by a diminution at temperatures above the

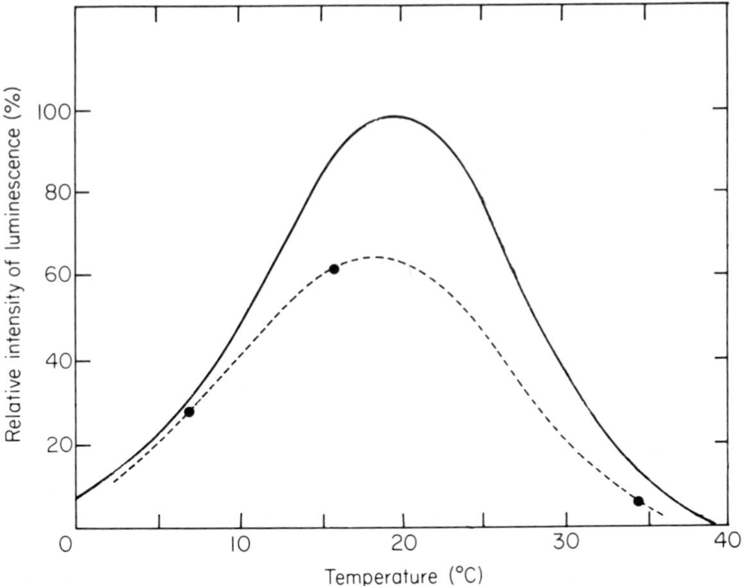

Figure 2.14 The rate–temperature curve for luminescence of the bacterium *Photobacterium phosphoreum*. The dotted line represents the near complete recovery of intensity after cooling the bacteria from 35°C, where it was almost totally inactivated. (After Brown, Johnson and Marsland, 1942.)

optimum. The points in Fig. 2.14 represent the intensity at 7°C and 15°C after a brief exposure to 35°C where intensity was but a small fraction of its optimal value. Although some irreversible destruction of the luminescence machinery was evident through the failure of luminescence to return completely to its former level after exposure to 35°C and cooling, it is clear that the inhibition at high temperature was largely reversible.

Complete reversibility depends on only very brief exposure to high temperatures. As a subject for study the luminescence of photobacteria is probably unique, in that its reaction rate can be measured instantaneously, so that the effects of extremely brief exposures to high temperature may be assessed. Other biological processes, such as growth, require much longer for rate measurements, so that irreversible inactivations are usually unavoidable.

To summarize, the net rate of an enzymatic process over the complete biological temperature range depends upon at least three reactions; first the catalytic reaction itself, secondly the reversible inactivation of the enzyme and thirdly the irreversible inactivation of the enzyme. The net rate due to the first two reactions can be described according to absolute reaction rate theory (Johnson *et al.*, 1974) by

$$k = \frac{cT\exp(-\Delta H^{\neq}/RT)}{1 + \exp(\Delta H_i/RT)\exp(\Delta S_i/R)} \tag{2.33}$$

where c is a constant, T the absolute temperature, and ΔH_i and ΔS_i are the change in heat and entropy, respectively, for the reversible inhibition reaction.

That simple enzymatic and some more complex biological processes conform to this theoretical treatment is convincingly shown in Fig. 2.15. The calculated curve for the intensity of luminescence of a luminous bacterium agrees tolerably well with the experimental data both above and below the maximum. Above 37°C and below 15°C the agreement is less good, and it seems that some other process(es) is involved. Similar impressive agreements have been obtained for the rate of replication of *E. coli*. In the latter case the fit is very close and it is likely that the process is, indeed, based upon two rate-limiting steps. In the former case the fit is close up to 45°C, but above this temperature the rate drops precipitously. In cases such as this the treatment is not entirely adequate and additional reaction schemes must be included. Unfortunately, this results in equations with so many variables that curve-fitting as a procedure to define the underlying mechanism loses its value.

The kinetics of the irreversible inactivation process can be studied in a similar way as for normal rate processes. For example, the loss of enzymatic activity at high temperature can be followed as a function of exposure time by assaying the residual activity (i.e. the activity remaining after exposure) at a non-inactivating temperature. A graph of log (residual activity) against exposure time usually yields a straight line, the slope of which provides a

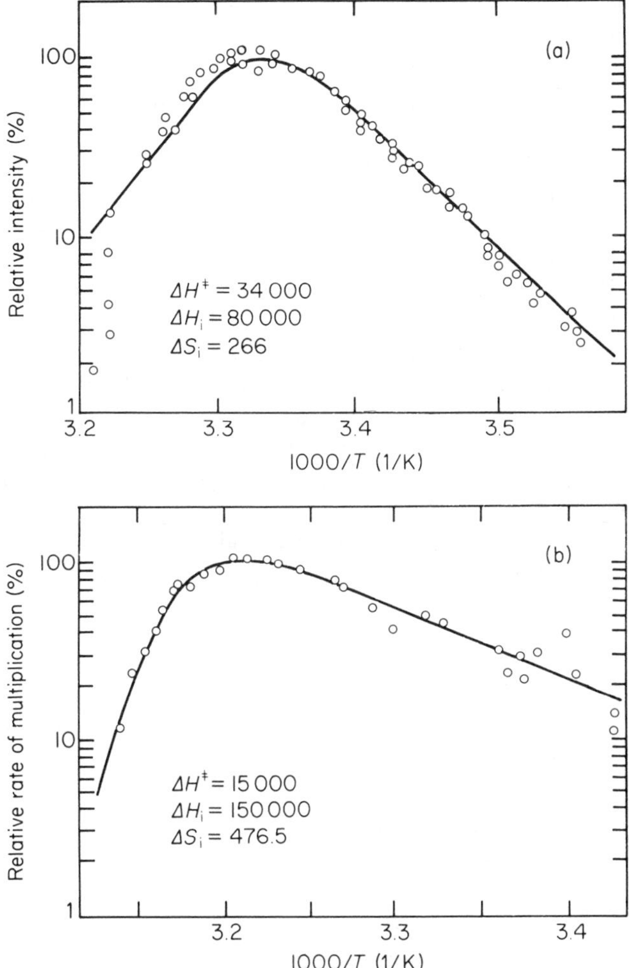

Figure 2.15 Arrhenius plots for (a) luminescence in the photobacterium *Achromobacter fisheri* and (b) for the rate of replication in *E. coli*. The smooth curves were calculated according to Equation 2.33 using the constants indicated. (After Johnson *et al.*, 1974.)

measure of the rate constant for inactivation. In some instances a biphasic graph is obtained, which suggests that at least two sequential or concurrent reactions are involved (see, for example, Cossins and Bowler (1976)).

2.8 Temperature optima of biological processes

The term 'optimum' has been used in two widely different senses. Sometimes it refers to conditions in which the animal is 'most comfortable', the criterion usually being such phenomena as longevity or animal condition. More usually it is defined as the condition at which a process proceeds at a maximal rate. Optimal temperatures for phenomena as diverse as enzymic reactions, longevity, intellectual performance and growth have been described in the literature. However, although there may be large differences between the thermal optima of different organisms which may be correlated with their natural habitat, thermal optima are usually of limited value for characterizing the temperature relationships of biological phenomena.

This is because at temperatures above the normal range where the optimum occurs the rate of a process is often not constant, but declines progressively with time as the destructive influences of temperature take effect. In the hypothetical example shown in Fig. 2.16 the rate remains constant with time at temperatures up to 25°C. At 30°C, however, although the rate is initially higher than at 25°C it slowly declines until it is less than at 20°C. At still higher temperatures the decline in rate is more dramatic. If the rate–temperature curve is constructed by taking rate measurements at time t_1, the optimal temperature would be about 40°C whilst at time t_2 the optimal temperature would be only 30°C. The optimal temperature

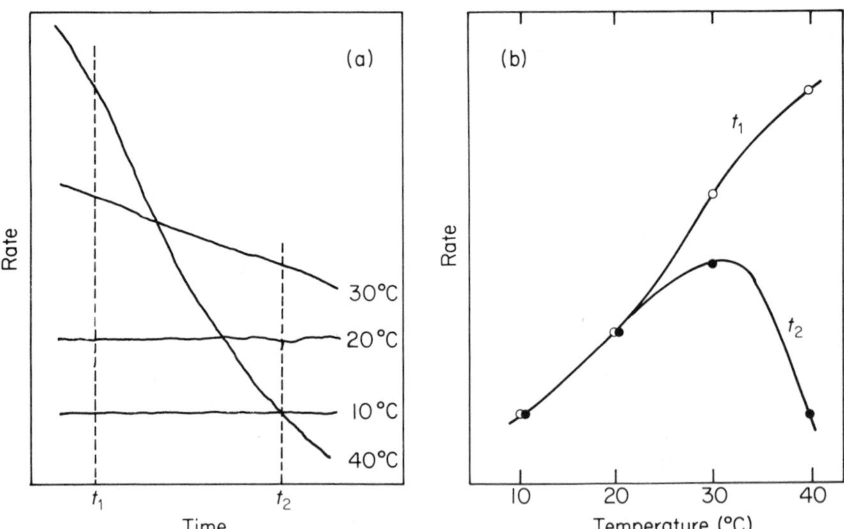

Figure 2.16 A hypothetical graph illustrating the effects of exposure at high temperatures upon the rate of a biological process. (a) The time-course of the rate at different temperatures. At the lower temperatures the rate is quite stable whilst at the higher temperatures the rate progressively declines as the damaging effects of temperature take effect. Rate measurements made at times t_1 and t_2 yield the rate–temperature curves shown in (b). The 'optimal' temperature so obtained decreases with exposure time.

therefore decreases with increasing exposure to the various temperatures, and the specific values of temperature optima have no special significance outside the experimental procedure with which they are determined. In addition, the optimal temperature for the different activities and processes of individual animals are not necessarily the same, so that in quoting an optimum the specific process for which it was determined must be mentioned.

There are some instances where thermal denaturation is not responsible for the reduction of rate at high temperatures and where the 'optimum' is a more meaningful concept. For example, the aerobic scope is the difference between the active and standard metabolic rates. At high temperatures both rates may continue to increase, but if the standard rate increases more rapidly than the active rate then the difference becomes progressively smaller.

Low Q_{10} and temperature-independent processes

The classical view of the effects of temperature upon biological processes is that metabolic rate has a Q_{10} of approximately 2 over the 'normal' range of temperatures. Nevertheless, a number of exceptions to this general rule have been reported. These range from small plateaux in an otherwise

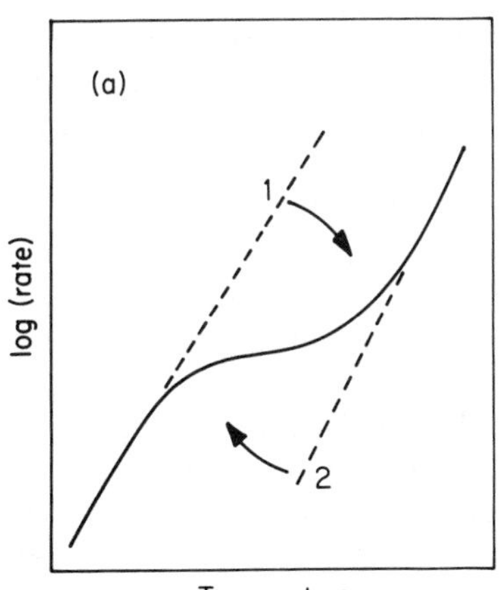

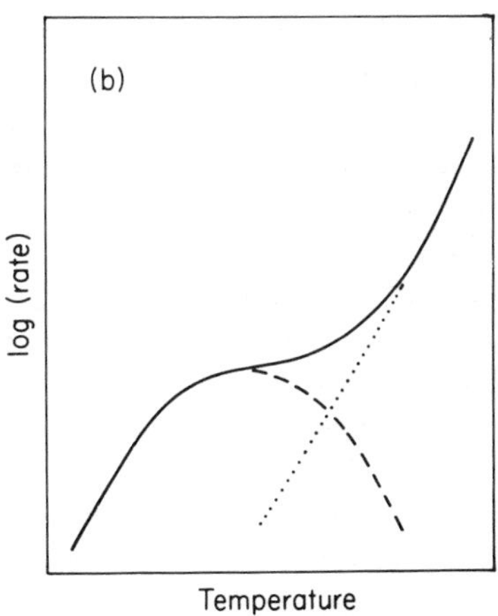

Figure 2.17 Diagrams illustrating the general form of the rate–temperature graphs in which plateaux occur and how they may be interpreted. (a) How this plateau can be produced by a change from one rate–temperature curve to another over a discrete range of temperatures. (b) An alternative idea of two separate rate–temperature curves which overlap to produce the plateau. (After Weiser, 1973.)

steadily ascending rate–temperature curve, to extended regions of temperature independence which occupy most of the normal temperature range.

The rate–temperature curve shown in Fig. 2.17 is typical of that found for aerobic metabolism in a number of studies (Weiser, 1973). The curve generally rises over the lower temperature range ($Q_{10} = 2$ to 3) and then flattens out ($Q_{10} = 1$ to 2) before increasing again at higher temperatures. The high Q_{10} encountered at low temperatures may enable a rapid acceleration of metabolism as temperature rises following a bout of cold torpor, and conversely, permits an equally rapid slow-down as temperature is reduced. The middle phase often occurs over the same temperature range as that when the animals are maximally active. Because the metabolic rate is effectively depressed compared with that expected by extrapolation of the lower curve, this can be viewed as a mechanism for moderating energy expenditure: a process termed 'metabolic homeostasis'. Alternatively, this plateau can be explained as the superimposition of two phases, a cold-active phase and a warm-active phase. The extent of the plateau depends upon the degree of overlap (Weiser, 1973). The former interpretation is supported by the fact that such curves are seen most distinctly in those animals that need to economize with their energy reserves.

Whilst this adaptive interpretation may be ecologically correct, an important question is whether these unusual curves result from departures of cellular and biochemical processes from the general principle of temperature dependence. Strict temperature dependence only occurs when the conditions in which the process takes place remain unchanged – as, for example, during a simple chemical reaction. In biological processes there are a number of potential sources of interference to this condition, ranging from cellular control processes (alternate metabolic pathways, feedforward or feedback inhibition) to systemic control influences (hormones, nervous and behavioural drive). Thus irregular rate–temperature curves are not, in themselves, incompatible with the Van't Hoff rule.

Unfortunately, most studies of metabolic rate do not successfully control the activity level in the experimental subjects, so that it is not possible to define unequivocally a rate for any particular process. Routine rate is generally measured which is largely under behavioural control, rather than a standard metabolic rate which more closely reflects the underlying physiological processes. In those few studies where activity level has been quantified and a standard rate has been estimated by extrapolation to zero activity, it is clear that standard metabolism does have a lower temperature dependence than active metabolism (Halcrow and Boyd, 1967).

A good example of this more general observation has been provided by Aleksiuk (1971) in the garter snake *Thamnophis sirtalis*. The rate of oxygen consumption in snakes obtained from a cool temperature region (Fig. 2.18) showed a normal logarithmic relationship with temperature above 15°C. However, below this temperature the relationship becomes non-linear.

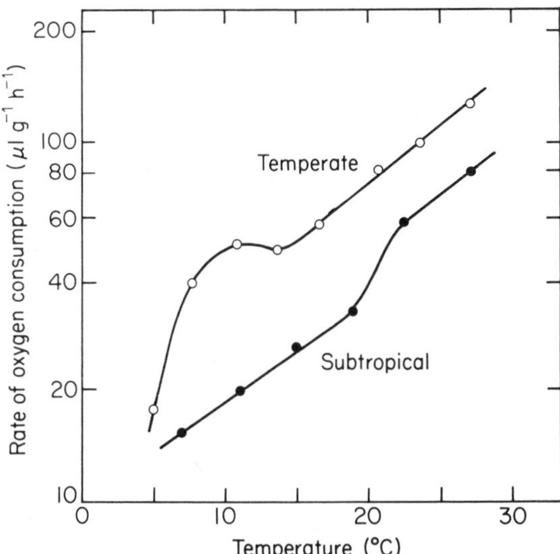

Figure 2.18 The rate–temperature curves for oxygen consumption in cool-temperate and subtropical populations of the garter snake *Thamnophis sirtalis*. The temperate population display an unusual plateau over the lower temperature range, which may be of adaptive significance. (After Aleksiuk, 1971.)

Oxygen consumption increases with a *decrease* in temperature between 15°C and 11°C, and then continues to decreases with further cooling. Aleksiuk (1971) interpreted this upward shift of the lower part of the curve as an instantaneous physiological compensation which offsets the depressive effect of lowered temperature upon metabolism. Thus the *in vivo* metabolic rate at 10°C is almost as high as that at 17°C. Below 10°C the Q_{10} is very high and promotes a profound depression in energy metabolism in the cold.

The adaptive interpretation offered by Aleksiuk was supported by two key observations. First, garter snakes from tropical populations showed a conventional rate–temperature curve (Fig. 2.18), thereby eliminating procedural artefacts. Secondly, the life-history of the cold-temperate population closely matched the laboratory experiments (Aleksiuk, 1976). Thus, body temperature is normally regulated at 29–30°C, but if this decreases activity is maintained down to 17°C. Below this temperature the animals seek shelter and become inactive, but digestion, growth and gestation are maintained. Below 10°C garter snakes hibernate.

At present the physiological basis of this response is not known although similar upward shifts in metabolic rate at low temperatures was consistently observed during experiments with isolated hearts and liver (Aleksiuk, 1971, 1976). If correct, this represents a truly remarkable physiological adaptation which is associated with life in a cool climate where the opportunity for normal thermoregulatory behaviour is strictly limited.

A much more extreme case is provided by sedentary intertidal invertebrates which suffer particularly rapid and large fluctuations in body temperature with the tidal cycle. Newell (1966) and Newell and Northcroft (1967) have noted that the rate of oxygen consumption in some intertidal animals is often not constant, but alternates between at least two phases. In one phase, when the animal shows active respiratory movements and locomotion the rate of oxygen consumption showed a Q_{10} of approximately 2. In the second phase, in which oxygen consumption was extremely low and during which the animals appear quiescent, this rate was largely

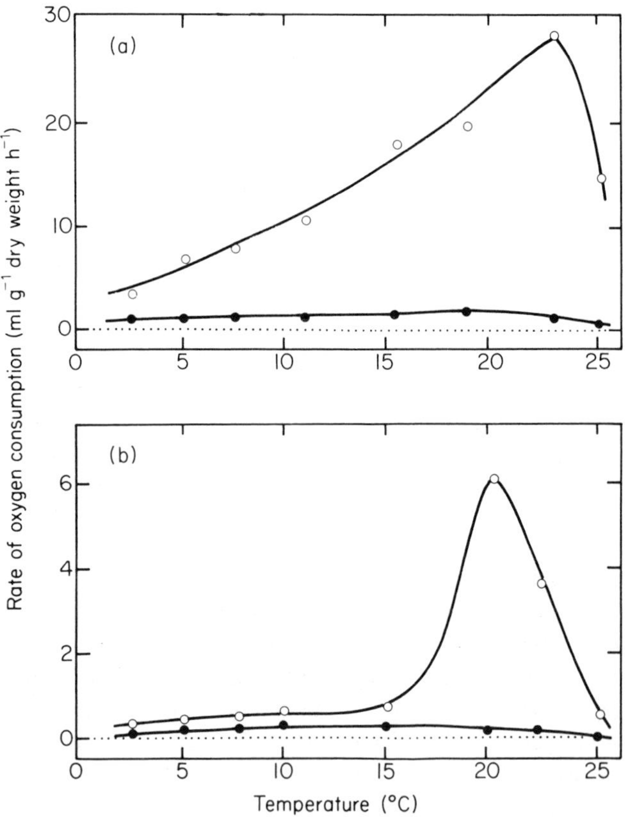

Figure 2.19 The rate–temperature curves for oxygen consumption of (a) the intertidal anemone *Actinia equina* and (b) the winkle *Littorina littorea*. Values at each temperature were estimated from graphs of rate against weight for animals of a given weight. The open symbols represent so-called 'active' respiration and the closed symbols represent 'standard' respiration. Note the extremely low levels of 'standard' respiration. (After Newell and Northcroft, 1967.)

independent of temperature (Fig. 2.19). Newell and Northcroft (1967) identified the 'quiescent' rate with maintenance metabolism and, in calling for reinterpretation of normal temperature dependence, concluded that this was held constant despite rapid fluctuations in temperature. The adaptive potential of this property is obvious, particularly as such short-term fluctuations in rate could not be mitigated by the more long-term process of acclimatization. Significantly, the temperature range over which temperature independence was observed was not fixed, but varied during different seasons of the year to match the range of environmental temperatures experienced by the animals (Newell and Pye, 1970) or during which gonadal growth occurred (Barnes and Barnes, 1969).

Although this interpretation is teleologically satisfying and has gained a degree of acceptance, it is by no means unequivocal (Tribe and Bowler, 1968; Davies and Tribe, 1969). The controversy rests on the precise physiological status of the animals during each of the respiratory phases, and the relevance of laboratory measurements to field conditions. Intertidal invertebrates normally sustain periods of anaerobic respiration, particularly during exposure to air, and the oxygen debt must subsequently be repaid. This interdependence of aerobic and anaerobic respiration make the separation of respiration into sequential phases highly questionable (de Zwann and Wijsman, 1976). Thus, the so-called 'quiescent' phase of Newell and Northcroft (1967) probably represents a period of more intense anaerobic respiration, particularly as the very low rates of oxygen consumption recorded during this phase may be insufficient even for maintenance purposes. Indeed, the factorial aerobic scopes (i.e. ratio of active to standard rates) observed in these studies (up to 20) are significantly higher than is generally observed (2–5).

The use of minimal respiratory rate as a measure of standard metabolism is quite arbitrary, and thus is suspect. In any case, for many poikilothermic invertebrates there is no 'basal' metabolic rate as the term is conventionally understood in vertebrates (McMahon and Russell-Hunter, 1977). During times of exposure seaside intertidal gastropods employ predominantly anaerobic respiration whilst withdrawn into their shells. At other times feeding and respiratory mechanisms are continuously active, and the gastropods show continuous and slow locomotion. This routine rate is thus equivalent to the standard rate.

In view of these severe interpretative problems it is surely premature to abandon the conventional ideas of the rate effects of temperature upon the biochemical processes in intertidal invertebrates. The apparent temperature independence observed by Newell may be explained by the fact that oxygen consumption in the quiescent phases bore no relationship to standard metabolism. However, this is not to say that the greatly reduced rate of oxygen consumption during the quiescent phase is without adaptive significance, since it undoubtedly leads to considerable saving of food resources and to protection from desiccation during exposure. However, to claim that the basic biochemical processes in these animals possess a large degree of temperature independence is not fully justified on the available evidence.

The precise interpretation of non-standard rate–temperature curves is often extremely complex and we should not uncritically accept plateaux as evidence of homoeostatic responses. A good example is shown in Fig. 2.20, which is a particularly complex rate–temperature curve for mitochondrial respiration of blowfly flight muscles (Davison, 1971). The curve for ADP-stimulated respiration may be conveniently divided into four phases: (1) at low temperatures the rate increases dramatically with increasing temperature whilst (2) over the middle range of temperatures there is virtually no change in rate with temperature; (3) at slightly higher temperatures the rate declines but (4) at the highest temperatures the rate again increases dramatically. It is commonly concluded that phase 2 is a cellular example of

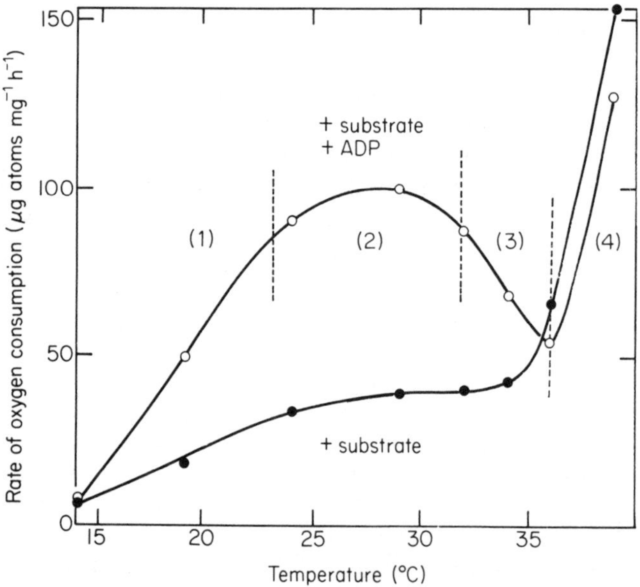

Figure 2.20 The complex rate–temperature curves for isolated mitochondria of blowfly flight muscles. Respiration in the presence of both substrate and ADP shows a complex form which may be interpreted as a plateau of adaptive value. However, there is a clear decline at higher temperatures (i.e. during phase 3) in the ability of ADP to stimulate respiration above that observed in the presence of substrate alone. This stimulation by ADP is a clear indicator of mitochondrial integrity, which is apparently severely disrupted at high temperatures. The plateau in phase 2 thus represents an 'optimum' rather than temperature independence. (After Davison, 1971.)

the homeostatic mechanism described in intertidal organisms by Newell and his colleagues. However, the decline in activity during phase 3 is associated with a dramatic reduction in the operating efficiency and integrity of the mitochondria, since the rate of oxygen consumption is no longer stimulated by addition of ADP in the normal way. Phases 1, 2 and 3 therefore represent the normal biphasic rate–temperature curve for ADP-stimulated mitochondrial respiration whilst phase 4 represents the dramatic temperature-dependent increase in respiration that occurs in the absence of ADP. Phase 2 cannot be considered to be any more independent of temperature than any process that shows an optimum. At the sub-organism level of organization there are very few unequivocal examples of processes which possess any degree of temperature independence. Newell and his colleagues have claimed that the mitochondria of intertidal organisms and other animals possess complete temperature independence over a wide range of temperatures, particularly when substrate concentration is low. This has not been confirmed by other studies and the original techniques of Newell have been subject to criticism (Tribe and Bowler, 1968; Davies and Tribe, 1969).

2.10 Conclusions

The mathematical analysis of rate–temperature curves is useful for two reasons. First, the derivation of temperature coefficients aids the quantitative description of the relationship and enables predictive calculation at temperatures other than those at which the process was measured. However, from a mathematical point of view, biological processes are rather diverse. Whilst one formula may hold reasonably well for one process, such as rate of development or rate of heart beat, it usually does not hold for many others. Thus, empirical formulae are rather limited in their application, and may even fail to apply with exactness to every example of a specific process. This is not surprising since these processes are highly complex, with diverse physical and chemical components interacting to give the observed relationship. Indeed, so complex are such processes as development, that it is surprising that they can be described at all in simple mathematical terms. The most useful general statement remains the Van't Hoff approximation of $Q_{10} = 2 - 3$. Virtually all exceptions to this so-called rule involve the regulatory control of the process as exemplified by the plateau in the rate–temperature graphs for routine oxygen consumption rather than any reduction in the fundamental temperature-dependence of the underlying biochemical processes. Perhaps the most convincing temperature-independent processes are the periodicity of biological rhythms or biological 'clocks'. Clearly, a $Q_{10} = 1$ is an absoulte necessity if the clock is to have any usefulness.

The second use of rate temperature analysis is to provide an understanding of the phenomena which underlie the process in question. Here, the theoretical formulations have proven to be of limited use to biologists. The equations provided by Arrhenius and by modern rate theory apply with rigour only to simple chemical reactions, and to enzymatic processes where a well-defined, rate-limiting step may be expected. For more complex processes one might expect that many reaction steps will influence the overall rate, perhaps over different parts of the normal temperature range. Nevertheless, the rates of some complex processes agree well with the Arrhenius equation, at least over a limited range of temperatures, and this may justify a belief in a single rate-limiting step. In principle, this rate-limiting step may be identified by the similarity of its activation energy with that of the overall process – though because of the variability of the activation energy of even simple processes with variations in experimental conditions, this has not been realized in any practical way.

3 Body temperature in bradymetabolic animals

Heat exchange between an animal and its environment obeys the physical laws of heat transfer, that is by radiation, convection, conduction and by the evaporation of water. These exchanges are complex, and most animals have evolved behavioural and physiological mechanisms to exert control over them. The relative importance of the various physiological and physical factors involved in the establishment of an animal's body temperature (T_b) are not easy to understand or quantify. This is because microclimate conditions, and the responses of the animal, are complex and subject to unpredictable variation.

Physical changes in environmental conditions as well as changes in physiological processes will modify not only the extent, but also the pattern of heat exchange with the animal, if the T_b is to remain constant with time, then the rate of heat gain must balance the rate of heat loss. The heat gained from metabolism always makes a positive contribution to T_b, although exchanges by conduction, convection and radiation may either add to or take heat away from the body. The likely exchanges of heat between the animal and its environment are summarized in Fig. 1.3. Evaporative heat loss is, of course, only available to terrestrial animals.

Some animals, notably birds and mammals, have evolved complex regulatory mechanisms to stabilize T_b, at a specific core temperature, depending on species, between 37°C and 41°C. However, most other animals have no such physiological control over T_b. Nevertheless, they must still function in a variety of environments and display the full range of their biological activities.

Animals have been divided by physiologists into two groups. Those cold to the touch were called cold-blooded, and those warm to touch were called warm-blooded. These terms are subjective and unsatisfactory, as they provide no indication of the source or method of maintenance of body temperature. The terms *poikilothermic* and *homoiothermic* are more usually applied to describe those two groups, respectively. Both terms have their roots in Greek, *Poikilo-* means various or many, and is applied to those animals whose body temperatures are labile and follow changes in ambient temperature, T_a (thermoconformers). It should be stressed that T_b need not equal T_a, but it will vary with it. *Homoio-* means to become alike, and is used to describe those animals that maintain a relatively constant temperature even though T_a might vary (thermoregulators). These two terms have a utility, and the latter is generally confined to the members of the two taxonomic groups, the birds and mammals, in which homoiothermy is maintained during the resting state.

More recently Cowles (1962) has proposed a different set of terms, based upon the primary source of heat used to establish T_b, and which emphasize the thermoregulatory properties of the animal. This new terminology arose because of the obvious inadequacy of existing terms when considering, for example, the physiological ecology of reptiles. Some reptiles, in field conditions, have a fairly constant elevated T_b. They achieve this by basking in the sun and from shuttling between sun and shade, and so did not satisfy the definition, poikilothermic.

The terms introduced by Cowles (1962) were *ectotherm* for animals whose T_b is determined by heat sources external to the body (i.e. solar radiation or a warm substrate) and *endotherm* for animals whose T_b is determined by heat derived from cellular metabolism. This terminology has became widely used owing to the greater appreciation of the wide diversity of thermoregulatory strategies that have been adopted within the animal kingdom.

Ectotherms are characterized by a relatively low level of resting metabolism (*bradymetabolism*) so heat production is low, and this is coupled with a high thermal conductance. As a consequence, metabolic heat is rapidly lost to the environment and T_b is determined by the thermal properties of the environment.

Cowles (1962) also separated ectotherms into two rather distinct groups according to the external heat source used. *Heliotherms* are those animals which rely upon basking in sunlight for thermoregulation, whilst *thigmotherms* derive their body temperature solely from the conductive medium in which they live. Thermoregulation in the latter group, which largely consists of aquatic and fossorial animals, is restricted to movement to preferred water or soil temperatures or away from excessive temperatures. Heliotherms, by contrast, are mainly terrestrial forms in which the main conductive medium is air. Conductive heat exchange is consequently reduced and solar radiation is able to exert a dominant role. As we shall see, postural control and shuttling between the shade and isolated areas can allow these animals to control T_b with remarkable precision. In that heliothermy requires a rapid conduction of heat from the integument to the rest of the body, then insulation at the body surface is clearly not appropriate.

Endotherms have a relatively high level of metabolic heat production associated with mechanisms for its conservation and this results in a T_b

elevated above T_a. This state is at its most sophisticated in the birds and mammals. These animals display high rates of basal metabolism (tachymetabolism) coupled with a low thermal conductance, and significantly they have developed complex neural mechanisms that are involved in the regulation of heat gain and heat loss thereby ensuring the stability of T_b.

The utility of Cowles' scheme has the emphasis placed on the energetics of each thermal strategy for this has profound implications for life-style, energetic status, life-cycle strategy etc. for the organism.

The interpretation of this terminology is compounded by those species (e.g. moths) that display ectothermy at rest but are endothermic in flight. Thus, thermoregulatory terminology has become increasingly difficult to apply, this is because our appreciation of the variety and complexity of thermoregulatory relationships displayed by animals has not been matched by the development of unambiguous semantics.

No natural taxonomic division exists between ecto- and endothermic organisms. Most invertebrates, fish, amphibians and reptiles are ectotherms, but endothermy is not restricted to birds and mammals; insects, fish and reptiles display endothermy. The terminology in this book follows that recommended by Bligh and Johnson (1973).

3.1 Thermal inertia – the contribution of body size to the stabilization of T_b

Animals of large body weight are slow to warm and cool. This is because tissues have a low thermal diffusivity. In fact, heat is only effectively distributed through the body of an animal by the circulatory system. Large body mass helps to stabilize core temperatures and this *thermal inertia* may allow metabolic heat to contribute significantly to T_b, even in animals with a low rate of heat production.

The simplest model for thermal exchange between an animal and its environment is for an animal that cannot use metabolic heat to contribute to its T_b and can be expressed as

$$\frac{dT_b}{dt} = \frac{hA}{Mc}(T_a - T_b) \tag{3.1}$$

where h is the thermal conductance ($W\,m^{-2}\,°C^{-1}$) of the insulating layer (integument), A is the surface area of the body and M its mass, T_b is the temperature of the core and c is the specific heat of the body ($J\,g^{-1}\,°C^{-1}$). If the animal is placed at an ambient temperature lower than T_b The change in core temperature can be followed with time. This can be plotted in the form of a cooling curve, of log $(T_b - T_a)$ against time, and a straight-line plot should be obtained with a slope of $-0.4343hA/Mc$, from which a value for the thermal conductance of the animal can be derived.

However, this simple analysis of cooling curves for animals makes a number of assumptions. First, it assumes the environment to be isothermal with an infinite heat capacity, but it is known to be more complex with conductive, convective and radiative factors involved in setting the environmental temperature of the animal. Secondly, a single value for h may not be applicable because heat transfer also depends upon environmental factors, such as wind speed. Thirdly, the end-point of the cooling curve may not be T_a, but rather the equilibrium temperature, as metabolic heat production or evaporative cooling may contribute to T_b.

Bakken (1976) has considered this complex problem and suggested that the following relationship describes the heat transfer of an ectothermal animal better:

$$\frac{dT_b}{dt} = \frac{-1}{\tau} T_b - \left[\left(T_e + \frac{H_m}{k_0} \right) \right] \tag{3.2}$$

where H_m is the net metabolic heat production (i.e. heat production minus evaporative heat loss), T_e is the operative environmental temperature, k_0 is the overall thermal conductance and τ is the time constant equal to the time required for 63% $(1 - 1/e)$ of the total response to a sudden temperature change to take place.

The concept of operative temperature scales was introduced by Gagge (1940). It is an important idea because it represents, in a single measure, the thermal environment experienced by an animal. The value used is obtained by deriving coefficients that relate radiant and convective heat exchanges to mean skin temperature and mean radiative and air temperatures. It contains no factor representing heat transfer by evaporation. It is a complex term that includes both organism and environmental components of heat transfer.

Bakken (1976) widened the use of this concept to include conductive heat-transfer coefficients to the ground. This generalized environmental temperature for an animal is called the *operative environmental temperature*. The difference between T_b and T_e is the net thermal gradient between the animal and its environment. Thus, if the animal produced no heat internally and lost no heat by evaporation, then the body would equilibrate with the environment and T_b would equal T_e. Indeed, this is the theoretical basis for the use of physical models to measure T_e. Walsberg and Weathers (1986) have recently compared the sensitivity and accuracy of the use of taxidermic mounts with painted spheres in the estimation of T_e.

Figure 3.1 shows diagrammatically a plot of log $(T_b - T_{air})$ as a function of time when T_b^{eq} equals and differs from T_{air}. When metabolic heat production or evaporative cooling contributes to the heat transfer, the plot deviates from the expected straight line. The time constant can be obtained from this plot, as the slope equals $-0.4343/\tau$. The thermal time constant is a

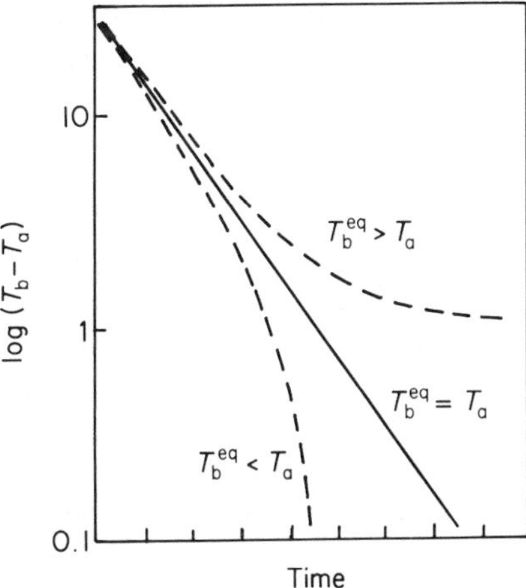

Figure 3.1 Plots of log ($T_b - T_{air}$) as a function of time when equilibrium body temperature (T_b^{eq}) is equal to, greater than and less than T_{air}. When metabolic heat production or evaporative cooling contribute to T_b the plot deviates from a straight line. In these cases the curvature of the plots prevent accurate determination of the time constant (τ) and overall heat transfer coefficient. (After Bakken, 1976.)

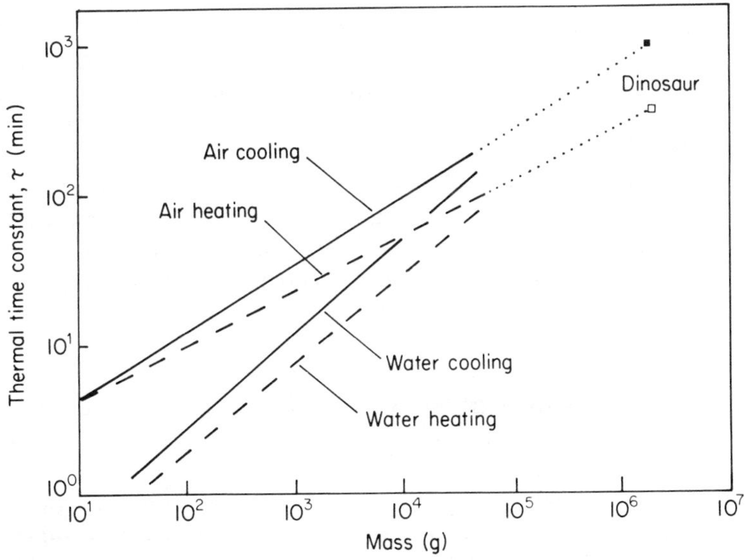

Figure 3.2 Relationships between the thermal time constant (τ) and size in lizard shaped reptiles. The time constants scale with mass over four orders of magnitude, for both heating and cooling, in water and in air. Lizards warm more quickly than they cool. Extrapolation of the curves in air to the mass predicted for large dinosaurs suggest that the mass of these animals would confer considerable thermal stability of T_b. (After Grigg *et al.*, 1979.)

particularly useful concept, as it is independent of the magnitude of the change in environmental temperature step.

Most of the practical examples available have been derived from work on reptiles and Fig. 3.2 shows that the value of the thermal time constant is very dependent upon mass. The value for a 10 g lizard cooling in air is only about 4 min, but for a 100 g lizard this lengthens to 30 min. Very small animals, such as an insect, would have values of the order of seconds rather than minutes. The greater thermal inertia of large lizards contributes considerably to their stable T_b. Another point of interest is that the slope of the curve obtained for heating is less than that obtained for cooling, both in water and air. This fits with the general observation that lizards warm quicker than they cool, and this implies that physiological control over heating and cooling takes place, presumably a function of the cardio-vascular system.

The thermal time constants for heating and cooling scale with mass over four orders of magnitude, and so provide an excellent predictive relationship. It is interesting then to extrapolate this relationship to the 2–3-tonne mass reported for dinosaurs when τ for cooling would be 15 h and for heating 6.3 h, but for a 20-tonne dinosaur the calculated values would be about 50 h and 15 h, respectively. Such cases suggest a considerable thermal stability in T_b would be achieved simply as a consequence of mass, so for these animals, even if they were bradymetabolic, the dissipation of metabolic heat would be a greater thermal problem than heat conservation.

When significant metabolic heat production occurs, and is conserved so that the equilibrium T_b is above T_a, the difference in temperature is usually referred to as the excess body temperature (T_x).

In animals that regulate T_b thermal relationships are very complex, as the animal is not a passive agent. For example, T_x may vary with T_a, but also short- and long-term physiological adjustments may result in a change in the value of thermal conductance, and heat production may also vary with T_a.

3.2 Selection of a preferred body temperature in ectotherms – an adaptive behavioural response

Ectotherms which experience seasonally varying temperatures have a number of strategies available to them. One physiological mechanism is to change cellular physiology and biochemistry in a process called acclimatization. As is explained in Chapter 5, a low-temperature-acclimatized animal generally has a higher metabolic rate than the same animal when it is warm-acclimatized. Such adjustments are thought to compensate for the direct effects of a low T_a on metabolism. Acclimatization is time dependent, taking days or weeks to achieve, and so is less useful to animals that experience large daily fluctuations in temperature, rather than seasonal changes.

The ability to select a particular T_a, suitable for the maintenance of an appropriate T_b, is another strategy available. This form of adaptive behaviour is shown by many ectotherms, and can be readily demonstrated in the laboratory. Several methods have been commonly used. They rely on allowing the animals under test to show a preference when given a choice of environmental temperatures. The simplest system is to establish a gradient of temperatures, hot at one end and cold at the other. The animals are introduced into the gradient and their distribution is recorded at intervals. Precautions must be taken to ensure that the floor of the gradient is not hotter or colder than the medium above, and that the animals could not be responding to other non-thermal gradients such as light, humidity or oxygen tension. The temperature where the animals are most frequently observed is usually designated as the *preferred temperature*. Such an experiment was carried out by Norris (1963) to determine the temperature selected by *Girella nigrans*, and the result is shown in Fig. 3.3.

The second method uses a simple choice chamber with two different temperatures to select. A series of such tests permits the preferred temperature to be determined, usually not as a single temperature, but rather as a range of temperatures within which the escape response does not occur. The advantage of this method is that the temperatures used are easier to control accurately, and so are easier to reproduce.

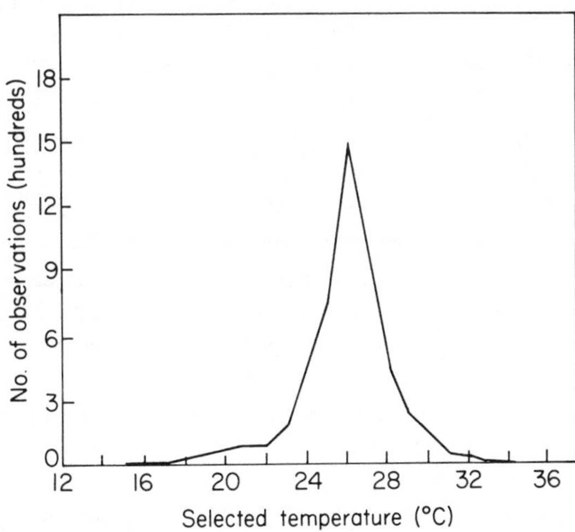

Figure 3.3 The selected temperature of 39 unacclimatized *Girella nigrans*. Fish were placed in a thermal gradient and, after a period of adjoinment to the chamber, recordings were made of the position of the fish every 6 s. The fish selected temperatures close to 26°C and strongly avoided temperatures only 2–3°C away from the mean selected temperature. (After Norris, 1963.)

The results obtained from laboratory gradient and choice-chamber experiments suggest that temperature selection is a widespread behavioural response to environmental temperature. Magnuson, Crowder and Medvick (1979) have discussed this in relationship to exploitation of the habitat, and it is clear that temperature selection is a significant adaptive strategy.

Laboratory studies show that a common feature of gradient studies is that the distribution of animals is usually skewed, with a marked tail-off towards the cold and a sharp cut-off at the warm ends of the gradient. It is likely that this reflects the behavioural avoidance of potentially damaging temperatures at the hot end, but at the cold end the skew may result from cold-suppression of locomotory activity.

In some species the preferred temperature is not fixed, but can be modified by acclimation or is shown to vary seasonally. Norris (1963) showed that intertidal fish *Girella* when acclimated to 11°C selected 13.7°C, but when acclimated to 23°C selected 20.7°C. The situation is complicated, in that the fish choose 26°C after exposure to naturally fluctuating conditions, a higher preferred temperature than can be obtained following acclimation in the laboratory.

Reynolds and Casterlin (1979) have shown that bluegill sunfish in a gradient will gradually change their selection to increasingly higher temperatures, and so move towards a '*unique preferred temperature*'. This gradual upwards shift in preferred temperature, to a temperature only just below the upper lethal limit, clearly has adaptive value. Movement of a cold-acclimated animal directly to such a high temperature may well prove lethal, but the gradual change allows some acclimation to occur. Presumably the unique preferred temperature is selected as it offers 'optimal' conditions for function.

Shuttle-box experiments have also been used with fish to determine their discriminatory ability with respect to temperature. In the hot side the water is heated until the fish leave for the cool side, which is then cooled until the fish leave for the warm chamber again. Turn-around temperatures as narrow as 4–5°C have been recorded in such experiments. In these experiments the preferred temperatures are those in which the fish do not receive thermal aversive stimuli. This was demonstrated in a remarkable experiment using goldfish (Rozin and Mayer, 1961). Fish were trained to cause a small fall in temperature of a heated water bath by pressing a lever to allow the entry of cold water. They stopped working the lever when the temperature fell below 33.5°C. In this way they were able to maintain their T_a between 33.5°C and 36.5°C for most of the time. These temperatures are several degrees higher than those selected by the fish in a gradient, and Rozin and Mayer suggest that the range 33.5–36.5°C is the highest in which the fish are comfortable and receive no thermal aversive stimuli. The highest temperature accepted by the fish is also called *maximum voluntary tolerated temperature* (MVT). This kind of thermoregulatory behaviour in aquatic ectotherms suggests that a high degree of thermal constancy of body temperature may be achieved in nature. The interpretation of behavioural experiments of this type has been discussed in a valuable article by Roberts (1979), which suggests behavioural thermoregulation in many ectotherms consists only of avoidance activity.

The terrestrial environment presents a greater variety of thermal niches, so terrestrial ectotherms are able to take greater advantage of environmental heat exchange to stabilize T_b than aquatic ectotherms. Insects and reptiles are the most widely studied, and preferred temperatures in insects are known to be affected by such factors as previous thermal history, stage in the life history and state of nutrition. Thermal preferenda in those terrestrial species that have been studied closely correlate with habitat temperatures (May, 1979), and consequently also correlate with measured field T_b when this is known (Cloudsley-Thompson, 1971), (see Fig. 3.4).

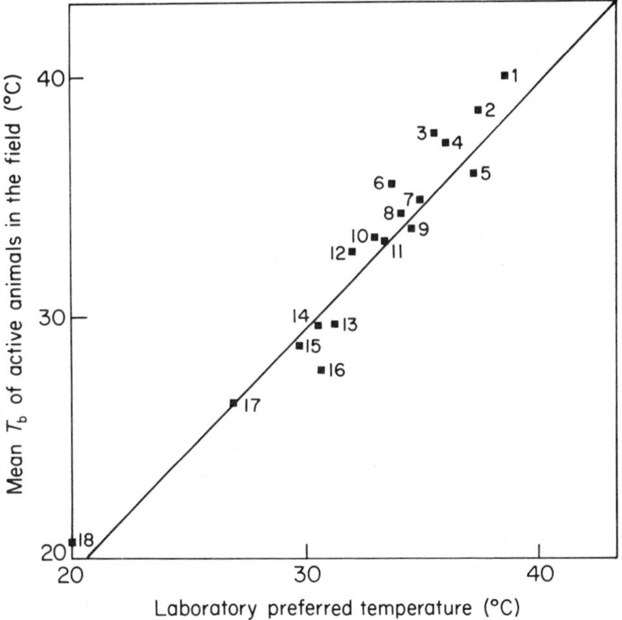

Figure 3.4 Relationship between preferred body temperature and observed body temperature of reptiles active in the field. In general a close agreement is found between laboratory preferred T_b and field T_b. (Data from Cloudsley-Thompson, 1971; Avery, 1982.) 1, *Dipsosaurus dorsalis*; 2, *Uma notata*; 3, *Sceloporous graciosus*; 4, *Varanus griseus*; 5, *Agama stellio*; 6 *Sceloporous occidentalis*; 7, *Lacerta sicula*; 8, *Eumeces obsoletus*; 9, *Conopholus pallidus*; 10, *Eumeces fasciatus*; 11, *Lacerta muralis*; 12, *Amblyrhynchus cristatus*; 13, *Coluber constrictor*; 14, *Lacerta vivipara*; 15, *Crotalus vividis*; 16, *Heloderma suspectum*; 17, *Pituophis catanifer*; 18, *Anniella pulchra*.

In spite of the many laboratory studies on the behavioural selection of preferred temperature, little is understood of the mechanisms involved. In particular, the question arises of how thermal stimuli are monitored and transduced into a locomotory movement to another temperature. The work on *Girella* implies that a thermostat of sorts exists in fish, and that it can be reset by a change in acclimation temperature. Both the temperature of the brain stem and that of the peripheral receptors are important in the thermoregulatory responses in fish. This is shown by lesions to the preoptic region of the brain which disrupt the response, and also that selective heating of the brain leads to the fish choosing a lower ambient temperature in a gradient. It is clear that some central nervous system integration of sensory information is involved in the activation of movement to an alternative set of thermal conditions (Prosser and Nelson, 1981).

3.3 Body temperature in field conditions

Information gained from laboratory experiments in gradients can provide only limited information about thermoregulatory powers in nature. This is because the ability to stabilize T_b is dependent not only upon the capacity of the animal to make physiological adjustments and/or behavioural responses, but also upon the actual thermal conditions obtaining in the environment.

Owing to the ease of measurement, more is known about terrestrial than aquatic ectotherms in field conditions. In insects the principal methods used are postural control of warming by solar radiation, microclimate selection and circadian activity patterns (May, 1979). Many species of grasshoppers and locusts climb vegetation and bask in the early morning sun, orientating their bodies perpendicular to the sun's rays to maximize heating. When warm enough they may face the sun to shade much of their body and avoid overheating. In the hottest part of the day migration into the shade may well occur. Evidence from butterflies, dragonflies, tenebrionid beetles and locusts suggest that this behaviour is widespread and effective in stabilizing T_b.

Edney (1971) has studied thermoregulation in the desert beetle *Onymacris rugatipennis*, which thermoregulates by shuttling into and out of the shade. Relatively few other insects are reported to use this behaviour, probably because of their small size and consequent low thermal inertia, which would make such behaviour inefficient. This beetle was found to have a T_b about 12°C higher than the air temperature. A similar beetle, *O. brincki*, displays an interesting dependence of T_b on the surface reflectivity. The black thorax was some 3–4°C higher than the white abdomen.

In some habitats the animals may be subjected to dangerously high levels of insolation. The surface of tropical deserts is a good case in point, and desert invertebrates face a severe thermal problem. A variety of different

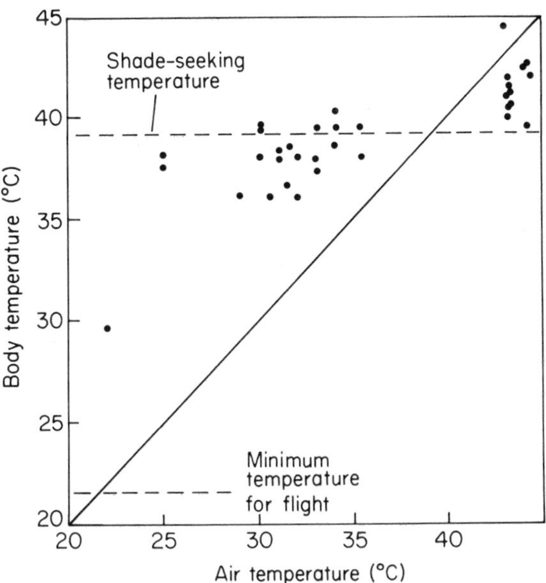

Figure 3.5 Body temperature of the desert cicada *Diceroprocta apache* as a function of field air temperature. Insects maintain a T_b of close to 40°C by basking at low air temperature, but when it exceeds 40°C they move into shade. (After Heath and Wilkin, 1970.)

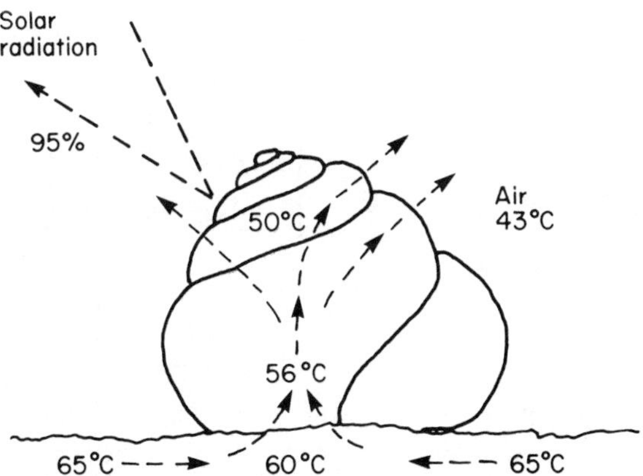

Figure 3.6 Temperature of a dormant *Sphincterochila* resting on a loess surface. Two factors are important in preventing the snail from reaching lethal temperatures. The high reflectance of the shell to solar radiation and convective exchange with the air, and the retention of an air pocket in the large whorl of the shell, which insulates against heat gain from the substratum. (After Schmidt-Nielsen *et al.*, 1972.)

adaptations have been evolved to avoid overheating, and these are nicely introduced in a book on desert organisms by Louw and Seely (1982).

The cicada of the Arizona desert has been studied by Heath and Wilkin (1970). This insect maintains a body temperature close to 39°C, over a range of air temperatures, by microclimate selection (see Fig. 3.5). They select perches in the cooler boundary layer on the shaded sides of stems, and in this way they can sing when conditions are too extreme for predators. The desert snail *Sphincterochila boisseri*, studied by Schmidt-Nielsen *et al.* (1972), has a more acute problem in that it is found dormant exposed to the full solar radiation in the Negev desert. Figure 3.6 shows that the substratum on which the snail is found may be at 60°C, however the snail's tissues are never above 50°C. This is probably because the shell has a high reflectance and some 90% of incident radiation is not absorbed. Heat gain from the substratum is limited as the snail withdraws into the upper whorls

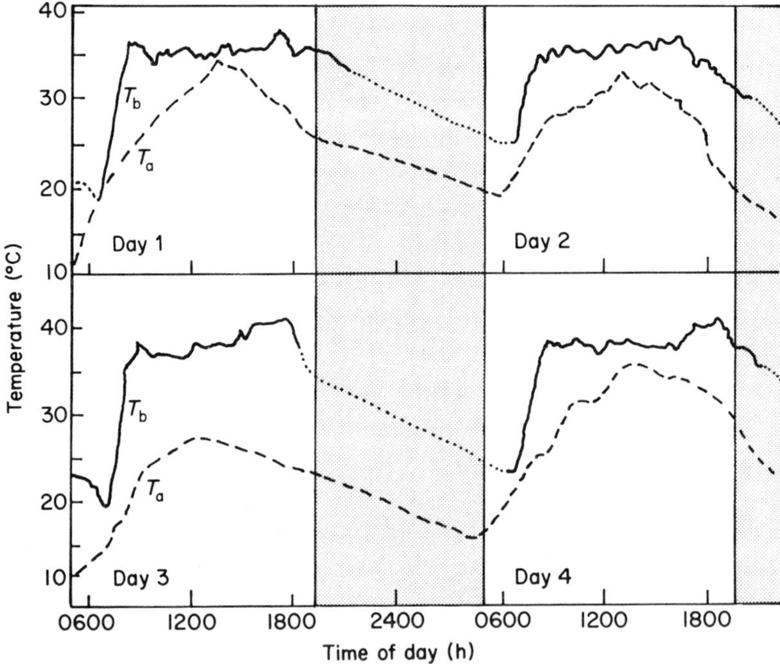

Figure 3.7 Body temperature of a free-ranging monitor lizard (*Varanus varius*) in its natural habitat. Body temperature was monitored by radiotelemetry (——). Air temperature was measured 5 cm above the ground (– – –). Times when no records were taken are indicated (..............). T_b rises quickly to the regulated level (35°C) by basking when the lizard leaves its overnight hide in a hollow tree. At night T_b falls slowly in the absence of solar radiation. Note how T_b is regulated to the same level on four successive days even though T_{air} is markedly different. (After Stebbins and Barwick, 1968.)

leaving an insulating air space in the lowest whorl of the shell. Heat is lost by convective cooling as air temperature is only 43°C, and this would balance the heat gain by radiation and conduction.

By far the best-documented thermoregulatory behaviour is that displayed by reptiles. In the absence of a radiant heat source a reptile behaves as a true poikilotherm, with T_b closely following T_a. The same animal is distinctly different in the presence of radiant heat, and it will maintain a fairly constant T_b. Figure 3.7 shows this clearly for a monitor lizard. During the day the lizard maintains its T_b between 34°C and 37°C irrespective of air temperature. T_b falls with air temperature at night but lags behind it, largely owing to the thermal inertia of this large lizard. Figure 3.7 also shows that heating and cooling rates are very different, an expression of the different values obtained for the time constants is shown in Fig. 3.2.

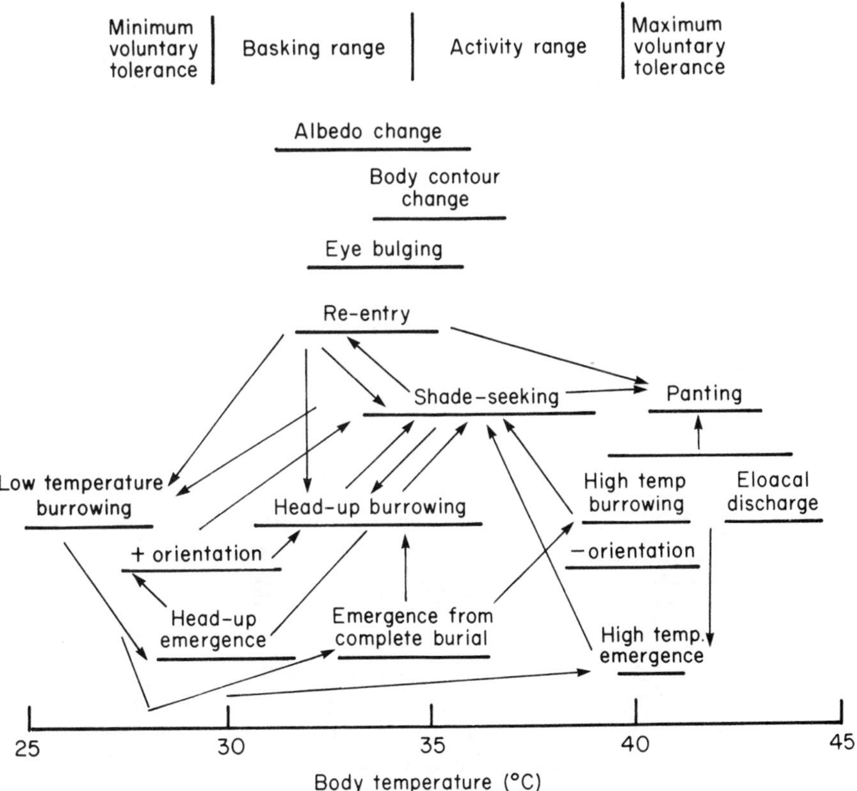

Figure 3.8 Thermoregulatory patterns in *Phrynosoma coronatum* as a function of body temperature. The lizards are active over a narrow range of body temperatures (34–39°C) and bask at lower T_b. Shade-seeking and burrowing are the major aversive behavioural responses. (After Heath, 1965.)

Thermoregulation in reptiles is at its simplest just shuttling between sun and shade. Such behaviour suggests that the animals possess on–off thermostats that switch the behavioural response from sun- to shade-seeking or vice versa. The actual mechanisms are rather more precise and show proportional control. That is, the behavioural thermoregulatory response can be graded to match the deviation of T_b from that preferred. When T_b is low, postural adjustments will maximize the area exposed to the radiation, but as T_b rises towards the preferred temperature the body can be orientated to reduce the radiant input and slow heating. Only when T_b rises further will shade-seeking occur. Postural changes can also alter conductive exchange with the substratum. Figure 3.8 shows the complexity of these behavioural responses to temperature in the horned lizard (Heath, 1965).

There is good evidence that physiological as well as behavioural mechanisms are involved in the regulation of heat gain and heat loss in a number of reptiles. The best-documented work is that on the Galapagos marine iguana. By basking on the barren rocks of the islands this lizard can maintain a T_b of 37°C. The amount of solar radiation absorbed can be

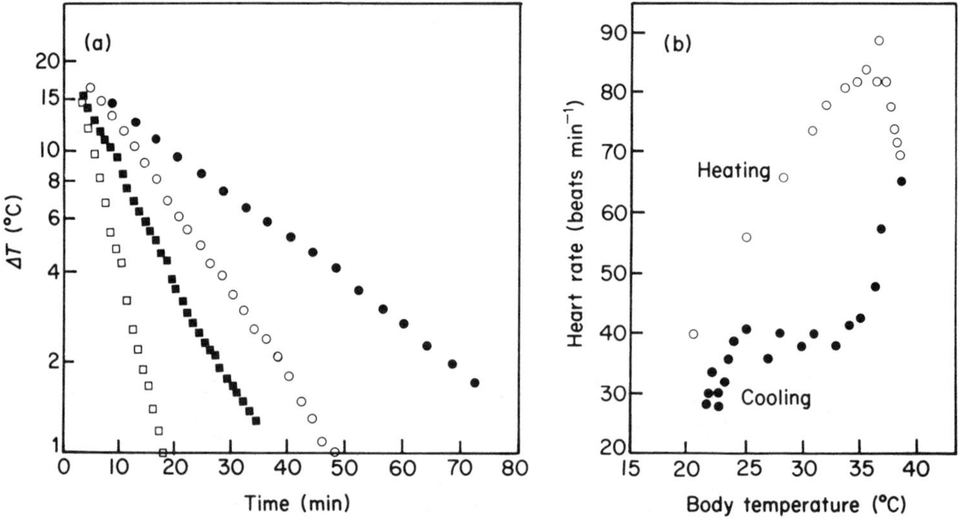

Figure 3.9(a) Heating and cooling rates of *Amblyrhynchus cristatus* in water and in air. ΔT is the difference between T_b and ambient temperature. In both air and water the rate of heating was about twice the rate of cooling. (b) Heart rate as function of body temperature during heating and cooling in water. A marked difference occurs in heart rate depending on whether the animal is warming or cooling. A marked bradycardia also occurs on diving. These points suggest cardiovascular adjustments are important in controlling heat loss when feeding in water, and in re-warming on land after a dive. ●, Cooling in air; ■, cooling in water; ○, heating in air; □, heating in water. (After Bartholomew and Lasiewski, 1965.)

altered by body shape and postural changes. When hot the lizard raises itself from the rocks to enable convective cooling from the inshore breeze. At night they display an unusual huddling behaviour, where groups of about 200 pile together and conserve body heat. The lizard feeds on marine algae in the cold waters of the Humbolt current around the islands. Bartholomew and Lasiewski (1965) have shown that this lizard conserves body heat on entering the water by peripheral vasoconstriction and a reduction in heart rate. Both responses reduce the flow of warm blood to the periphery, and core temperature is preserved at the expense of peripheral temperatures. On the return to land heart rate increases and peripheral vasodilatation occurs, so rewarming by basking rapidly brings T_b to the preferred level. A clear hysteresis occurs between the rate of warming and the rate of cooling, with the former happening at about twice the rate of the latter. This is shown in Fig. 3.9 for the marine iguana, but similar data exist for other reptiles and suggest that the circulatory control over heat gain and loss is widespread.

3.4 Ecological significance of body temperature – costs and benefits

The understanding that physiological performance in an ectotherm is optimal at a specific temperature (or over a narrow range of temperatures) is of profound ecological significance. It infers that T_b is the relevant temperature that ecologists should consider, rather than T_a. Many ectotherms probably operate at close to optimal performance over a fairly wide range of body temperatures, at least when given the opportunity for acclimatization to occur (see Chapter 5). What is not so readily apparent is the cost and consequences of operating at temperatures other than those giving optimal functioning. Huey and Stevenson (1979) have discussed the need to establish the range of temperature over which an organism can function normally when considering ecological and zoogeographical problems.

The concept of the ecological niche then should take account of the temperature–performance profile of the animal. Porter *et al.* (1973) have developed these ideas and have introduced the concept of *climate-space* to describe the niche better. In a terrestrial environment this refers to a four-dimensional space whose boundaries are set up by radiation, air temperature, wind and humidity. In an aquatic environment other factors such as oxygen tension and salinity are likely to be important. These factors seem to interact in a complex fashion to delimit a space within which the animal can function, for the physiological limits of performance are set by these factors. The climate-space of an animal may be larger than that observed under field conditions, since biotic factors, such as food availability, predation and competition, may result in the occupation of a smaller climate-space or niche. Porter produced a predictive model of the climate-space of the desert lizard *Dipsosaurus dorsalis*, and it is shown in part in Fig. 3.10.

In the field these lizards maintain a T_b of 38–43°C, so it is possible to construct a predictive model of their activity pattern. The animal must seek shade or enter its burrow when surface temperature exceeds 43°C. The model predicts that for most of the year the time available daily for activity, and so for feeding, is often as short as 2–3 h. Between November and March surface temperatures are too low for activity and the lizard lies torpid in its burrow.

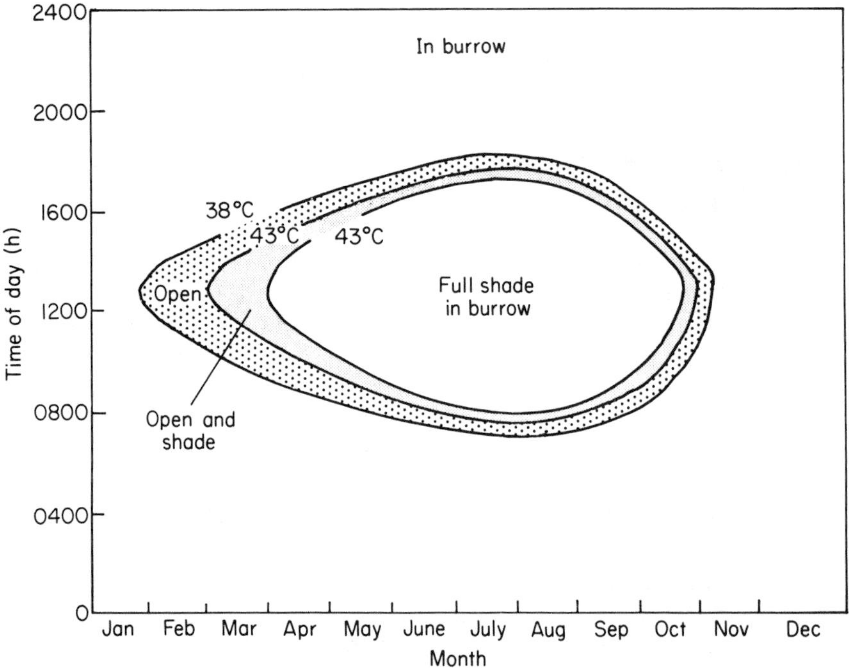

Figure 3.10 Predicted seasonal behavioural pattern for *Dipsosaurus dorsalis*. The model assumes a cloudless sky and uninterrupted sunshine. Field measurements of T_b show the lizard to be active with a T_b between 30°C and 43°C. Lizards are active on the surface only when they can behaviourally regulate T_b within those limits. At other times they seek shelter within the burrow. The model predicts four patterns of behaviour: active in the open in the day; active in the shade in the day; inactive in the burrow during the day when T_{air} is too high; and inactive in the burrow at night. The time available for essential activities on the surface is quite small (about 3 h) for most of the year. In spring and autumn this is around midday, but in summer it is confined to two periods at dawn and dusk. (After Porter *et al.*, 1973.)

Only in spring and in late-October would midday temperatures allow activity in the open. During April to September activity would be restricted to morning and afternoon sessions. Activity periods during summer can be extended if shade from bushes is available. The model constructed was tested against the measured T_b and field temperatures for this lizard, and provided a good fit. This confirmed the ability of this species to function with a T_b close to that which would be lethal, an adaptive strategy for an animal living in hot deserts.

These workers also constructed a similar model for a putative prey species of ant that was active only when the surface temperature was between 21°C and 33°C. Only in early spring would the ant fall prey to the lizard, for only then would both be active in the open at the same time. This suggests that if changes in seasonal temperature differentially affect the activity patterns of predator and prey species, the composition of the diet may be dependent upon seasonal temperature.

In another study, by Smith and Miller (1973), a predictive model of the thermal balance for two species of intertidal crab has been constructed taking a number of climatic and organism parameters into account. *Uca pugilator* lives on the open beach with little shade, except that it can retreat into its burrow; whereas *U. rapax* lives in the mangroves at a lower tidal position on the shore. This habitat difference was reflected in the behavioural responses made to seasonal and daily variations in temperature. The most effective means of avoiding damaging hyperthermia was to retreat into the burrow in *U. pugilator* or into shade in *U. rapax*. The open-shore species also showed a number of interesting additional responses. In full sun it blanched, so increasing its surface reflectance. It also wetted itself to increase evaporative cooling. It was calculated that without these additional facilities, *U. pugilator* would have often experienced close to lethal T_b in summer radiation. It was also able to raise its T_b above the air temperature by basking, but on overcast days its T_b was below the air temperature. Low lethal conditions were also avoided by retreat into the burrow where temperatures did not fall below about 16°C.

3.5 Is preferred T_b the temperature for optimal functioning?

It is believed that an animal will, given the choice, select a T_a that results in a specific T_b, or behaviourally regulates to establish a specific T_b, which provides for the most efficient functioning. This idea assumes that the preferred T_b and optimal T_b are coincident. There is some evidence that suggests that this may be so. For example, in the desert lizard the highest assimilation of ingested food is at 38°C, the preferred T_b in this species.

Thus, the adoption of a preferred T_b can be considered as a powerful adaptive strategy that maximizes fitness. According to this argument the preferred T_b is the one that maximizes the difference between the costs and benefits associated with selecting that temperature, and so produces maximal fitness within a particular niche. A shift in seasonal temperature may displace this relationship between costs and benefits, and the change in acclimatization state may be necessary to restore the balance fully or partially. If only partial redress occurs, then a change in preferred temperature may also be necessary.

In this respect it is interesting that many species will select a lower preferred temperature when there is a temporary food shortage. In rainbow trout this may differ by 4°C between fed and starved fish. It may be significant that the starved fish do not select the lowest temperature possible, but merely a lower temperature. This means that they are not simply minimizing energy utilization in choosing the lower temperature. This may be because often benefits accrue from the new temperature selected. For example, it may be that digestion and assimilation of the food available requires a relatively high temperature. It has been shown in fish (see Chapter 7) that the optimal temperature for growth falls when ration is restricted. This results from a complex interaction of the effect of temperature on food assimilation and energy utilization, but at a lower temperature less energy is expended on maintenance metabolism and a greater proportion of the reduced energy intake is available for growth. This may also contribute to the costs and benefits equation and explain why many ectotherms select lower temperatures when starved.

Many ectotherms face a relative absence of food in certain seasons. In these animals survival will depend upon the amount of energy that is stored during times of plenty, and the rate at which this store is being depleted when food is not available. For these animals a decrease in body temperature, with a minimum metabolic demand, will maximize the time of survival and ensure that the animal can replenish energy reserves when food is available again. There are many examples of temperate species that face this seasonal problem, e.g. frogs, insects and crayfish (see Table 5.2). They enter a period of torpor and remain quiescent, during which time metabolic rates are at a supposed low level until seasonal conditions become more favourable. It is likely that metabolic activity is set at the new low level by hormonal and/or central nervous system involvement. This type of acclimatization is discussed fully in chapter 5, and produces the benefit of the sparing use of energy stores, for even if environmental temperatures rise temporarily the metabolism remains below that predicted for the Q_{10} response. The cost of this strategy is that the animals have a reduced 'scope for activity', and so lose fitness with respect to their ability to exploit their environment and avoid predation.

For the vast majority of species there is a lack of detailed information concerning habitat micrometeorological conditions, and as a consequence, field T_b. This hampers our understanding of how temperature impinges upon the lifestyle of an animal. In the absence of such detailed information the determination of preferred T_b probably provides the best estimate of an animal's thermal characteristics.

3.6 Endothermic animals

Birds and mammals are most readily identified as endothermic animals. A unique feature of their thermoregulation is the precision with which core temperature is regulated, continuously, by physiological means. Some large insects, fish and reptiles also display endothermy.

Several mechanisms are common to all endothermic animals. First, all show increased heat production through metabolism. This is coupled with heat-conserving mechanisms and, usually, the development of insulation.

The primary source of heat available to aerobic cells is the oxidation of substrates by mitochondria. The complete oxidation of glucose in the cell yields the same quantity of energy as it would if burnt in a test tube. However, in the cell, only some of the energy is released as heat for some is conserved in the synthesis of ATP from ADP.

With *in vitro* preparations of mitochondria respiratory rate is dependent upon the supply of ADP and inorganic phosphate, and in the absence of ADP respiratory rates may be only 10% of the value in its presence. There is good evidence that mitochondria *in vivo* respond similarly to cellular levels

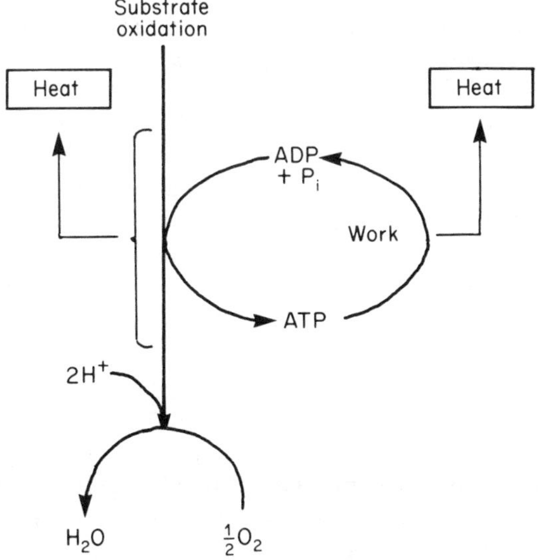

Figure 3.11 Sources of heat production in aerobic cells.

of ADP and ATP. Thus, high work rates in cells result in a high production of ADP from ATP and are associated with increased respiratory rates. This is summarized in Fig. 3.11.

The hydrolysis of ATP is usually coupled with the performance of work, and so relatively little heat will be released from ATP breakdown. However, in some cellular activities ATP hydrolysis is not coupled with work, so the enthalpy of ATP formation will be released as heat.

Several distinct mechanisms of heat production are involved in different tissues in endotherms, and specific reference will be made to each in context. (They are dealt with in detail by Hochachka (1974).) All are variations on the scheme shown in Fig. 3.11, and so potentially are available to all organisms. Whether this heat contributes to the maintenance of T_b will depend not only upon the oxidative capacity of the tissues, but also upon thermal conductance, which determines the rate of heat loss to the environment.

Endothermy is achieved at considerable energetic cost, so under a negative thermal balance T_b can be maintained only for as long as adequate nutrition is available.

3.7 Why set T_b between 30°C and 45°C?

The temperatures at which some endotherms regulate T_b, together with the preferred T_b or field temperatures of some ectothermal animals, is shown in Table 3.1. Examination of the table shows that most endotherms regulate to maintain T_b at a temperature that is above 35°C but below 45°C. This temperature range also covers that attained by ectotherms regulating behaviourally. What is significant about this particular range of temperatures is not clear, but it seems to have resulted from evolutionary selection on several separate occasions. That the specific temperatures regulated are not just a feature of endothermy, but occur amongst ectotherms regulating behaviourally, suggests that some fundamental property is associated with this set of temperatures.

Whether thermoregulation evolved to provide as high a T_b as possible or to stabilize T_b within narrow limits is not easy to distinguish. If T_b is to be maintained constant, then it is better to do this at a level above the highest T_a likely to be experienced, because the animal is in negative heat balance and can lose heat to the environment in a regulated manner by controlling radiation, convection and conduction exchanges. If it is in positive heat balance then it could lose heat only by the evaporation of water. This is obviously not a course open to aquatic animals, and for many terrestrial animals the conservation of water is a serious problem and would not permit its use in routine thermoregulation. This in part explains why temperatures above 25–30°C are used to set T_b, but other proposals have

Table 3.1 Body temperatures of some endothermic animals, and the measured field temperatures of some ectothermic animals whilst active.

Species	T_b(°C)	Reference
Centris pallida (solitary bee)	43–47	Chappell (1984)
Birds (range for various species)	39–44	Dawson and Hudson (1970)
Dipsosaurus dorsalis (desert iguana)	38–43	Porter *et al.* (1973)
Sphinx moths (13 species in flight)	38–43	Heinrich and Casey (1973)
Scarabaeus laevistriatus (dung beetle)	38–42	Heinrich and Bartholomew (1979)
Placental mammals (range for various species)	36–39	Altman and Dittmer (1968)
Dasyuroides byrnei (marsupial)	36–38	Smith and Dawson (1984)
Amblyrhynchus cristatus (marine iguana)	35–37	Bartholomew and Lasiewski (1965)
Apis mellifera (honey bee)	34–37	Heinrich (1976)
Marsupials (range for various species)	34–36	Tyndale-Biscoe (1973)
Didelphis virginiana (opossum)	33–37.5	Treagust *et al.* (1979)
Chelonia mydras (green turtle)	32–37	Standora *et al.* (1982)
Eumeces obsoletus (skink)	33	Schmidt-Nielsen and Danson (1964)
Podarcis muralis (wall lizard)	33–36.5	Avery (1982)
Thunnus sp. (tuna)	18–28	Stevens and Neil (1978)
Ornithorhynchus anatinus (platypus)	30.5–32	Griffiths (1978)
Tachyglossus aculeatus (echidna)	29–32	Griffiths (1978)
Anniella pulchra (fossorial lizard)	15–30	Brattstrom (1965)
Anurans (range for various species)	3–36	Brattstrom (1963)
Salamanders (range for various species)	– 2–27	Brattstrom (1963)
Diamesa sp. (chironomid active on glacial snow)	– 16–0	Kohshima (1984)

also been presented. The commonest supposes that benefits accrue from functioning at a T_b that is raised above T_a. The benefits concern the enhanced cellular activity at high T_b, since the higher rate functions for enzymes allow a more rapid and stronger response to changes in the cellular environment. The fact that cell temperature will be stabilized is also seen as a benefit, for the cell can evolve enzymes that are maximally efficient at that particular temperature. Both benefits are thought to improve cell performance. It is also likely that benefits would equally apply to neurological and hormonal functions, the speeding up of information processing and transfer would permit more complex integrative patterns to develop. This would affect organism behaviour as well as providing a more sophisticated and responsive physiology. Thus, the temperature range selected may be a compromise between having as high a T_b as possible but without incurring damage at high temperatures.

An alternative idea considers that liquid water is the medium in which cellular biochemistry takes place. The physical and chemical properties of water are likely to have profound effects on cellular activities, and many of these properties are temperature dependent. Most dramatic are the changes in state that occur at 0°C and 100°C. Between these extremes the properties of liquid water are also temperature dependent. For example, this relationship in the case of ionization, viscosity and osmotic behaviour is not linear but logarithmic, and thus the midpoint in the temperature scale is not 50°C but 37°C (i.e. 100°/2.71828). At 37°C many of the physical and chemical properties of liquid water are also close to being most typical of liquid water. Paul (1986) has recently drawn attention to the fact that the specific heat of water is minimal at 35°C. So thermal constancy can be maintained with generation and/or dissipation of minimal amounts of heat. The proposal is then that 35–40°C is the most suitable temperature for cellular chemistry and physics and this is why this temperature range has been chosen. To accept this uncritically ignores the many species that function with a T_b distinctly below 37°C without any apparent disadvantage (see Table 3.1).

3.8 Endothermic insects

Endothermy has evolved several times in the insects, and endothermic species are found in dragonflies, katydids, dipterans, bees, wasps, beetles, moths and butterflies (May, 1979). In most of these cases endothermy relies on the high heat production of the thoracic flight muscles. This is one of the most active aerobic tissues known, and as the flight mechanism is only about 25% efficient then flight is associated with the production of heat. It is only in some of the species with elevated thoracic temperatures that T_b is regulated. In small insects such as *Drosophila* and midges the metabolic heat

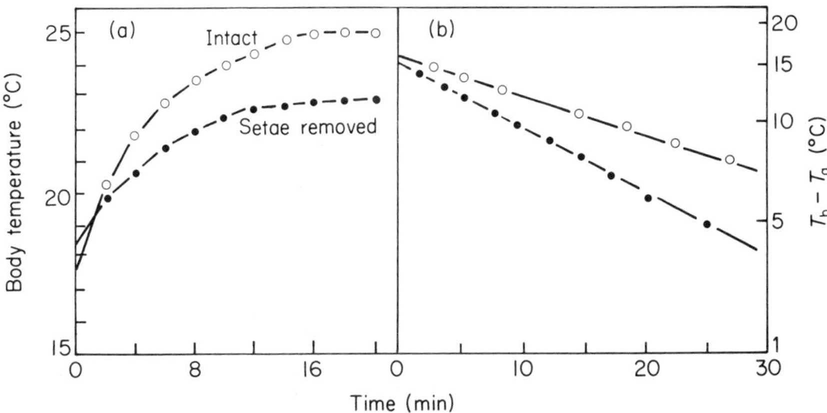

Figure 3.12 The effect of setae on heat exchange with the environment in Gypsy moth caterpillars. Caterpillars have long lateral setae and short dorsal setae. These act to reduce convective transfer without affecting radiative heat transfer. (a) Heating curves in the presence of a radiant heat source. A lower steady state of T_b occurs when the setae are removed. (b) Cooling curves in the absence of a radiant heat source showing greater rates of heat loss when the setae are removed. (After Casey and Hegel, 1981.)

of flight is dissipated by convectional cooling, and they fly with thoracic temperatures close to T_{air}. Only in insects larger than about 100 mg body weight is the thermal inertia great enough to conserve heat and so warm the thorax. Some of the large insects are also covered with hairs and scales that have insulatory value (see Fig. 3.12).

3.9 The need for 'warm-up'

A high flight-muscle temperature is a prerequisite for flight in larger species. This is because the lift required for free flight, above stalling speed, demands a minimum wing beat frequency and force. This is only met when the muscles are warm. As a high thoracic temperature is necessary for flight in some species, a warm-up mechanism is essential if flight is to be initiated at low T_a. Two mechanisms are proposed, a form of shivering and futile cycling.

In many species flight is preceded by low-amplitude wing vibrations which are associated with a dramatic rise in thoracic temperature (see Fig. 3.13) but with no equivalent rise in abdomen temperature. The rate of rise of thoracic temperature lies between 1 and 8°C min^{-1}, and is independent of T_a but is probably directly related to thoracic temperature. In asynchronous muscle during warm-up the muscle is thrown into tetanic-like contraction and there are no wing movements because of the

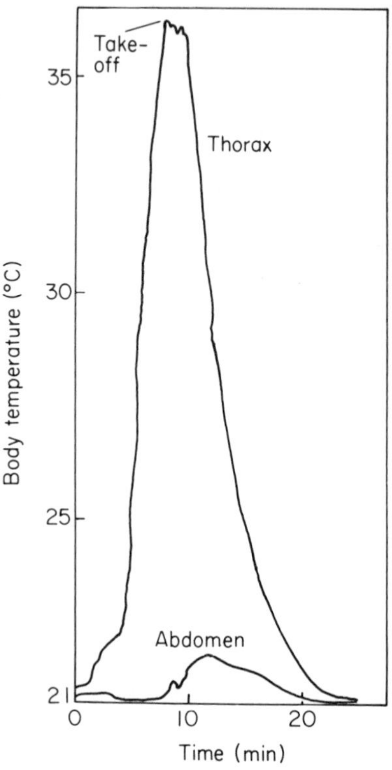

Figure 3.13 Thoracic and abdominal temperature during 'warm-up' in the dragonfly *Anax* at an air temperature of 20°C. There is a sharp rise in thoracic temperature, and take-off occurred when this reached 36°C. Thoracic temperature falls in an approximately exponential way after a period of flight. Abdomen temperature rises well after the start of 'warm-up' and may be the result of warm haemolymph entering the abdomen from the thorax. (After May, 1976.)

mechanical uncoupling that occurs. In synchronous muscle the wing upstroke and downstroke muscles are normally activated alternately, but during warm-up these pairs of muscles are simultaneously stimulated so that wing movements are small. Whether this type of warm-up requires the same neuronal elements as flight is not known. The metabolic rate at the end of warm-up is similar to that of free flight and so much metabolic energy is invested in this form of endothermy.

The other method of heat production proposed is futile substrate cycling. This is a form of non-shivering thermogenesis proposed to occur in bumblebee muscle. The process is shown in Fig. 3.14 (see also Hochachka, 1974). In glycolysis the reaction fructose-6-phosphate + ATP → fructose-1, 6-bisphosphate + ADP is catalysed by phosphofructokinase (PFK), a key regulatory enzyme. The back-reaction is catalysed by fructose bisphosph-

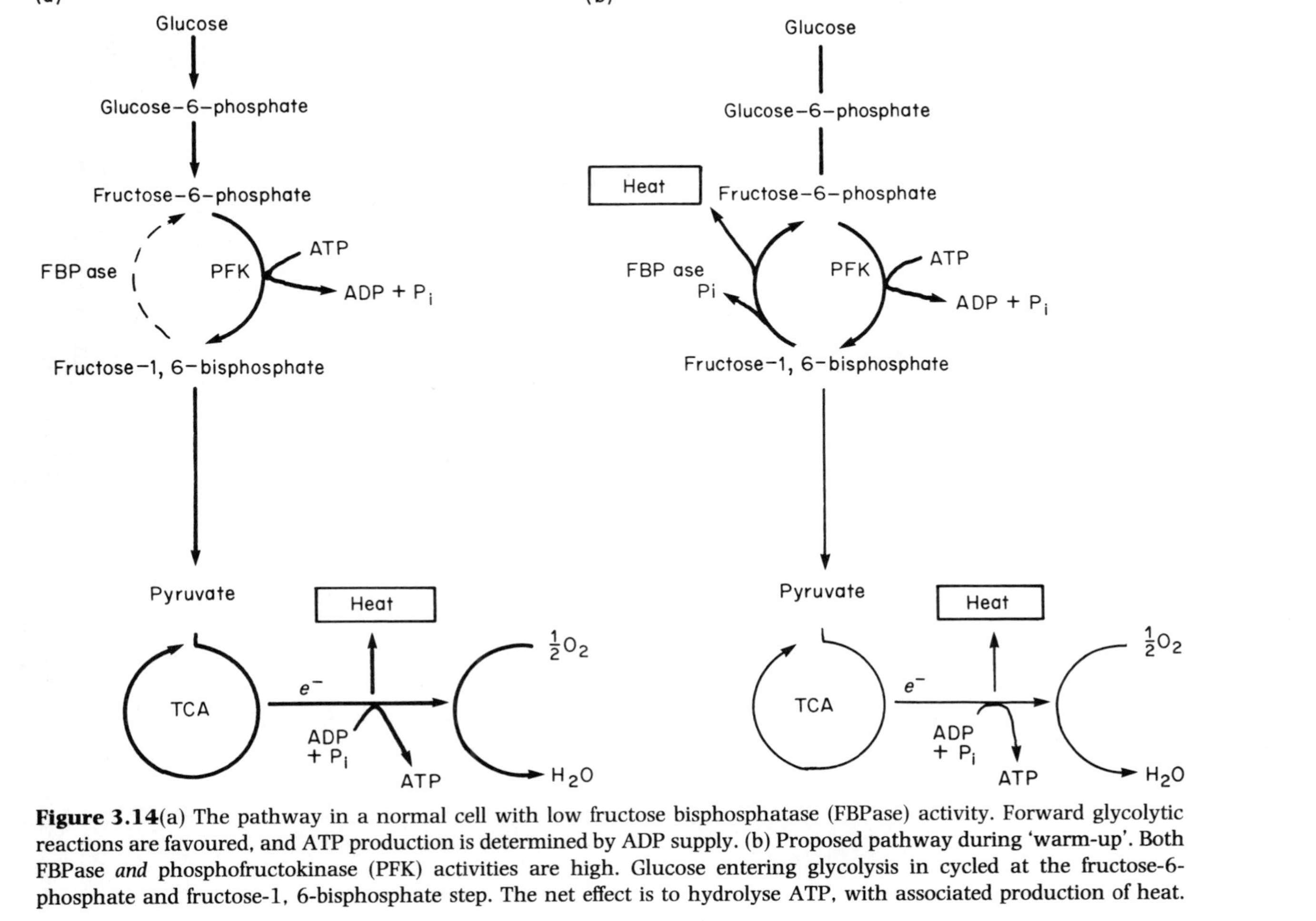

Figure 3.14(a) The pathway in a normal cell with low fructose bisphosphatase (FBPase) activity. Forward glycolytic reactions are favoured, and ATP production is determined by ADP supply. (b) Proposed pathway during 'warm-up'. Both FBPase *and* phosphofructokinase (PFK) activities are high. Glucose entering glycolysis in cycled at the fructose-6-phosphate and fructose-1, 6-bisphosphate step. The net effect is to hydrolyse ATP, with associated production of heat.

atase (FBPase). In most cells it is inappropriate to have both enzymes working together at high levels, and this is prevented by having the two enzymes differentially sensitive to the intracellular levels of AMP. The PFK is activated by intracellular AMP whilst the FBPase is very sensitive to AMP inhibition. In such cases the possibility for futile cycling between fructose-6-phosphate and fructose-1, 6-bisphosphate is limited.

It is reported that bumblebee flight muscle possesses high levels of both PFK and FBPase activity and, furthermore, that the latter enzyme is insensitive to AMP inhibition. In this case futile cycling at this step in glycolysis is possible, and it has been calculated that at maximum cycling rates the muscle could produce some 10 J min^{-1}, about half of which would result from the complete oxidation of fructose-1, 6-bisphosphate that slipped through the futile cycle. This is needed to generate sufficient ATP to allow the phosphorylation step catalysed by PFK.

It is thought that futile cycling in bumblebees is only effective at very low T_a, and at higher temperatures warm-up is by shivering. Clearly futile cycling must stop to permit both shivering and flight, for both require large quantities of ATP. It seems that the switch mechanism is the release of sequestered Ca^{2+} from the sarcoplasmic reticulum. This is required for the contraction of the muscle, but also switches off futile cycling. Bumblebee flight muscle FBPase is inhibited by raised intracellular Ca^{2+} levels, so the back-reaction is prevented and the fructose-1, 6-bisphosphate formed is completely oxidized through the TCA cycle.

It is likely that some fliers that are endothermic, such as bees, use shivering to maintain high thoracic temperatures between bouts of flying. In such resting insects metabolic rate is likely to be inversely related to T_a, which contrasts with the situation in ectothermal insects. It does mean that endothermic insects can forage at low ambient temperatures; although at some metabolic cost.

3.10 Thermoregulation in flight

Thoracic temperatures both after warm-up and on the initiation of flight are characteristic of the species and are also independent of T_a. Continuous flight will elevate thoracic temperature in large insects, but whether thoracic temperature is regulated during flight is not clear for all species (see Heinrich and Casey, 1973). In sphinx moths T_b in flight is stabilized even though T_a may vary, and it is thought in cases like this that regulation occurs. The necessity for regulation might also occur in flight at high ambient temperatures, particularly in species with well-insulated thoraces. Regulation of heat production maybe a feature in some dragonflies and butterflies where periods of gliding intersperse with flight activity.

The metabolic requirements for flight are unlikely to be set by thermoregulatory needs, but rather they will be determined by aerodynamic constraints. This suggests that thermoregulation is likely to involve the control of heat loss rather than of heat production. There is good evidence in moths and bumblebees that the uninsulated abdomen acts as a thermal window through which unwanted heat can be lost to the environment. Heinrich (1976) has shown that in bumblebees during warm-up, and in flight at low T_a, heat produced in the thorax is conserved by regulating the flow of haemolymph between thorax and abdomen. He has suggested that the control may be more sophisticated in some species with the possibility of counter-current heat exchange systems operating to prevent heat being lost from the thorax. When overheating is threatened the heat exchanger can be uncoupled and the extra thoracic heat load can be carried into the uninsulated abdomen for dissipation. In bumblebees this heat exchanger may lie in the petiole region between the abdomen and thorax, and is shown diagrammatically in Fig. 3.15. It is unlikely that insects would use evaporative cooling as a routine method of cooling, because their water reserves are too low.

The mechanisms that control thermoregulatory responses in endothermic insects are not clear. Shivering probably uses a set of neuronal elements similar to those that initiate flight. Hanegan and Heath (1970) showed that heating or cooling the thoracic ganglia of the moth *Hyalophora cecropia*, independent of thoracic temperature, affected the warm-up response. Cooling of ganglia in an otherwise warm insect can initiate warm-up behaviour and interrupt flight, so the pattern selected may depend on temperature. This would suggest that warm-up starts when thoracic temperature cools to below a low set-point. However, non-thermal

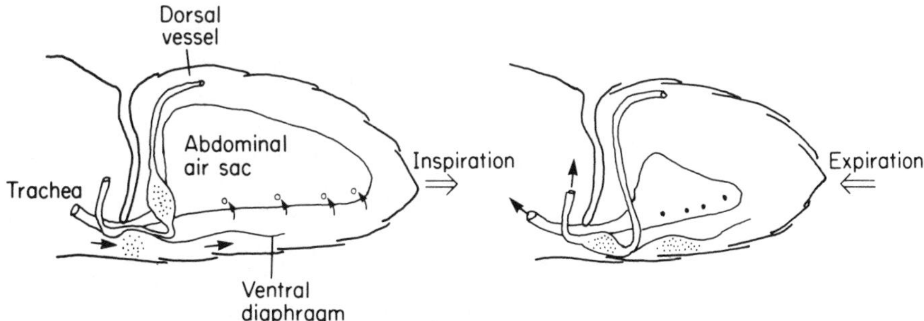

Figure 3.15 Diagram showing the possible air flow and haemolymph flow in the petiole region of the bumblebee at high air temperature. The ventral diaphragm acts as a two-way valve. At inspiration pulses of blood pass through the petiole region to the abdomen, whilst at expiration the pulses of blood pass into the thorax via the dorsal vessel. Thus, the opposing blood flows are uncoupled and no counter-current heat exchange can occur. The warm blood dissipates heat to the abdomen, from where it is lost to the environment. (After Heinrich, 1976.)

inputs must modify these responses since it would be energetically inefficient to initiate warm-up each time T_b fell below the low set-point. Thoracic temperatures above an upper set-point may well be the stimulus for causing circulatory adjustments that result in heat loss from the abdomen. In the sphinx moth *Manduca sexta* the circulatory response to overheating occurs in response to thoracic, not abdomen, temperature.

3.11 Thermoregulation in honeybee swarms

Heinrich has shown, in an elegant series of experiments, that honeybees thermoregulate not only individually during flight but also at rest in response to hive temperature. Recently he has also shown that swarms of bees thermoregulate to control swarm temperature. Large and small swarms of bees maintained the core of the swarm at 35–36°C over a wide range of T_a. The temperature of the mantle of the swarm did vary with T_a, and at 1–16°C the mantle was kept at between 15°C and 21°C, but at T_a higher than 16°C the mantle temperature was kept 2–3°C above T_a.

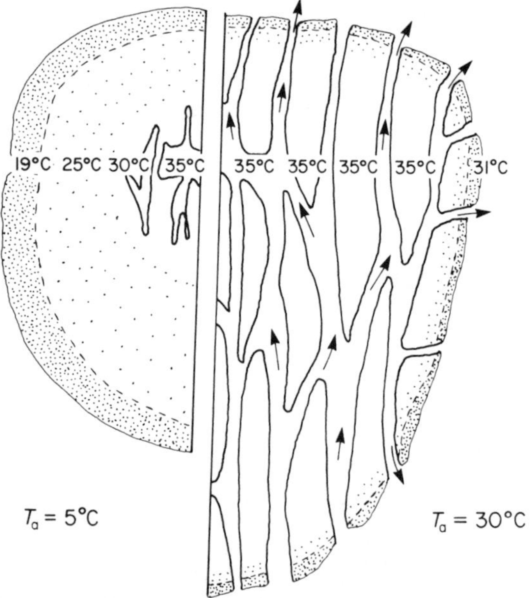

Figure 3.16 Diagram showing thermoregulation of a swarm of honey-bees at low T_a and high T_a. At low T_a the mantle bees are closely packed and a steep thermal gradient exists through the swarm. At high T_a the swarm is loosely packed with open airways, and only a small gradient exists. Dark stippling, mantle bees; light stippling, active thermoregulating bees; clear area, bees with resting levels of metabolism; arrows indicate avenue of heat loss. (After Heinrich, 1981.)

As is shown in Fig. 3.16 the swarm increased in volume as T_a increased. This is because large, open airways arise within the swarm at high ambient temperatures, and these allow convective cooling to occur. At low T_a a steep thermal gradient exists through the swarm and a denser packing of the bees occurs. This reduces individual heat loss and in consequence the heat loss from the swarm. In the contracted swarm the metabolism of the core bees is still sufficient to maintain the core temperature at 35°C, but the mantle bees may have to shiver to prevent individual T_b falling to a point below which they cannot maintain contact with the swarm. The mantle bees seem to have the primary role in swarm thermoregulation, for they close the open airways when T_a falls and maintain mantle temperature.

The ecological significance of this thermoregulation involves the readiness for finding potential new nest sites. If bees cool below about 15°C they will lose contact with the swarm and will not be able to fly. The majority of the swarm are at 35°C, and so are always ready to fly and exploit any new site discovered by 'scout' bees. In the swarm only the mantle bees have to use shivering to maintain individual temperatures, so precious food reserves are used only sparingly for thermoregulation. Southwick (1983) also reports that clusters of bees overwintering in the hive thermoregulate, and do so in a way that clusters of some 20 000 bees behave as if they were a 'superorganism'. In winter the oxygen consumption of a cluster is like that of a mammal of similar body weight (about 1.5 kg). Furthermore, like a homoiotherm, the cluster increases its metabolism if environmental temperature falls, rather than metabolism being lowered as expected for most other ectothermic insects.

3.12 Ecological costs and benefits of endothermy in insects

Endothermic insects are temporal endotherms, and their periods of endothermy coincide with periods of activity. Such small animals cannot afford the energetic cost of continual endothermy, and so the energy expended in thermoregulation must depend upon the availability of food. In the bumblebee, for example, the cost of foraging at low T_a may require some 2 mg sugar min^{-1}, and so it would only be worth foraging if food could be collected at that rate.

The clearest evidence for an advantage to the individual in being endothermic comes from studies on the African scarab beetles by Heinrich and Bartholomew (1979). A high T_b has been shown to give a positive advantage in the exploitation of fresh dung by these beetles. The speed with which a beetle can construct a ball of dung and roll it away is greater in beetles with temperatures of about 40°C. At lower T_b they are less successful and small endocoprid beetles can quickly enter the ball and render it useless.

Ball stealing also occurs, and the outcome of resulting intraspecific contests is also dependent on T_b, with the beetles with the highest T_b being the likely winners. This is because both the rapidity and the power of movement are positively correlated with muscle temperature. The beetles arrive at fresh dung with a T_b close to 40°C and they will try to maintain this temperature after landing by occasional shivering, since only when continuously warm will they be reasonably sure of success in competition.

Many of the interesting aspects of insect thermoregulation introduced above are extensively dealt with in a recent book by Heinrich (1982).

3.13 Endothermic fish

Several tuna fish species and some sharks are found to be several degrees warmer than the water in which they are swimming (Carey *et al.*, 1971). This is atypical of fish since the gill, which acts as an efficient gas exchanger, also serves as an equally efficient heat exchanger. So blood temperatures normally reach equilibrium with water temperature on passage through the gills.

Several factors contribute to the establishment of regional heterothermy in tuna. First, they have higher levels of heat production than other teleosts of the same size. Secondly, much of the heat produced in metabolism is conserved using elaborate circulatory heat-exchanger mechanisms. This, together with a relatively large size, contributes to the large thermal inertia shown by these fish (Stevens and Dizon, 1982).

The high metabolic rate results from the contraction of the red muscle that runs laterally alongside the vertebral column. This muscle is particularly rich in myoglobin and oxidative enymes, and is intensely aerobic. It is used in rapid cruise swimming, so is continuously active. It contrasts with the bulk white muscle, which fatigues, is more anaerobic and is used mainly in burst swimming.

The anatomical basis of heat conservation in the red muscle lies in the way in which the blood supplies this muscle. Blood is carried by four artery–vein pairs which run subcutaneously, delivering oxygenated blood into the muscle. This contrasts with the normal teleost pattern of radial supply from a central dorsal aorta. Venous blood leaving the muscle is also relayed to the subcutaneous vessels rather than to a central post-cardinal vein, as in most other teleosts. The significant feature of this tuna blood supply is the *rete mirabile* formed from an extensive elaboration of arterioles supplying and venules draining the muscle. These fine capillaries lie in close apposition to each other, and heat from the venous blood flowing out from the muscle is transferred back to the arterial blood flowing into the muscle in this elaborate counter-current heat exchanger.

In the large tuna species these rete are present as two slabs of vascular

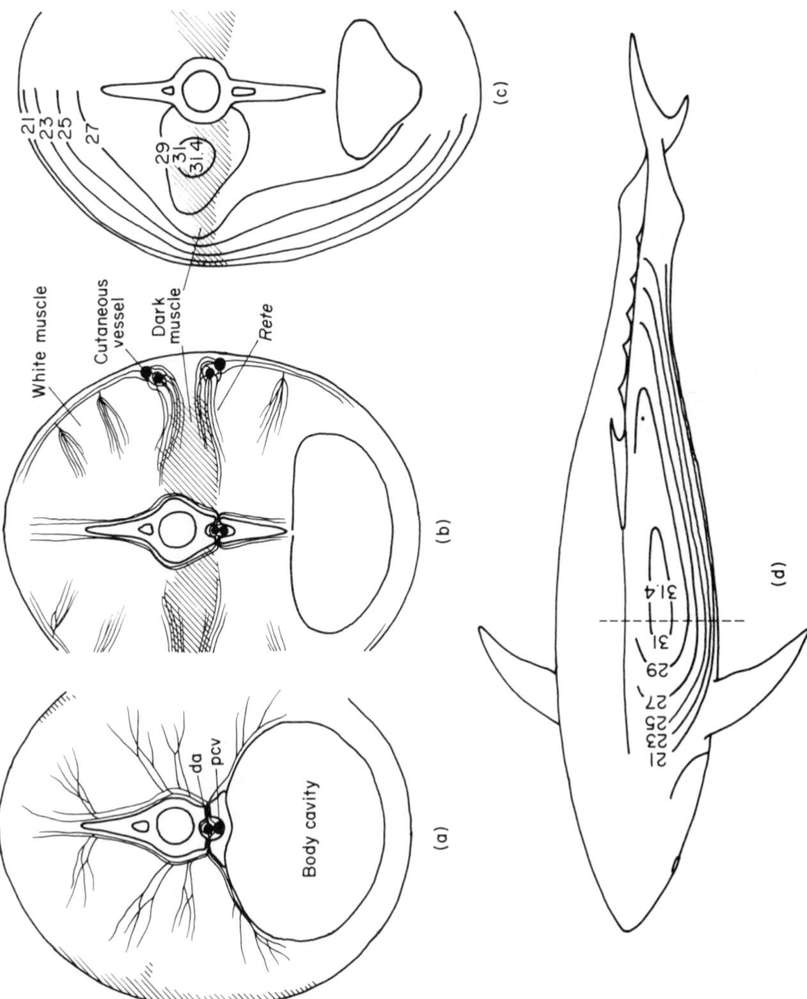

Figure 3.17 Blood supply to the muscle and thermal profiles of the muscle in the blue-fin tuna. (a) A typical teleost. Blood is supplied to the muscle by segmental branches of the dorsal aorta (da), and the venous supply is collected into the posterior cardinal vein (pcv). Little or no counter-current heat exchange is possible. (b) In the blue-fin tuna the dark muscle is supplied from four artery–vein pairs, two on each side of the fish. These break-up into numerous small-diameter vessels that form a rete in two slabs above and below the muscle. Heat exchange occurs between arterioles and venuoles, so conserving heat within the dark muscle. The white muscle has no rete but limited heat exchange can occur, and this muscle too is warm. (c) and (d) show the isotherms (in °C) for the blue-fin tuna. The warmest muscle corresponds to the dark muscle, but the white muscle is also warmer than the water. (After Carey, 1973.)

tissue dorsal and ventral to the red muscle. In the smaller species the most dominant rete lie centrally beneath the vertebral column. It is calculated that some 90% of the heat generated in the dark muscle is retained there by the action of these rete. The arrangement of the rete in tunas and also the supply of blood to the muscle in other teleosts as reported by Carey is shown in Figs 3.17 and 3.18. Figure 3.17 also shows the internal temperatures in an active tuna—and, as can be seen, a steep gradient of temperature exists with the red muscle being the warmest, but note too that the overlying white muscle is also warmer than water temperature. Rete also occur in association with the viscera and the brain–eye regions. These areas are also maintained several degrees above T_b.

There is no agreement, based on existing information, whether tuna can regulate T_b, or rather the excess body temperature (T_x). Figure 3.19 shows the temperature of tuna muscle as a function of T_a. In all cases T_x falls as T_a rises, but this is most marked in the large blue-fin tuna. As stated earlier, the

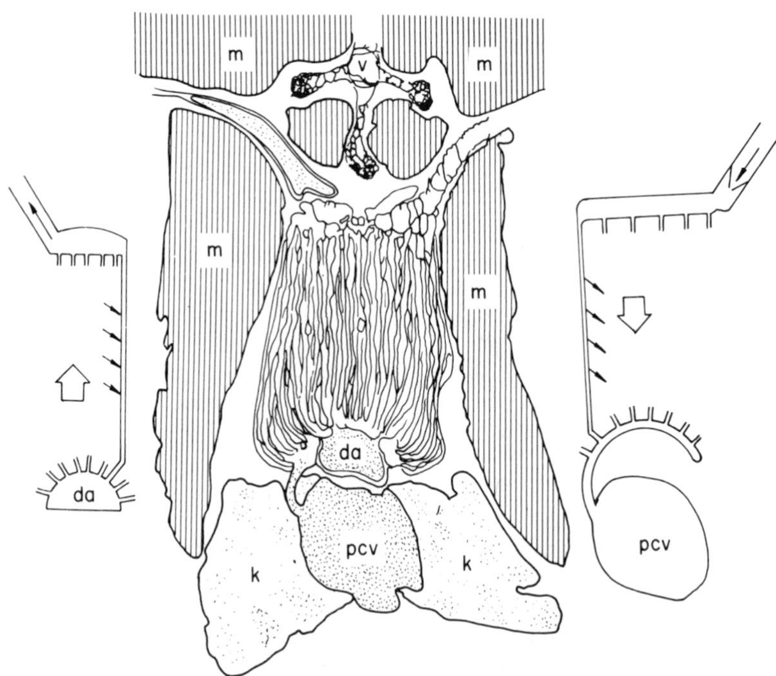

Figure 3.18 Diagram of the vascular heat exchange of the skip-jack tuna. The capillary bed lies centrally beneath the vertebral column. It is composed of arterioles and venules running in close opposition. Insets show the pattern of blood flow from the dorsal aorta (da) and into the posterior cardinal vein (pcv) – the small arrows indicate the direction of heat transfer. m, muscle; v, vertebra; k, kidney. (After Stevens *et al.*, 1978.)

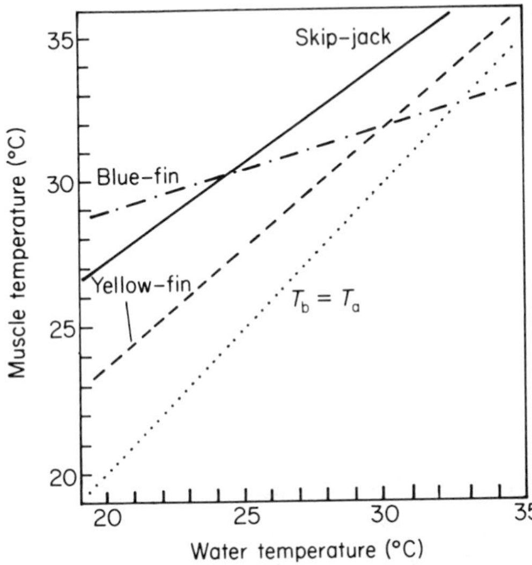

Figure 3.19 Muscle temperature of some tuna as a function of water temperature. Skip-jack and yellow-fin tuna appear to maintain muscle temperature at a fixed level above water temperature, and so have little thermoregulatory control. Larger blue-fin tuna have developed a powerful thermoregulatory mechanism using an efficient heat exchange between muscle and gill. (After Carey and Teal, 1969.)

thermal inertia, particularly in the large tuna, contributes to the stabilization of T_b, so rapid changes in T_a are not reflected by changes in T_x. This would allow tuna to thermoregulate behaviourally by selecting a T_a that results in an acceptable T_x at routine swimming speed. Swim speeds are not noticeably greater at lower T_a and, therefore, there is no evidence for increased heat production at low T_a to produce the larger T_x evident in Fig. 3.19. The question of how tuna avoid lethal overheating when swimming at higher T_a without reducing swim speed has been addressed by Dizon and Brill (1979). At a higher T_a tuna may recruit proportionally more white muscle into service, and as this muscle is devoid of rete the heat produced will be lost to the water across the gill. At a still higher T_a it may be necessary to introduce circulatory adjustments to increase the rate of heat loss from the red muscle so that overheating is avoided without affecting swim speed. The effectiveness of the rete could be reduced by shunting more blood past, rather than through, the heat exchangers. In these ways, through their thermal inertia, by behavioural selection of water temperature and also by physiological adjustments to circulation, it is possible for tunas to regulate T_b over a wide range of T_a both at routine and at burst swim speeds.

Several adaptive advantages arise from tuna endothermy. Most obvious is that it permits both the greater generation of muscle power and also higher rates of contraction and relaxation. In addition, recovery after violent activity is very short compared with non-tuna teleosts of the same size. This allows a shorter interval between successive bursts of activity. This is an advantage to a fish foraging in an environment that is patchily distributed with prey species. They may also benefit from the relative stabilization of muscle temperature, since muscle physiology and biochemistry can be optimized with respect to temperature. For example, the swimming power of tuna requires very high rates of oxygen uptake, and it is possible that the elevated temperature in the red muscle facilitates the transport of oxygen to the mitochondria via the rich supply of myoglobin present. Furthermore, the high visceral temperatures are thought to be adaptive for a fish with a small gut, since it will speed the rate of food processing and assimilation.

Thermal inertia contributes greatly to the thermal stability of tunas. They can, for example, penetrate below the thermocline into cold water without suffering the dramatic drop in T_b experienced by other teleosts of the same size. This may be a crucial factor for tuna muscle temperature in predator–prey interactions, and therefore swimming speed is rather independent of rapid changes in water temperature.

3.14 Endothermic reptiles

Mrosovsky (1980) has provided evidence, similar to that discussed for tuna fish, that large aquatic turtles also have a marked regional heterothermy. Active muscles during swimming possess the greatest T_x, which may be as high as 18°C. Less experimental evidence is available than for tuna, but it is clear, that two factors contribute to the excess temperature. First, metabolic heat production from the muscles operating the flippers is conserved by means of counter-current heat exchangers and, secondly, the animals have a marked thermal inertia, owing to their large size and insulative properties of the carapace and plastron. The value for k of 0.002°C min^{-1} per °C gradient for the leatherback turtle is much lower than values for typical fish and are close to the values obtained for air-breathing aquatic mammals. The advantage gained by this form of endothermy is that it may increase swimming ability and so be a factor in the long-distance migrations made at sea by these turtles.

The only other reptile that clearly shows endothermy is the brooding Indian python. This snake coils around its eggs and produces metabolic heat by the spasmodic contraction of its musculature. When T_a is below 33°C the animal can raise T_b by as much as 7°C to incubate its eggs. The metabolic cost of this endothermy is high, for oxygen consumption can rise to nine times the resting rate.

Generally, heat conductance in lizards is so large that metabolic heat is usually rapidly dissipated to the environment. In large lizards activity metabolism, coupled with a large thermal inertia, can contribute to an excess body temperature of 1–2°C, but of course this is only maintained as long as the animals are active.

3.15 Conclusions

Virtually all animals display responses directed towards temperature. Whilst ectotherms do not possess the physiological specialization for precise thermoregulation found in birds and mammals, they should not be considered as less suited or adapted to their ecological niches. Thus, ectotherms should not be considered as evolving towards endothermy or as failed endotherms, since ectothermy is itself a successful strategy which enables animals to occupy niches not available to birds and mammals.

Ectotherms have a resting metabolism that is considerably lower than that of birds or mammals, and this together with the low night temperatures means that the daily or annual energetic demand is considerably less. A lizard, for example, may expend only 5% of the energy per day that is required by a mammal of the same size. This reduces the need for high rates of food acquisition and allows specialist patterns of feeding to develop. In carnivorous ectotherms such as spiders and snakes, a sit-and-wait policy can be adopted in hunting, and the intensive foraging of birds and mammals is not required. Similarly, ectotherms can be fasted for extended periods without ill-effects, whilst endotherms would perish. Ectotherms also have a high efficiency, of up to 40% in some reptiles (Andrews and Asato, 1977), in converting ingested energy into biomass, since respiratory heat production is much reduced compared with endotherms. This high efficiency may be a factor in the flexible growth patterns seen in many ectotherms. In some species growth is slow but continues even in the adult form.

In endothermy heat produced by metabolism is lost across a surface area, so the ratio of surface area to the heat-producing mass is a critical factor, which is why there are no very small birds or mammals. Ectotherms, however, are not constrained by surface area/mass relationships, so may be very small in size and furthermore may display a wider range of body forms with a large surface area/mass ratio. This is evident in the greater range of morphological types seen in many ectothermic taxa, where an elongate worm, eel or snake body plan would not be possible for birds or mammals.

Whilst the understanding of thermoregulation of ectotherms has been revolutionized by an appreciation of the various strategies employed, it is clear that thermoregulatory behaviour in animals in the field cannot always be understood on physiological considerations alone. Thus, some reptiles are quite passive to ambient conditions whilst in others the regulated

T_b may be quite diverse and depend upon conditions in the field. A more useful framework for understanding these properties comes from relating the costs of thermoregulation to the benefits that accrue. The physiological benefits of achieving high T_b are obvious and have been shown for many physiological functions, including metabolic scope, neuromuscular performance and food digestion. These benefits are maximized when T_b is held at the optimal value throughout the day, and are reduced in proportion to the time not spent at that temperature.

There are also important costs associated with thermoregulation as displayed by ectotherms. These include energetic costs of two kinds; first the higher T_b will elevate metabolic rate (although this is surely related to the physiological benefit) and, secondly, the postural and locomotory adjustments will require energy expenditure. Shuttling and basking behaviour whilst lengthening the time available for activity can also interrupt the time available for other activities such as feeding and courtship. In some species the need to thermoregulate may lead to intraspecific competition for basking sites, with increased social interactions having an energetic cost as well as interfering with basking *per se*. Finally, exposure to potential predators is a cost, but one that is difficult to evaluate. A higher T_b should enable the prey to detect and escape the predator more easily. However, the exposed basking sites and shuttling movements may draw the attention of predators which may then force the prey to abandon thermoregulation temporarily.

Now, since the physiological benefits of ectothermy are largely independent of the environmentally related costs, it is clear that the physiological optimum body temperature does not always coincide with the ecologically optimal T_b (Huey and Slatkin, 1976). Thus, reductions in regulated T_b are likely to be related to an unacceptable 'cost' of achieving a higher T_b. This approach is now supported by an elaborate mathematical model which, in general, provides more satisfactory appreciation, not only of the diverse thermoregulatory strategies employed by animals, but also the factors which are important in shaping them.

Ectothermy, then, is a frugal energy strategy that allows the exploitation of many niches not available to birds and mammals. Nevertheless, some bradymetabolic animals display endothermy, but here the usual pattern is to switch between normothermia and hyperthermia. We shall see in Chapter 4 that in birds and mammals temporal endothermy is characterized by a shift in the opposite direction from normothermia to hypothermia.

The bradymetabolic species that display endothermy do so as a result of heat produced during activity, usually by specialized muscle groups. In fishes the heat produced is conserved by using complex vascular countercurrent heat exchangers. In insects only the larger species are endothermic, some of which have insulation to reduce heat loss. The temporal endothermy seen in insects is associated with the need for the flight muscle to be warm so that the stroke rate and power of the wings is sufficient to provide lift. The high associated energetic cost in these large insects must be matched by their ability to gather sufficient food of a high energy content. This is perhaps best seen in bees. The warm body temperature of tuna, compared with most fish, is thought to have a variety of adaptive advantages. Tuna live in an environment where food of high quality is patchily distributed. They are energy gamblers, investing large amounts of energy in continuous swimming at high cruising speeds with the expectation of large and sustained energy returns in prey capture. Having a warm body is part of that gambling strategy. The warm muscle allows sustained cruising speeds, the warm central nervous system must facilitate information-processing and the warm gut speeds digestion and food assimilation. It is also thought that this strategy enhances escape from predators, for tuna move from warm surface water to deeper cold water when chased by predators, their warm bodies enabling greater swimming speeds in cold water.

4 Body temperature in tachymetabolic animals

In endothermic insects, fish and reptiles only specific muscle groups produce heat in quantities sufficient to raise T_b significantly; in birds and mammals on the other hand, there is a general elevation in tissue metabolism. The biochemical basis of this high basal metabolic rate is not fully understood but, in part at least, it stems from a greater oxidative capacity of the cells (Akhmerov, 1986; Else and Hulbert, 1985), which of course results in a larger heat production (see Fig. 3.11). In addition to a high rate of heat production, which alone would elevate the equilibrium temperature, birds and mammals have decreased their thermal conductance so that heat loss is reduced.

In most birds and mammals T_b is fairly precisely regulated by balancing heat loss and heat production. This can be achieved only by the integration of the various processes involved. Sustaining a high T_b is metabolically costly, and in some species special mechanisms have evolved to permit a daily or seasonal relaxation of the strict maintenance of T_b. Seasonal acclimatizations also occur that allow the animal to adjust heat production and heat loss in an adaptive manner.

The birds and mammals have separate origins from the reptiles, and it is very probable that their respective forms of endothermy, each with a sophisticated thermoregulation, have evolved independently in these classes. The physiology of thermoregulation in birds and mammals is nevertheless sufficiently similar for them to be dealt with together.

4.1 Body temperature

That most tachymetabolic birds and mammals are homoiothermic (i.e. maintain an approximately constant T_b) is well documented, but the understanding of the processes by which homoiothermy is achieved is still only partial. Heat is produced internally by cellular activity, and is lost through the body surface; consequently axial and radial temperature gradients exist in the body. Skin temperatures are more variable and

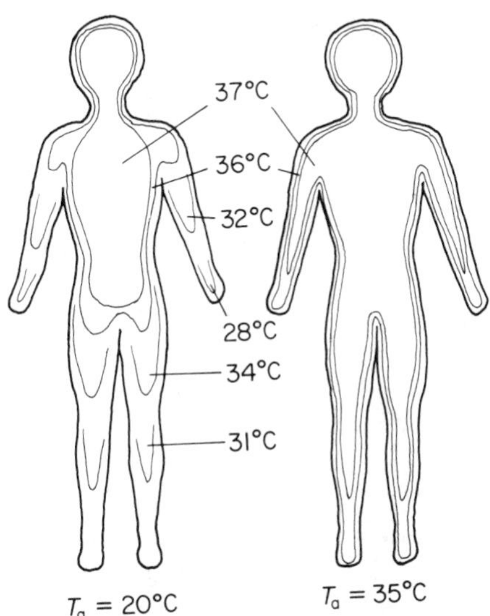

Figure 4.1 Body temperature isotherms of man in a cool and warm environment. Note the axial thermal gradients in the limbs in the cold and the extension of the area of the core in the warm. (After Aschoff and Wever, 1958.)

generally lower than internal temperatures. Therefore, no single temperature can be considered as representative of the T_b of either a bird or a mammal. In fact, only the core, consisting of the thorax, abdomen and brain, has an approximately constant temperature. The shell or periphery, which can constitute some 20% or more of body volume, has a more variable temperature, which largely depends upon the prevailing T_a and the extent of the physiologically controlled peripheral blood flow. This is illustrated in Fig. 4.1 for man.

Skin temperature is also greatly influenced by the extent of the thermal insulation provided by hair and feathers, the greater the insulation is, the less dependent T_{skin} is upon T_a. It is important to appreciate that at the extremities even muscle temperature may differ significantly from the core temperature (Fig. 4.1).

Core temperature in mammals is generally somewhat rather lower (34–38°C) than that found in birds (40–41°C), and it is usual to take orifice temperature as being representative of core temperature. The concept of a mean T_b has also proved useful, and is determined from the relationship

$$\text{mean } T_b = (T_{rectum} \times 0.67) + (T_{skin,mean} \times 0.33) \qquad (4.1)$$

This assessment of mean body temperature is of value experimentally because a change in mean body temperature is definite evidence of a change in body heat content.

4.2 Distribution of heat within the body

Heat production is not uniform throughout the core, and some tissues, particularly muscle, liver and brain, have a higher capacity for heat production than others such as skin, bone and depot fat. Heat produced at these sites is distributed throughout the body via the forced convection of the circulatory system. When T_a is less than T_b, blood flowing to the periphery is cooled there and the control over peripheral blood flow is a major determinant of the rate of heat loss from the core to the environment. Consideration of two extreme cases will illustrate this point. When core temperature is high and harmful hyperthermia threatens, it is necessary for large volumes of blood to flow through skin capillaries. Skin temperature may well approach core temperature, so heat transfer to the environment will be greatly increased. Conversely, when threatened by low ambient temperatures and a possible hypothermia, heat loss must be reduced. Peripheral blood flow is greatly reduced and skin temperature will fall, so reducing heat loss. Heat transfer from the core to the skin is minimized because convective transfer by the blood is reduced and only the relatively poor conductive transfer through the tissues occurs. Thus, control over core temperature is effected by using the shell as a variable heat sink and conductor of heat flow from the skin to the air.

Both central and peripheral thermoreceptors are involved in the control of blood flow. For example a cold environment will cause a fall in skin temperature, which is sensed by skin thermoreceptors. Local vasoconstriction occurs, perhaps as a direct effect of the temperature fall, but this vasoconstriction can be reinforced by centrally mediated signals. Vasoconstriction results in the effective shut-down of blood to skin capillaries and the opening of subcutaneous arterio-venous shunts. The shut-down cannot be complete for the skin still requires O_2 and nutrients (Fig. 4.2). Heat loss is greatly reduced by returning venous blood to the core via deep veins running in close proximity to the arterial blood supply, permitting counter-current heat exchange to conserve heat by warming of the returning venous blood (Fig. 4.2). Then the thermal gradient between blood in skin capillaries and the air is reduced.

Vasodilation of skin capillaries occurs by the closing of arterio-venous anastomoses, so arterial blood flows more rapidly through skin. T_{skin} rises and heat loss can be further facilitated by returning blood via superficial veins.

In mammals and birds the core temperature is not constant, but fluctuates through a daily cycle. The extent of the variation found is species-specific. In sheep it is only 1 °C whereas in camels and the African buffalo it is wider, being about 3°C, (Bligh and Harthoorn, 1965). Similar species-specific daily variations in T_b have been described in birds by Simpson and Galbraith (1905). The position of the maximum and minimum temperature

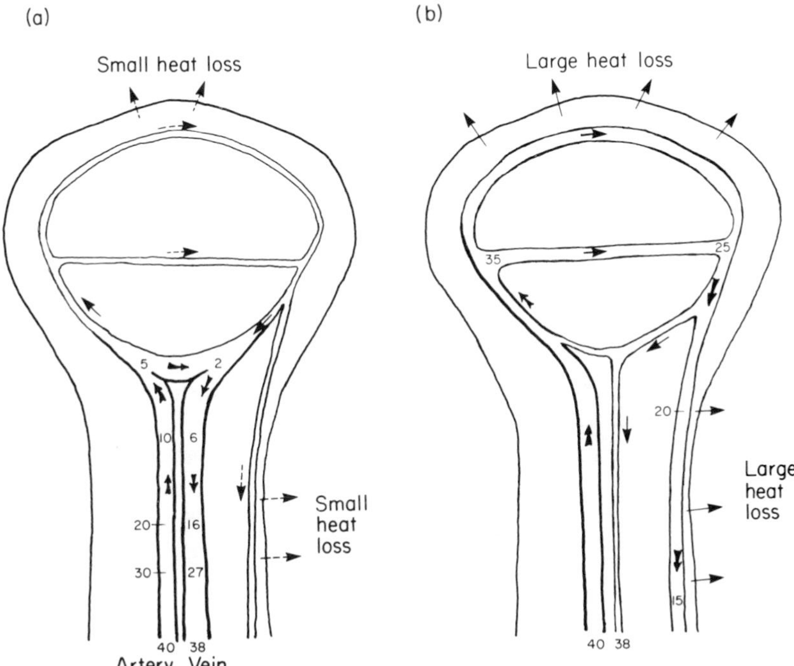

Figure 4.2 Pattern of blood flow through a limb. $T_a = 0°C$, temperatures in °C. (a) Heat conservation. The central artery–vein pair act as a counter-current heat exchanger, with heat transferred from the warm outgoing blood to the re-entering cool blood from the periphery. Note the peripheral circulation of blood is low, so heat loss from the extremity is reduced but its temperature will fall. (b) Heat loss. The arterio-venous shunt is now closed, so blood is diverted away from the central vein. Increased blood flow occurs in the peripheral circulation and is returned via surface veins. Heat loss is greatly increased as there is no longer any effective heat exchange between arterial and venous blood. The thickness of the arrows indicates the amount of blood flow or heat loss. The numbers indicate the likely temperature at that site.

recorded depends upon whether the species is diurnal or noctural. In those species active in the daytime the maximum occurs in mid- to late-afternoon, and the minimum occurs in the early hours of the morning. In nocturnal species the timing of the maximum and minimum T_b is reversed. In man the 24-h rhythm persists in the absence of environmental clues, but becomes out of phase with the sleep–wake cycle. The rhythm is considered to depend upon an endogenous clock mechanism. How such rhythms are set up is not understood, nor is it known how they relate to the mechanisms that determine the set-point temperature. Thus, there are species-specific differences in the exactitude with which T_b is controlled, some species being relatively thermolabile. The extent to which these differences in the thermostability of the core are due to ambient conditions is not clear. However, in some tropical species of large mammal, notably the camel, it is clear that the ability to allow a wide diel variation in T_b is an adaptation to heat stress (Schmidt-Nielsen *et al.*, 1957).

4.3 Metabolism and ambient temperature

Changes in T_a have a pronounced effect on the metabolic rate of tachymetabolic endotherms; the relationship is shown for a typical mammal in Fig. 4.3, and should be contrasted with that obtained for an ectotherm shown in Fig. 2.16. In explaining the complex relationship it must be realized that the mammal maintains a constant T_b over the range of T_a, unlike the ectotherm. Thus, if T_b is to be kept constant at about 37°C whilst T_a falls, then heat production must increase to offset the increased heat loss

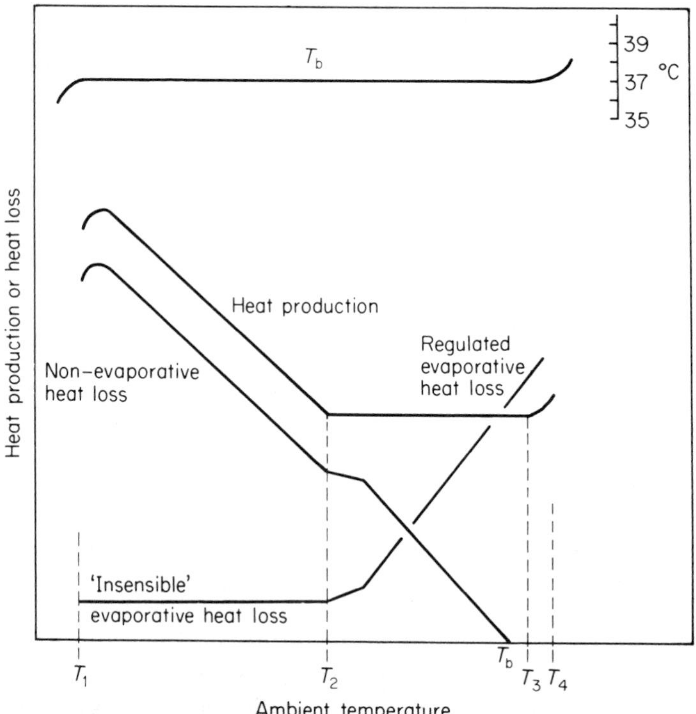

Figure 4.3 Diagram showing relationship between heat loss and heat production as a function of ambient temperature. T_1, ambient temperature at which hypothermia occurs; T_4, ambient temperature at which hyperthermia occurs; T_2 and T_3, lower and upper critical temperatures, respectivley, bounding the thermoneutral zone (TNZ).

at the lower T_a. This is the reason for the nearly linear increase in metabolic rate as ambient temperature falls between T_2 and T_1 in Fig. 4.3.

It is also usually found that metabolism is at a minimum over a narrow range of T_a (between T_2 and T_3) just below T_b. This is called the *thermoneutral zone* (TNZ), and within this zone the metabolism is shown to be independent of T_a and conductance is minimal. However, the relationship is more likely to follow a curve with a minimum rather than a plateau. This occurs because several complex factors, involving both heat-production and heat-loss mechanisms, with different temperature dependencies, interact to produce the TNZ. The zone can be conveniently considered as the range of T_a over which the animal can make metabolically inexpensive thermoregulatory adjustments to maintain T_b even though T_a may change. These adjustments might include alteration to peripheral blood flow, piloerection or fluffing of feathers and postural changes. Below T_2 heat production and metabolism must rise if T_b is to be maintained; T_2 is called the *lower critical temperature*. Below this temperature heat is lost from the animal mainly by non-evaporative means, and the slope of the line for heat production against T_a gives a measure of the thermal conductance of the animal. That is, the net rate of transfer of heat per degree Celsius difference between T_b and T_a.

A simple model for thermal exchange between an animal and its environment is given in Equation 3.1. This equation predicts that the rate of fall in temperature of the animal is proportional to the difference in temperature between its surface and the environment, and is derived from Newton's law of cooling. In birds and mammals T_b does not fall with T_a, so the application of the law is not strictly valid. For an animal with a constant T_b, heat transfer from its core to the surface is largely by forced convection and by conduction. It is described by:

$$H_c = -h\frac{A}{l}(T_b - T_s) \qquad (4.2)$$

That is, the rate of heat loss (H_c) is inversely proportional to the thickness of the shell (l) and directly proportional to the thermal conductivity of the shell (h), the surface area (A) and the temperature difference between the core and the surface.

Heat loss from an animal to its environment (H_T), however, is a more complex phenomenon with natural convection and radiation being of prime importance. Paradoxically this is best described by a more simple relationship:

$$H_T = C(T_b - T_a) \qquad (4.3)$$

where C is the thermal conductance, a composite term containing elements of conductive, convective and radiative exchanges.

As metabolic heat production (MR) must balance heat loss, this equation can be rewritten as

$$\mathrm{MR} = C(T_b - T_a)$$

or

$$C = \frac{\mathrm{MR}}{T_b - T_a} \tag{4.4}$$

that is, the slope of the line below T_2 in Fig. 4.3. Thermal conductance is expressed as cm^3 O_2 consumed g^{-1} body weight h^{-1} $°C^{-1}$. This is contrary to established practice, where it is expressed per unit surface area rather than mass (see Section 3.1). The volume of oxygen can be converted into joules by using the calorific equivalent of oxygen.

As temperature rises above the lower critical temperature (T_2) evaporative heat loss makes an increasingly important contribution to the overall heat loss. This is, of course, because the other avenues of heat exchange depend upon the temperature difference between the body and the environment, and as this gets smaller the heat transfer falls. When T_a rises above T_b the animal will be in a positive heat balance with its environment, and heat will be gained by conduction, convection and radiation. Indeed, the only way that heat can then be lost is by evaporation. At temperatures above the TNZ metabolic rate rises. This increase in MR as T_a rises has two components: a rise in the mean body temperature as the rate of heat loss fails to match that of heat production has the physical effect of increasing the rate of metabolism and an increase in metabolism associated with the muscular activity of thermoregulatory panting. At an even higher T_a the inability of thermoregulatory measures to equal the rate of heat gain results in a progressively increasing core temperature, which will ultimately cause death. Regulation can also become ineffective at low T_a. In this case the rate of heat production will not equal that of heat loss, and eventually the core temperature begins to fall. As mean body temperature falls there will be a progressive reduction in the rate of metabolism or heat production, ultimately resulting in lethal hypothermia.

4.4 Heat production

Metabolism provides the heat required to maintain T_b above T_a in birds and mammals. The magnitude of this heat production depends on species, T_a, activity, acclimatization state and on size. This last factor varies in a predictable way, and has been shown experimentally to have the relationship

$$\text{heat production} = kM^n \tag{4.5}$$

where k is a constant that depends upon body form and n is an exponent

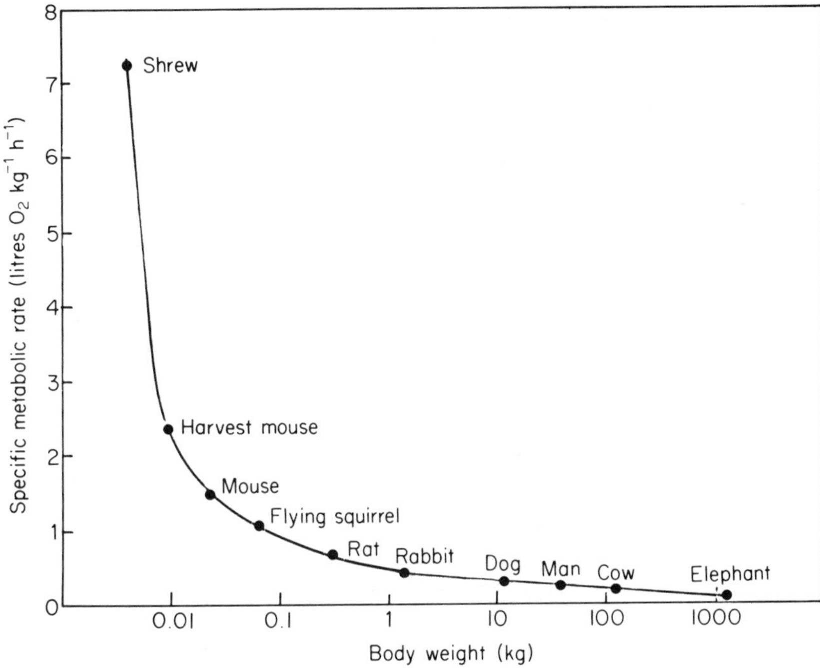

Figure 4.4 Semi-logarithmic plot to show the relationship between body weight and weight specific metabolism. Note the exponential increase in energy expended per kg body weight as body weight falls. (After Schmidt-Nielsen, 1983.)

that varies between 0.75 and 0.82. M is either body mass or volume.

This equation shows that heat production, and therefore metabolic rate, increases at a lower rate than size. In the shrew weight-specific metabolism is some eight times higher than in a mouse, and would be about 200 times greater than that of an elephant. This is shown graphically in Fig. 4.4, and as can be seen a sharp rise to an asymptote occurs in weight-specific metabolism as low body weights are reached. This effectively creates a lower size limit for homoiotherms, the basis of which is their high surface area/volume ratio. This relationship is shown for a hypothetical spherical homoiotherm in Fig. 4.5. As body weight is reduced the mass of heat producing tissue falls, but this is accompained by a relative increase in the area over which heat is lost. This relative increase in thermal conductance requires that a high weight-specific metabolism if T_b be maintained.

It is usually suggested that this limit is set by the ability of the animal to obtain and process its food, or even by the cellular biochemical reactions that produce heat. High metabolic rates also need fuelling with oxygen, and it may be the inability to supply oxygen to the tissues that is limiting. Shrews and hummingbirds have high blood oxygen-carrying capacities

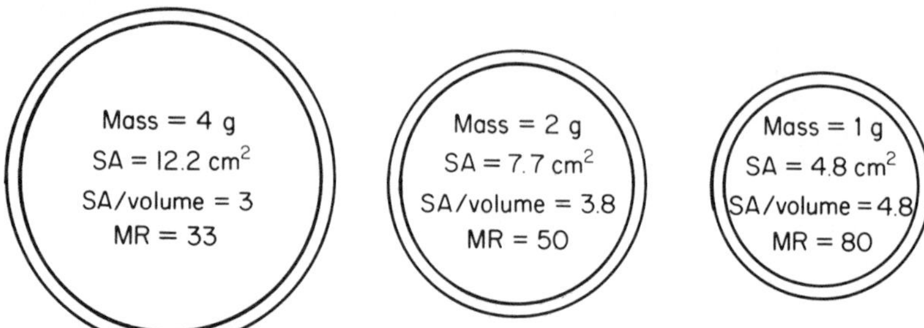

Figure 4.5 Diagram showing the relationship between mass, surface area (SA) and metabolic rate (MR) in a homoiotherm. The animals are assumed to be spherical.

and relatively large hearts with a very high beat frequency. The cardiovascular system in these vertebrates could be close to its physiological design limits (Schmidt-Nielsen, 1984).

Very large mammals may experience the problem of heat dissipation, particularly during exercise, owing to their low surface area/volume ratio. This may be why large tropical mammals tend to be hairless, and in the case of the elephant possess large external ears. These add some $2\,m^2$ to its surface area without any significant increase in weight, and so may serve a thermoregulatory function. Many arguments have been raised to account for the observed power function in Equation 4.5 for the relationship between MR and body weight (see Schmidt-Nielsen, 1984). One of these suggests that the different surface area/volume ratios between large and small animals is important. Thus, as body size falls the mass of heat-producing tissue falls but the relative area over which heat is lost increases. However, the same relationship between size and oxygen consumption is found in ectotherms. Heat loss is not a factor of those animals, so it is likely that some general, unexplained principle exists other than that offered for heat production and heat loss.

There are three major mechanisms of heat production available in addition to basal metabolism: voluntary muscular activity, and shivering and non-shivering thermogenesis. The last two are the principal methods used in thermoregulation, and will be considered more fully.

4.4.1 *Shivering*

Shivering, an involuntary tremor of skeletal muscle, is a powerful means of increasing heat production that is used by both birds and mammals. The rate of heat production can be raised during shivering to some two to five times that of basal metabolism, depending upon the magnitude of the increases in tremor and the number of motor units involved. Shivering can be initiated by either a fall in core temperature without a fall in skin temperature, or by a fall in skin temperature. It is controlled by somatic motor nerves and not by the autonomic nervous system. The control centre appears to be in the posterior hypothalamus, and the stimuli are sent to the muscles by neurons that pass via both mid- and hindbrain, and lateral areas of the spinal cord. This pattern of stimulation of the motor units is different from that in normal motor activity. The synchronous contraction of small groups of motor units occurs but out of phase with, and alternate to, motor units in antagonistic muscles, so no gross movement occurs.

Shivering rather than an increase in locomotion is used for thermoregulation, despite the fact that exercise can result in a 20-fold increase in basal metabolism, because exercise has a number of thermoregulatory disadvantages. Movement disturbs the pelage so that its insulatory efficiency is reduced. Locomotion also causes an increase in peripheral vasodilatation, and so the heat produced during exercise is not so efficiently retained for thermoregulation. Shivering and exercise are also not additive, as shivering stops as soon as exercise begins.

4.4.2 *Non-shivering thermogenesis (NST)*

Non-shivering thermogenesis is the production of heat by processes that do not involve muscle contraction. It was first demonstrated as a cold-induced increase in metabolism which persisted even when muscle contraction was blocked by curare (Hsieh and Carlson, 1957). Heat produced from the basal metabolism is, of course, a form of non-shivering thermogenesis, as it does not involve muscle contraction. However, this is not a specifically thermoregulatory form of NST, even though basal metabolism contributes to the heat-balance equation, and adaptive changes in basal metabolism occur under the influence of thyroid hormones. Thermoregulatory NST is associated with the calorigenic action of noradrenalin, the administration of which, in mammals, leads to an increase in metabolism and T_b. Figure 4.6 shows the effect of the infusion of noradrenalin (NA) on oxygen consumption of warm- (25–28°C) and cold- (4°C) acclimated rats. In the cold-acclimated group the extra oxygen consumed as a consequence of the NA infusion is equivalent to about 1.5 kJ of heat production, the same NA infusion in warm-acclimated animals caused only about 0.4 kJ of extra heat production.

There is still some doubt about the sites of NST. Brown adipose tissue (BAT) is known to be the principal site of NST, but it is not clear whether liver and muscle are also involved in the metabolic response to noradrenalin. The BAT is a specialized thermogenic tissue which differs from depot fat in appearance and function. The cells are filled with small lipid droplets and numerous mitochondria. The tissue is also richly supplied with blood

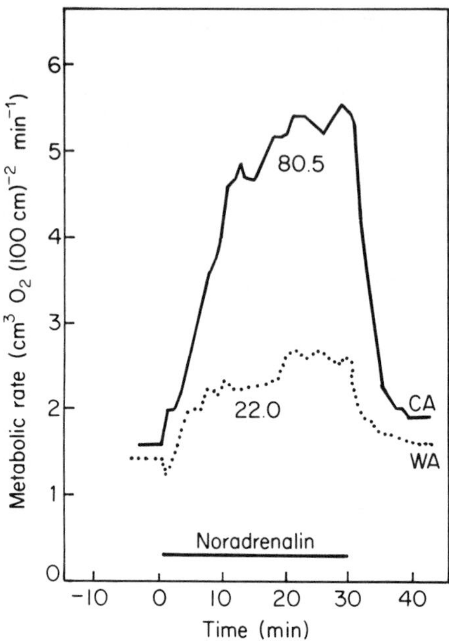

Figure 4.6 Oxygen consumption of warm- and cold-acclimated rats during noradrenalin infusion. The values 80.5 (cold-acclimated, CA) and 22.0 (warm-acclimated, WA) is the total O_2 consumption consumed during the 30-min infusion. (After Himms-Hagen, 1970.)

vessels and is innervated by the autonomic nervous system. It has a high metabolism when stimulated by noradrenalin. The quantitative contribution, to total metabolism in mammals, made by BAT has been reviewed recently by Foster (1984). As will be seen in Fig. 4.32, in cold-acclimated rats BAT is responsible for about 60% of the increased thermogenesis, but for only 30% in warm-acclimated rats. The small increase in heat production shown by skeletal muscle, heart, and rib-cage and diaphragm is a consequence of the increased mechanical work of the heart and respiratory musculature, and to the increase in metabolic rate of all tissues owing to the elevation of T_b during NA infusion. The blood supply to BAT is shown to increase dramatically during cold exposure or NA infusion. Oxygen consumption of BAT was also linearly related to blood flow, and both were linearly related to the oxygen consumption of the whole organism. This suggests that the NST response is largely due to BAT.

Birds are generally believed not to possess BAT, and this is probably the reason why their metabolism is not stimulated by noradrenalin. However, the demonstration of NST in birds is equivocal, since it is suggested that the hormone glucagon might activate a non-shivering increase in heat production, although no discrete site for this NST has been shown, but Oliphant (1983) reports BAT-like tissue in two bird species.

Brown adipose tissue is found in young mammals, where it is the most important site of thermoregulatory heat production in the newborn. However, it is gradually lost so that most adult mammals possess little or none. Those adults that retain extensive BAT tend to be of small body size or hibernators. It occurs in discrete masses or pads situated mainly in the neck, subscapular and thoracic regions. It may constitute some 1–5% of body weight. The pattern of the vascular drainage of the pads suggests that the brain, anterior spinal cord, heart, lungs and kidneys preferentially benefit from this localized increase in heat production.

Non-shivering thermogenesis is subject to two controlling influences. One influence, presumably hormonal, may result in hypertrophy or atrophy of BAT. For example, in some species, during development BAT is lost and the capacity for NST is progressively lost. How this is controlled is not understood. During acclimatization to cold BAT may hypertrophy, and the capacity for NST increases. This will be dealt with more fully on pages 143–5 see also Fig. 4.32.

Short-term increases in NST, in response to cold stress are induced by the efferent control of the sympathetic nervous system and by its release of noradrenalin, which acts on specific receptors on the BAT cells. The response time is very short and the elevation in metabolism can occur within 1 min of the stimulus. Noradrenalin binds to a fat-cell plasma membrane receptor site and the binding causes a rapid lipolysis, releasing free fatty acids that are the substrate for the thermogenic respiration of mitochondria. The rate of activation, transport and oxidation of these fatty acids is very high (see the recent review by Nicholls and Locke, 1984).

In normal mitochondria the rate at which protons are extruded by the respiratory chain equals their rate of re-entry via the proton translocating ATP synthetase. This proton re-entry is thought to be the driving force for ATP synthesis and the two processes are tightly coupled. When the demand for ATP is low, proton re-entry is reduced and the proton gradient across the mitochondrial membrane enlarges. The increase in the gradient slows further proton extrusion and so slows respiration. Thus, if the proton gradient can be discharged through a route other than the ATP synthetase, respiration will continue at a high rate with the liberation of large amounts of heat (see Fig. 3.11).

Brown adipose tissue possesses a powerful special mechanism that allows rapid oxidation of substrates without it being coupled with ATP synthesis. At first it was thought that BAT mitochondria existed in an uncoupled state, but this is shown not to be the case; uncoupling is coincident with tissue activation by noradrenalin.

The BAT mitochondria possess a unique mechanism that permits normal functioning, with ATP synthesis when the cells are unstimulated, but when stimulated with noradrenalin they become uncoupled, dissipating the proton gradient and releasing heat. Nicholls and Rial (1984) describe that these mitochondria contain a 32 000 dalton protein that binds purine nucleotides, and in so doing the coupling of the mitochondria remains tight. If the binding of these nucleotides is prevented, then coupling is lost and the proton gradient is dissipated. The protein has been called *thermogenin* and is thought to form a specific channel through which the proton gradient can be discharged without ATP synthesis. Thus, the binding of nucleotides to this protein is part of the regulatory control over the heat-producing function of this cell. The stimulation of BAT by noradrenalin causes a rapid lipolysis and release of long-chain acyl coenzyme A molecules. These are proposed to dislodge bound purine nucleotides from thermogenin, uncoupling the mitochondria and stimulating heat production. A scheme, proposed by Cannon and Nedergaard (1983), is shown in Fig. 4.7 for the proposed regulation of thermogenin activity in both short-term and long-term responses.

In marked contrast with shivering, NST and exercise are additive heat sources. Indeed, it seems that in many species with the capacity for NST the shivering reflex is inhibited as long as NST can provide sufficient heat to maintain spinal-cord temperature. This suggests that NST may be more effective or economical as a source of heat than shivering during cold exposure.

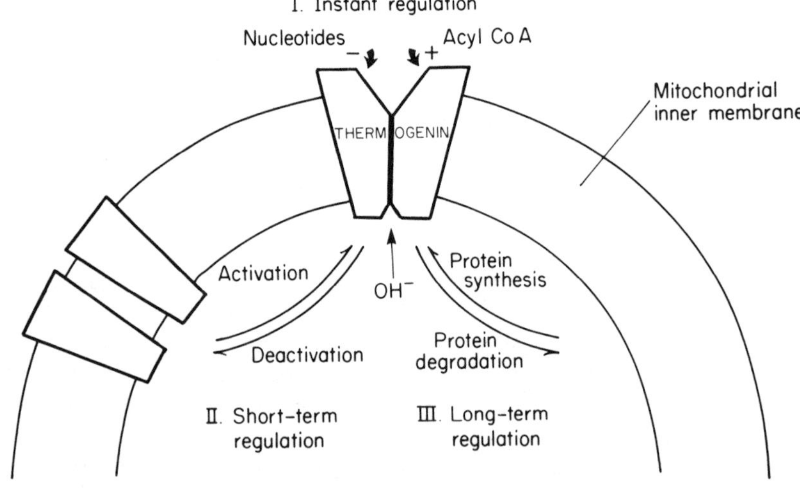

Figure 4.7 Regulation of thermogenin activity in BAT mitochondria. Three forms of regulation may occur. I, Instant. Negative modulation by GDP, GTP, ADP and ATP; positive modulation by Acyl CoA. II, Short-term. Interconversion between putative active and inactive forms. III, Long-term. Alterations in the amoung of thermogenin present by control over synthesis and degradation. (After Cannon and Nedergaard, 1983.)

4.5 Heat loss

In the tachymetabolic endotherm heat is being continuously produced internally and lost to the environment. Because T_b remains stable it must not be supposed that the heat flow from the organisms is uniform, as this will depend upon many external factors such as T_a, wind speed and relative humidity, as well as on regulatory factors over which the animal has control. In the simplest terms the total heat loss from an animal is given by Equation 3.1 and is discussed in Chapter 3.

Heat is transferred from the core to the surface by non-evaporative means, but from the surface to air both evaporative and non-evaporative transfer can take place. Heat transfer by evaporation is governed by different laws than by non-evaporative means, and so heat loss from the core to the surface and from the surface to the environment must be computed separately.

An analogy can be drawn between heat flow and the flow of electric current, i.e.

$$\text{current} = \frac{\text{voltage}}{\text{resistance}}$$

and

$$\text{heat flow} = \frac{\text{temperature gradient}}{\text{resistance}}$$

or

$$\text{heat flow} = \frac{\text{vapour pressure gradient}}{\text{resistance}}$$

The same mathematical rules apply, so when considering simultaneous heat flows that occur in parallel, as for evaporative and non-evaporative heat loss from the skin, the total heat loss is obtained by summation.

Conductive heat exchange between birds and mammals and their environment tends to be the least important avenue. Its rate depends on the difference in temperature between the skin and the environment and, of course, the area of contact between the two, and the thermal conductivity of the substratum. The larger the area of contact and the larger the gradient are, the greater the heat transfer will be.

Convective heat exchange is much more complex, as it depends upon a variety of environmental and organism characteristics (see Chapter 1). In still air the rate of transfer is simply dependent on the temperature gradient between skin and air. If air flow rises, then forced convective cooling occurs and this is more rapid since the exchange of air molecules at the surface does not rely simply on passive processes. Under forced-flow conditions the rate of heat transfer is a complicated function of the form and surface composition of the animal, and its description is beyond the scope of this book. Suffice it to say that a small animal has a greater heat loss per unit area than a large animal for a given temperature gradient.

In nature it is the net, rather than the absolute, radiative exchange that is important; that is, the difference between the radiation entering and leaving a surface. Air is a poor absorber of heat, so radiative exchanges between a bird or a mammal and its environment takes place with solid objects. The pelage lessens heat loss by radiation as the surface of fur or feathers is usually at a lower temperature than the underlying skin. Heat radiating from the skin is therefore absorbed within the pelage. Heat loss by radiation therefore tends to be greater from naked skin, but will of course depend upon skin temperature, which in turn depends on the control of the blood flow to the skin. A heat transfer model for an animal in the form of an analogue to an electric current is shown in Fig. 4.8.

In natural conditions behavioural seeking of shelter is a powerful component of an animal's response to the cold. Shelter reduces the thermal gradient from core to environment and, correspondingly, reduces the radiant and convective loss of heat. What may be particularly important is shelter from cold winds that may, in the open, penetrate the pelage and reduce its insulative value. The nest environment significantly reduces

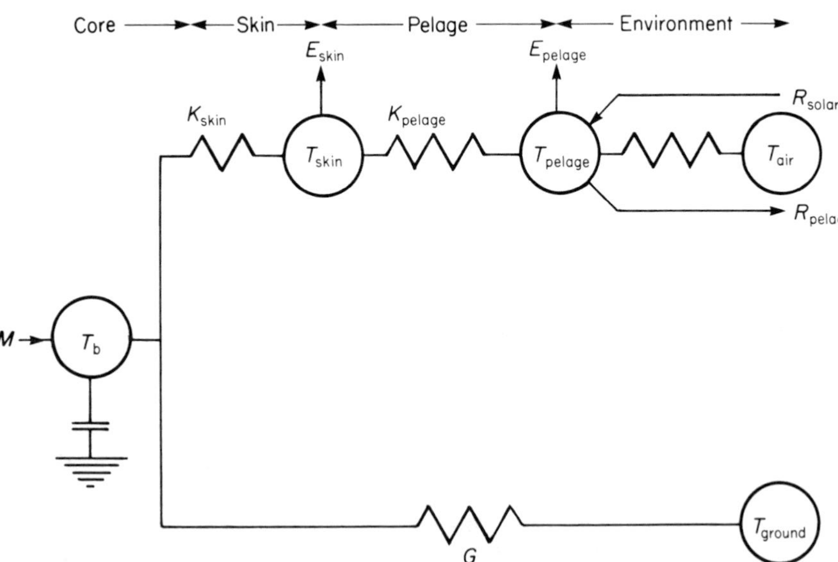

Figure 4.8 Electric-circuit analogue of heat transfer model of an animal. The circles represent interface temperatures. The resistance symbols represent conductance (K) between adjacent interfaces. E, evaporative heat loss from a surface; G, the conductance to the ground from the core and R, radiative exchanges. The capacitor symbol represents the heat storage capacity of the core. (After Bakken, 1981.)

wind speed near the nestlings, and so reduces convective heat loss. Reduction in the ratio of surface area/mass by huddling has been shown to be the most important factor affecting heat loss.

The use of a single derived value for the ambient temperature experienced by an animal, the operative temperature (T_e), was proposed by Gagge (1940), see also Equation 3.2.

In effect this is a one-dimensional model in which the animal exchanges heat by conduction, convection and radiation with an environment having uniform air and radiative temperatures. It takes no account of evaporative heat loss. Bakken (1981) has criticized this calorimetric approach because it ignores the directional nature of wind and solar radiation, and the consequent differential thermal loads experienced. He has produced a two-dimensional model with greater predictive qualities, and readers are referred to the article, and to Chapter 3, for further details.

4.5.1 *Evaporation*

When ambient temperature exceeds T_b in the absence of evaporative heat loss the animal will not only gain heat from its surroundings, but will also retain metabolic heat. If a rise in T_b is to be avoided, heat can only be lost by the evaporation of water from its surface. Effective cooling will only occur if the surface from which evaporation is taking place is richly served with blood vessels, for then the blood will be cooled and the cooled venous blood returned to the heart and then redistributed about the body.

We have seen that the driving force for evaporation is the difference in vapour pressure between the skin surface and air (Chapter 1). The rate of evaporation also depends upon the wettedness of the surface and upon the movement of humidified air away from the skin. At higher wind speeds humid air close to the skin will be replaced quickly by drier air, enhancing the evaporation rate.

Heat loss from the core to the surface is a complicated process involving forced convection by blood circulation and tissue conduction. Heat loss from the surface to the environment involves natural convection, radiation and evaporation and these are complex functions of posture, piloerection etc.

Rate of heat loss by evaporation (H_{evap}) is most simply seen as the product of the mass of water evaporated (m) and the latent heat of vaporization of water (l):

$$H_{evap} = lm \tag{4.6}$$

Two avenues of heat loss by evaporation can occur: by loss from the body surface (cutaneous) and loss from the respiratory tract. In determinations of H_{evap} both must be computed. Heat loss by evaporation is governed by the difference between the vapour pressure of the skin (P_{skin}) and of the air (P_{air}), and the area of skin wetted:

$$H_{evap} = h_e(P_{skin} - P_{air})A \tag{4.7}$$

where h_e is a coefficient that includes the latent heat of vaporization and values for wind speed and direction. H_{evap} is expressed in $W\,m^{-2}\,mbar^{-1}$.

All mammals lose water by diffusion through the skin (insensible water loss) and the lining of the respiratory tract. This basal loss is not regulated and, as shown in Fig. 4.3, is relatively independent of T_a. Regulated water loss occurs by two routes: sweating and panting.

4.5.2 *Sweating*

In mammals, active thermoregulatory (sensible) water loss occurs through the skin by way of sweat glands. These are of two morphological types, atrichial (without hair) and epitrichial (associated with hair). However, as far as thermoregulation is concerned the only important criterion for a functioning sweat gland is whether it produces a hypotonic fluid, which in response to a heat stress is discharged onto the surface of the skin.

Thermoregulatory atrichial sweat glands are apparently confined to some primates, with their greatest development in man. In many other mammalian species the epitrichial glands have a thermoregulatory secretory function, and in some species their activity matches that of the atrichial glands of man. Very high rates of evaporation by sweating can be achieved by cattle, horses and man, but in other mammals (e.g. pig, sheep, rodents and carnivores) sweating is weakly developed or absent. Figure 4.9 shows records of sweat-gland activity of some mammals at 40°C. In the pig the epitrichial glands show no secretion, whilst in some sheep and goats intermittent synchronized bursts of secretion occur, but with little evidence that the rate increases with rising T_b. In horse and man, and to a lesser extent in cattle, heating produces a continuous release of sweat, and an output that increases in parallel with the heat load. Those sweat glands involved in thermoregulation responses are stimulated by an increase in the activity of their sympathetic innervation.

The thermoregulatory efficiency of sweating has been best studied in man. Individuals born without sweat glands cannot regulate T_b in warm ambient conditions or during physical work. The amount of sweat produced in man is related to thermoregulatory requirements; however, the thermoregulatory effectiveness of sweating depends on the ability of the secreted fluid to evaporate from the skin surface. This is most efficient in a hot, dry environment. In humid conditions sweat may simply run off the surface without evaporating, and it is recorded that at high humidity heat stroke can occur at 32°C. This should be contrasted with the ability of a naked man in the hot (120°C), dry conditions of a sauna to avoid a damaging rise in T_b. However, in man high rates of sweating ($> 1\,l\,h^{-1}$) may

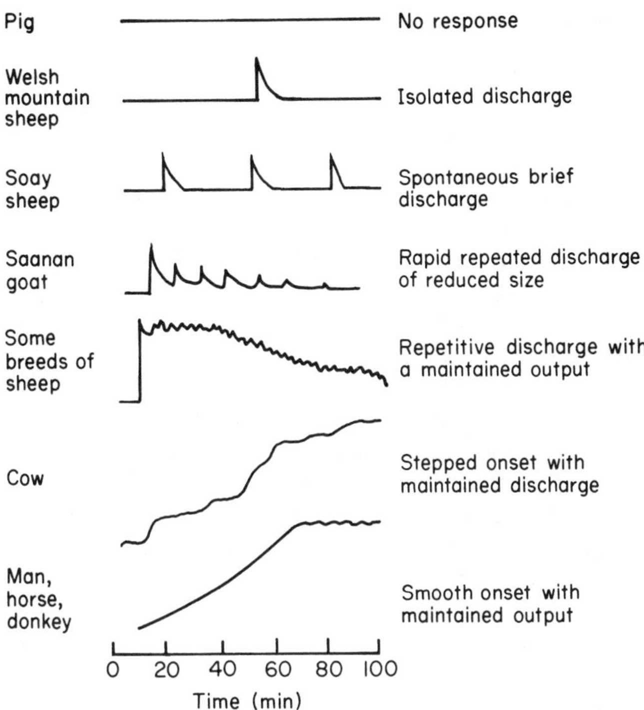

Figure 4.9 Sweat-gland activity in various mammals exposed to an ambient temperature of 40°C. These data were collected from a variety of observations in different laboratories, so it is not possible to represent the fluid discharge quantitatively. The discharge of sweat in isolated bursts probably results from the simultaneous contraction of the myoepithelium which surrounds the glands. The pattern produces an occasional discharge over a wide area of skin, but probably performs no thermoregulatory function. The other pattern of secretion is of a continual discharge in response to a thermal stimulus. The discharge of sweat occurs by displacement from the lumen as more fluid is secreted from the gland. Some myoepithelial contractions may occur but these are not the dominant method of discharge. Some intermediary patterns of secretion also occur. (After Bligh, 1967; Allen and Bligh, 1969.)

cause exhaustion of the glands when sweat production ceases and hyperthermia ensues.

In mammals there tends to be an inverse relationship between the importance of sweating and panting as those species, such as horse and man, that have whole-body sweating do not pant; whereas in poorly-sweating species, such as dogs or cats, a high capacity for panting occurs. Birds also lack sweat glands and may pant when heat stressed.

4.5.3 *Panting*

All birds and mammals lose some heat as a result of evaporation of water from their respiratory passages. This is because inspired air is cooler and less humid than expired air. Heat is lost from the evaporatory surface in both warming and humidifying this air. Some heat (and water) is conserved during expiration when the warmed and humidified expired air meets cooler respiratory surfaces, and some heat and water is given back as condensation occurs. However, in many species an increase in the heat load can lead to an increased respiratory rate and a corresponding increase in evaporative heat loss. This avenue of heat loss is the dominant one during heat stress in many birds and mammals.

In contrast with sweating, thermal cooling by panting is metabolically expensive, as it requires muscular activity to create the increased flow of air across the moist mucous membranes. A further disadvantage is that the increased ventilation during panting can lead to an increased removal of CO_2 from the blood, causing a detrimental respiratory alkalosis. However, the greatly reduced tidal volume of air during panting prevents plumonary hyperventilation, so the risk of respiratory alkalosis and a disturbance in blood acid–base balance is much reduced. This emphasizes the point that if the ventilation of respiratory surfaces is used in thermoregulation it must not interfere with respiration.

Panting, as it occurs in carnivores, is rather characteristic of panting in mammals, a topic reviewed by Schmidt-Nielsen *et al.* (1970). It consists of a series of shallow, rapid respiratory movements, driving air across the moist mucosal surfaces. The mechanisms have been well studied in the dog. At the start of panting respiratory rate switches to a new frequency that coincides with the resonance frequency of the respiratory passages. This minimizes the metabolic costs of panting, since less energy is required to cause the displacement of a system that is oscillating at its natural resonance frequency. In the dog also, a unidirectional flow of air occurs, with inspiration through the nasal chamber and expiration through the mouth. During inspiration the air is moistened, taking heat from the blood flowing through this nasal region. The humified and warmed air is then exhaled through the mouth. This separation of inspired and exhaled air is calculated to permit an extra $50\,J\,l^{-1}$ air of heat to be lost to the environment.

In some species of bird gular fluttering is used, rather than panting. The gular region lies in the floor of the mouth and the upper oesophagus. The skin in this region is highly vascular and its surface is moist (Fig. 4.10). During fluttering the birds open their beaks agape and rapidly move the gular region up and down. The air moving across this region is humidified and this cools the blood running through the skin. As the gular skin is thin, little energy is required to drive this cooling mechanism. An additional

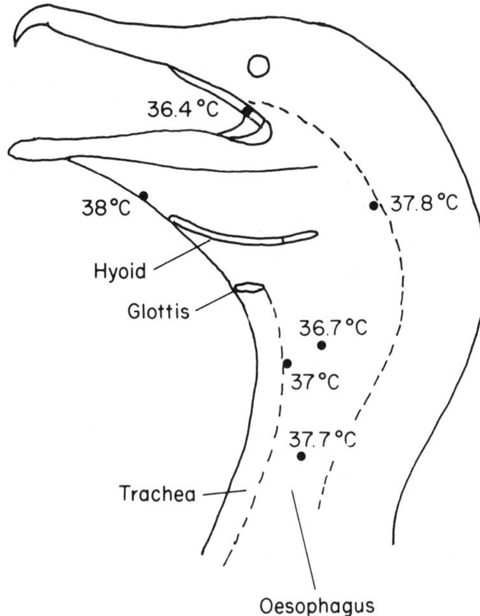

Figure 4.10 Temperatures of the evaporating surface of the buccal cavity and oesophagus during gular fluttering in the cormorant. (After Lasiewski and Snyder, 1969.)

advantage is that it also removes the possibility of respiratory alkalosis.

In many birds gular fluttering or panting occurs at a fixed frequency that corresponds to the resonant frequency of the gular apparatus or the respiratory tract and thoracic cage. In other species, however, respiratory frequency of fluttering rates increase in a step-wise fashion with increases in T_a and heat load.

4.5.4 *Saliva-spreading*

In some small mammals and marsupials the spreading of saliva on the fur is another means of inducing evaporative heat loss. This is less effective than sweating, since the fur must be wetted to the skin before significant body heat can be lost. Studies on rats show that an increased production of saliva is a thermoregulatory response, occurring when T_b is elevated to 38.5–40°C. Hainsworth and Strickler (1970) report that this secretion of saliva, like sweating and panting, is controlled by the hypothalamus, as when the hypothalamus alone is heated saliva flow from the submaxillary and parotid glands occurred. The thermoregulatory function of the saliva spreading is obvious because as T_b falls the flow of saliva is stopped.

4.6 Thermal insulation

Insulation serves to impede heat flow between the animal and its environment, so the ability to alter the insulation is a powerful thermoregulatory tactic. In fact, it is the mechanisms involved in reducing heat flow from the core to the environment that show the most powerful adaptation in most tachymetabolic animals. This means that it is more convenient to deal in the resistance to heat flow than to the heat flow itself. To draw on the analogy of Ohm's law again, heat loss could be expressed as follows

$$H_{\text{loss}} = T_1 - T_2/I \tag{4.8}$$

where I is the specific insulation and is given by

$$I = 1/K$$

The SI unit of insulation is °C m^{2}W^{-1} which is the inverse of that for thermal conductance. The *clo* is a widely used and convenient unit of insulation, devised originally to help military personnel to understand the insulative value of clothing worn by combatants. A clo unit corresponds to the value of the insulation given by clothing worn by a sedentary worker indoors and has a value of 0.155°C m^2 W^{-1}.

Insulation values are equivalent to electric resistances and are additive, to give a total insulation value, only when they act in series, as for example in the case of tissue insulation and pelage insulation, when

$$I_{\text{total}} = I_{\text{tissue}} + I_{\text{pelage}} \tag{4.9}$$

Evaporative heat loss, which is governed by different laws to non-evaporative heat loss, is not involved in the question of insulation.

The most effective animal body tissue for insulation is adipose tissue, because of its low thermal conductance. It is a living tissue, so the actual value of its insulation depends heavily on the blood flow through it. In cold ambient conditions peripheral circulation is restricted and a large thermal gradient is then created between the skin surface and the core, since skin temperature then falls dramatically to a value not much greater than that of the environment, as is shown in Fig. 4.11. The subcutaneous fat layer is most pronounced in non-furred animals, and in aquatic mammals where the thermal gradient may be some 35°C. In hot ambient conditions, however, adipose tissue has little insulative value. This is because peripheral vasodilatation occurs and warm blood then passes through the adipose tissue to the skin, the temperature of which will be close to the core temperature.

The insulation afforded by the pelage is additional to that of the tissues of the shell. Hair or feathers act to trap a layer of air close to the skin, and so impede the convective loss of heat. This trapped layer of air is the most effective part of pelage insulation, since the thermal conductivity of the

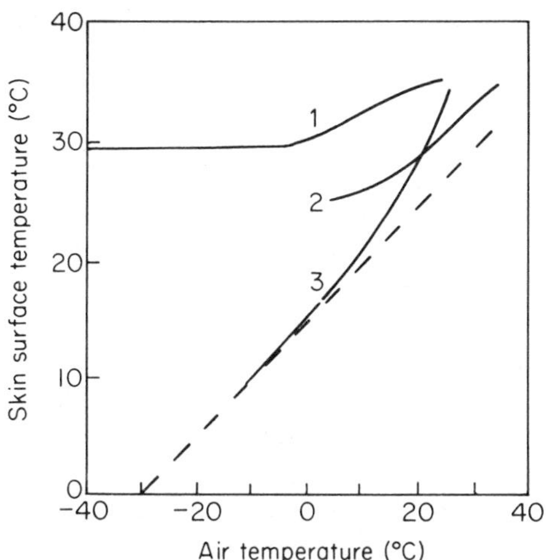

Figure 4.11 Relationship between skin surface temperature at different air temperatures and insulation in mammals from different climates. 1, Thick-furred Arctic animals; 2, Californian pigs; 3, Alaskan pigs. Note that the insulative value of the pelage keeps skin temperature high, even at very low ambient temperatures. In pigs, however, the lack of pelage, together with the insulation of the adipose tissue, means that the skin surface temperature falls with falling ambient temperature. This is more pronounced in the cold-adapted Alaskan pigs. (After Irving, 1956.)

compressed coat of a mammal is several times larger than that of an equal thickness of still air. Even the coats of Arctic mammals with a conductivity of about 50 mW m^{-1} are higher than that of still air, as shown in Table 4.1.

Aschoff (1981) has drawn attention to the fact that minimal conductance is not constant but varies with the activity pattern of the animal, and can be 50% greater in activity than during periods of rest. Aschoff points out this can lead to serious miscalculations for both energy expenditure and

Table 4.1 Thermal insulation of some animal coats compared with still air.

Animal coat	Insulation (°C m^2 W^{-1})
Still air	0.36
Red fox	0.26
Lynx	0.24
Husky dog	0.22
Merino sheep	0.22
Cheviot sheep	0.12
Cattle	0.07

conductance values of animals. His study considered a large number of birds and mammals and, on average, mammals have a 35% higher minimum conductance than birds of the same size, which probably reflects the greater insulative value of feathers than hair.

Thus, the density and length of the fur or feathers are important factors determining the insulative value of the pelage. Mammals with sparse coats are more likely to be wind-chilled as well as losing more heat by radiation from the skin surface. Piloerection or the fluffing-up of feathers increases the depth of the trapped air layer, and so increases insulation.

The importance of the form of the coat on the transfer of heat is shown in Fig. 4.12. The open coat of the Awassi sheep allows air to penetrate the coat and most of the solar radiation received is lost by convection through the coat. On the other hand, Merino sheep have a dense coat and the incident radiation is absorbed by the surface of the fleece, which may be at as high a temperature as 85°C. This heat is lost by re-radiation, since little convective exchange of heat occurs within the coat. The coat can also be used to reflect solar radiation, as is the case in the camel and Brahman cattle; this requires a smooth white surface. The form of the coat can be adaptive in insulation,

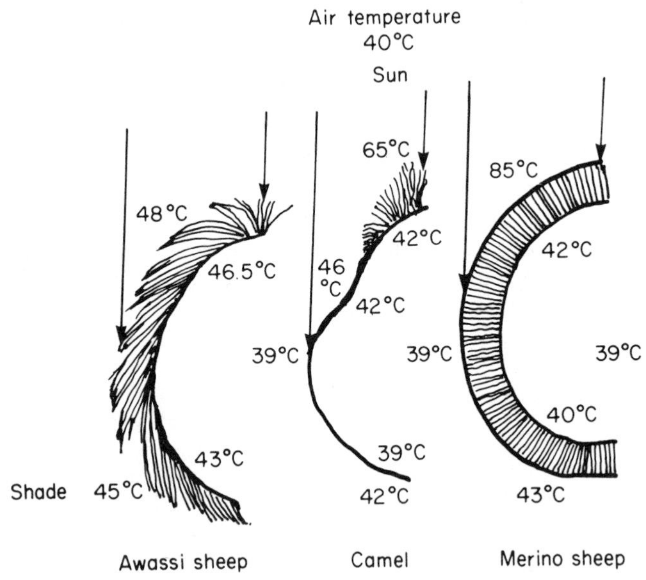

Figure 4.12 The effect of coat form on heat transfer. Awassi sheep; the open coat allows radiation to penetrate, so the mid-layers of the coat are hot. The skin is also relatively hot. Camel; the short coat reflects solar radiation and sweating is used to keep the skin cool. Merino sheep; the dense coat absorbs radiation at its surface and warms, heat is re-radiated from the wool tip. The skin remains cool, but a steep thermal gradient (about 40°C) exists across the coat. (After McFarlane, 1964.)

not only to conserve metabolic heat, but also to avoid heating at high levels of insolation.

The amount of insulation carried by a bird or a mammal is related to its size as well as the climate experienced. Larger animals carry more pelage per unit area. Figure 4.13 shows this relationship for a number of terrestrial mammals, and it clearly results from the need to compromise between having a coat that provides the most effective insulation without impeding movement. The relatively lower thermal conductivity of the coat of a large mammal is a contributory factor in its lower weight-specific metabolism (see page 104) and lower critical temperatures (T_2), see Fig. 4.3. This size–thermal conductivity relationship may be the biological basis for *Bergman's rule*, which states that in closely related mammals or birds the forms living in cold climates will be larger than those forms living in warmer conditions. The argument assumes that it will be energetically less costly in

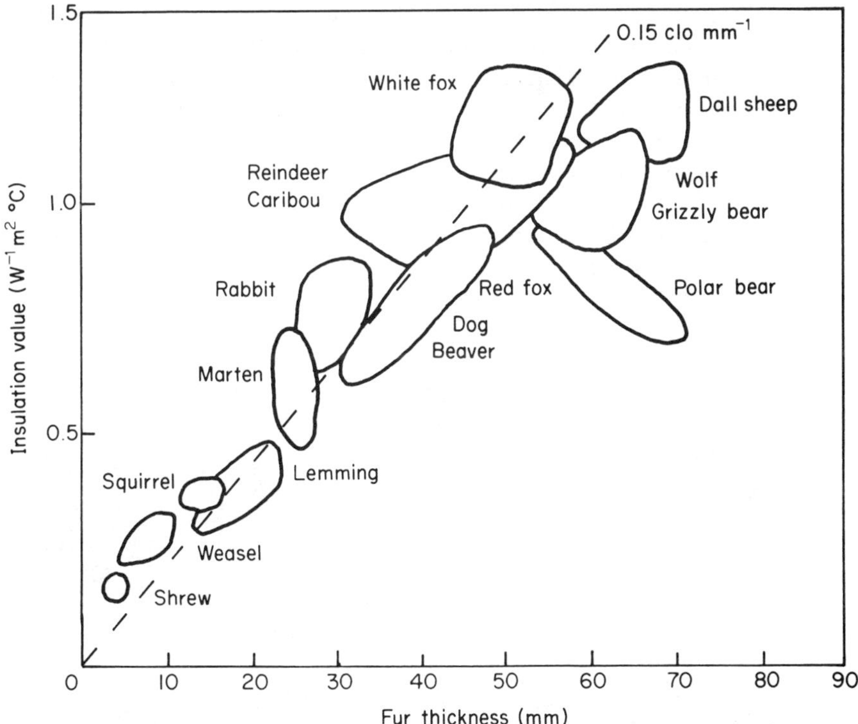

Figure 4.13 Insulation value in relation to fur thickness in a series of mammals. The smaller mammals have much less insulation than the larger ones; indeed, the insulative value of the fur of small mammals overlaps with that for many tropical species. In animals larger than the fox there is no correlation between size and insulation value of the fur. (After Scholander *et al.*, 1950a.)

a cold environment for thermoregulation if the animal is large. It assumes that thermal relationships have been dominant over other physiological and behavioural adaptations to climate during endotherm evolution, which is perhaps an oversimplification. In fact, not all birds and mammals living in the cold are large. Bergman's rule cannot simply be explained as an adaptive strategy for energy conservation, because a large animal will have a greater overall energy requirement than a small animal, notwithstanding the differences in weight specific metabolism. The large size may be related more to the greater ability to store energy, which is perhaps of more importance in a harsh environment.

4.7 Regional heterothermy

It is not possible to insulate the whole body surface, since there are times, even in extreme cold, when it is necessary for the animal to lose heat rapidly. It is usually the extremities that are relatively bare of insulation and so function at reduced temperature (Fig. 4.14). It may be that these areas require to be relatively free from insulation to permit effective function. Alternatively, the insulation of narrow cylinders (such as a bird's leg) may be ineffective because the added insulation would of necessity be very thin and might not be sufficient to overcome the increase in surface area it would bring. This suggestion is based on the 'critical radius concept' of engineering heat transfer, which shows that with cylinders and spheres of smaller than a critical radius a layer of insulation might increase rather than

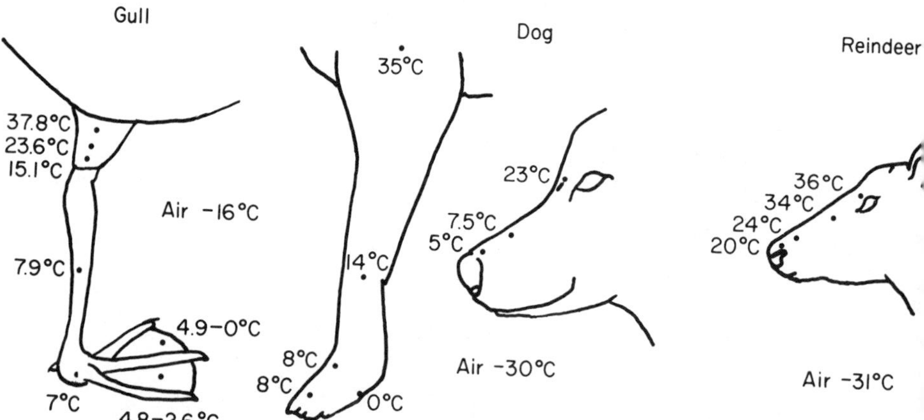

Figure 4.14 Temperatures at the extremities of three Arctic animals. Core temperatures were 38°C. A marked regional heterothermy occurs and the temperature of the extremities may be close to 0°C. The anatomical basis of this regional heterothermy can be seen in Fig. 4.2. (After Irving, 1964.)

decrease the rate of heat exchange. The application of this concept in homoiotherms has recently been discussed by Turner and Schroter (1985). It is of course more effective in heat conservation to reduce the length of the exposed extremity. This may well be the biological basis for *Allen's rule*, that animals in colder climates have shorter extremities than similar animals from warm climates. A good example, is the reduced ear-length in arctic compared with temperate or desert lagomorphs.

Heat loss from the extremities is controlled, in many endotherms, using counter-current blood flow as a heat exchanger. The anatomical basis for the system is shown in Fig. 4.2, and the working is shown diagrammatically in Fig. 4.14. During heat conservation, blood is carried out to the limb in central arteries but little passes through the peripheral circulation. Instead, the arterial blood is shunted into central veins that are closely apposed to the supplying artery. In this way the outgoing blood warms the blood returning to the core, so core heat is conserved. As peripheral tissues need to be supplied with nutrients and oxygen, the peripheral circulation cannot be closed completely, so some heat is lost. Peripheral circulation is also opened up at times to prevent the occurrence of cold-injury and frostbite. When heat loss is necessary arterial blood can be carried into the peripheral circulation, but more is also returned via superficial veins, so preventing heat conservation in the central counter-current system. This allows a large heat loss to occur, as may be necessary after intense bouts of activity.

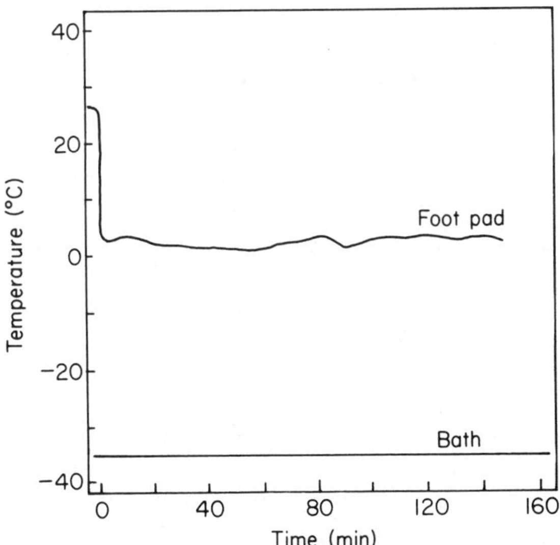

Figure 4.15 Temperature of the foot pad of a wolf on immersion in a cold bath at −35°C. Rectal temperature remained constant at about 38°C and air temperature was 18°C. Foot pad temperature shows an initial dramatic fall, but it is held above 0°C for the 2 h of the immersion. (After Henshaw *et al.*, 1972.)

Henshaw *et al.* (1972) have shown how effective the specific regulatory control of blood flow is to the foot pad of Arctic carnivores. In the wolf, immersion of the foot in a bath at −35°C causes blood flow to the foot pad to fall, but skin temperature was always held above 0°C (Fig. 4.15). Vasodilative drugs caused a massive increase in blood flow and foot pad skin temperature rose to 15°C. Anatomical studies showed that the foot pad possessed a massive arteriovenous plexus through which blood flow to, and therefore heat loss from, the foot pad was controlled. The supply of nutrients to the cutaneous tissues occurs through a more superficial but fine-structured plexus.

It is clear that the peripheral tissues in the extremities are adapted to function at low temperatures (see Fig. 4.14). This point is well illustrated by the cold sensitivity of the tibial nerve that runs to the feet of the gull. A section of nerve in the upper, feathered part of the leg will not conduct nerve impulses below about 12°C, whereas the nerve in the unfeathered leg will still conduct impulses at 3°C.

4.8 Thermoregulation

4.8.1 *The concept and models*

Core temperature in birds and mammals is regulated at a fixed and approximately stable level characteristic of the species. While the constituent cells of the body can tolerate considerable deviations from this regulated level, the deviation of core temperature is evidence of either a pathological condition or of intolerable thermal stress, which could have fatal consequences. Such an accurate and persistent regulation of body temperature is usually explained in terms of a central function that integrates afferent information about local temperatures from core and peripheral thermoreceptors. On the basis of this information the mechanism activates effector systems by which the rates of heat loss or production are varied. It is also usual to propose that the CNS has the property of a thermostat, the function of which is based upon a comparison of a signal representative of the regulated variable (core temperature) with a reference signal representative of the set level of regulation. If such a system relied solely on a deviation in core temperature to initiate a corrective response, because of the large thermal inertia of the body large changes in heat content of the body would have to occur before a corrective response was activated. There would therefore be considerable fluctuation in core temperature. By providing an early warning of changes in the thermal relationships between organism and environment, afferent information from peripheral thermoreceptors largely eliminates the need for a change in core temperature before thermoregulatory effectors are activated and so provide anticipatory responses.

Such systems are commonly employed by engineers, and it is not surprising therefore that explanations of biological thermoregulation have been based on such cybernetic concepts and so-called 'control theory'.

The two main models produced to explain mammalian thermoregulation are shown in Fig. 4.16. They differ principally in that one is based on

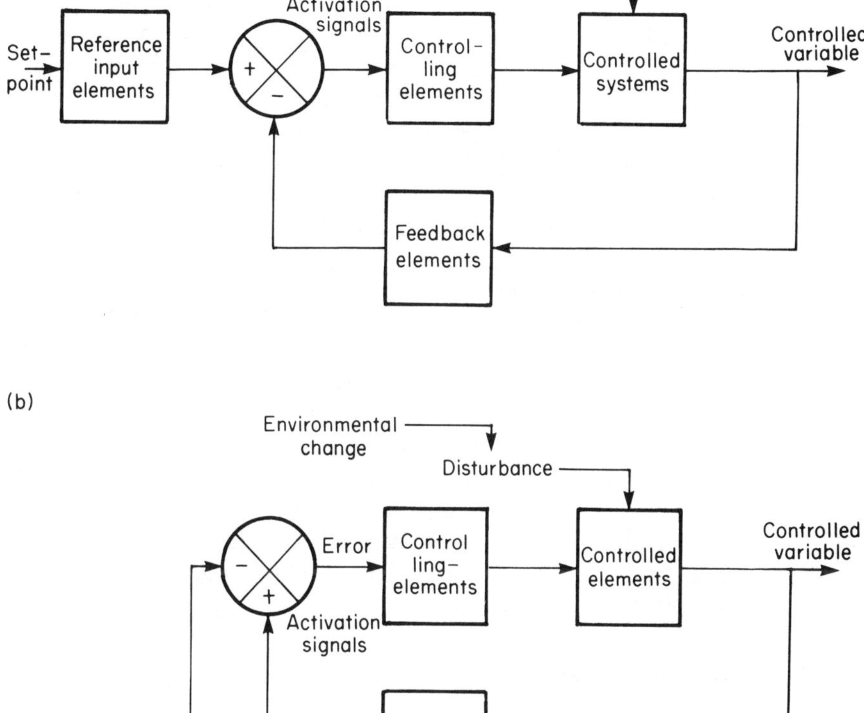

Figure 4.16 Two models for the operation of thermoregulation in a mammal. (a) System with a reference or 'set-point'; (b) system with feedback control but no reference.

orthodox control theory, and requires a stable reference signal to create a set-point, while the other does not. In the first model information on the controlled variable (core temperature) is continuously monitored and is fed back into the regulator, where it is compared with the stable reference signal. The difference between the two signals, the error, is used to activate and control the correction effectors (controller elements) which thus adjust the rate of heat production and/or of heat loss until the error becomes zero. Note that in such a system the set-point can be changed by altering the reference signal.

In the second model there is no stable reference signal. It requires the feedback to a comparator of the two opposing or reciprocal signals about rises and falls in the controlled variable. The error signal is now generated by comparing the two outputs: when they are of equal strength, they cancel each other out and the product, the error signal, is zero. When they are of unequal strength (i.e. there is a thermal imbalance) an error signal is produced which activates the controlling elements in the appropriate fashion to eliminate the error signal. In this model the 'set-point' can only be reset by changing the gain or sensitivity of the feedback elements.

A feature common to both models is that it is the intensity (a temperature) rather than the flow (heat) of the controlled variable that is detected or monitored. Other models have been proposed in which it is heat flow that is detected, and heat flow that is also controlled. These models are thought to be more unlikely to reflect the actual mechanisms than the models that detect and control temperature (intensity). This is because systems that use flow detection are subject to error as a result of 'drift'. Furthermore, thermoregulatory systems in birds and mammals actually use an inbalance in heat flow to correct deviations of T_{core} from the normal, and the flows (heat loss and heat production) are balanced again only when the deviation is redressed.

4.8.2 *Mechanisms*

The value of these analogue models is that they offer hypotheses on how the system might work. However, they still require a biologically equivalent structure(s) to be demonstrated. It is predicted that the 'controlling' structure will be one or more neural networks lying in the CNS. Whatever the mechanism, and the nature of the central controlling system, the models must possess the following properties:

(1) Responsitivity, so that fluctuations in T_{core} are narrow and steady-state values can be quickly re-established;
(2) Ability to integrate information from both peripheral and central thermosensors, whether it be reinforcing or conflicting;
(3) Ability to permit proportionality between input information and output responses;

(4) Capability of generating the thermostatic levels T_{core} characteristic of fever, torpor, hibernation and the diurnal cycle.

Whilst much is known about the mechanisms of heat production and loss in birds and mammals, the composition and working of the central regulator, or of what is regulated, is still in dispute. Thus, the identity and content of the 'black boxes' in Fig. 4.16 is unresolved. However, the following points can be made. It is clear that on–off switching systems exist for thermoregulatory effector mechanisms. As a consequence sensory information is integrated and interfaces with, the relevant effector systems. The data shown below in Figs 4.19 and 4.20 illustrate that this is the case for heating and cooling systems. In some cases, at least, these central integrators must possess an override facility, allowing the central activator to override peripheral activation. Whether it is necessary to invoke a central thermostat as a sort of reference system is also an open question. If such a system does occur, then the 'set-point' may reside in temperature-insensitive, endogenously active neurons acting as stable signal generators, or it may arise from the differential activity–temperature (Q_{10}) characteristics of two populations of CNS thermosensitive neurons.

In addition, some mechanisms must be included that allow behavioural responses to be integrated with the autonomic systems that control heat production and loss. The remainder of this chapter will consider some of the properties of these systems.

4.9 Thermoreceptors

These are thermal sensors which are widely distributed in the body. They are present in the skin, in the hypothalamus and in spinal cord in relatively high concentrations, as well as in other sites. They are generally considered to form two populations – cold- and warm-sensors – that respond dynamically to change in temperature but also respond statically to sustained temperature.

Figure 4.17 shows the static or sustained impulse traffic from discharge of several skin thermoreceptors plotted against sustained levels of skin temperature. Both plots are approximately bell-shaped, but have peaks at a lower temperature for the 'cold' sensors. The proportional relationship between the firing rate (impulse traffic) and temperature shown by cutaneous receptors is important, since it could be providing a basis for the graded response of thermoregulatory effectors to a change in tissue temperatures. Recordings made from single neurons in the hypothalamus and spinal cord of birds and mammals show that some are thermosensitive. That is, they respond to a change in local temperature with changes in their discharge frequencies. Some have a high positive Q_{10}, the firing rate increasing as local temperature rises and are considered to be warm-

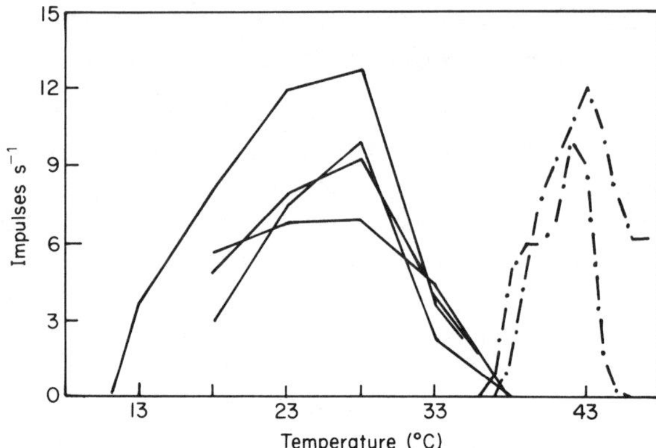

Figure 4.17 Static sensitivity curves for 'cold-receptor' (——) and 'warm-receptor' (—·—) preparations from scrotal skin. The cold-receptors had maximal activity at 28°C and were silent at 38°C. The warm-receptors were maximally active at about 42°C. There was no simultaneous activity for the two receptor types. (After Iggo, 1969.)

sensitive neurons, whilst others have $Q_{10} < 1$, their firing rate increasing as local temperature falls, and are considered to be cold-sensors.

An overlap in the activity patterns of these two groups of hypothalamic thermosensitive neurons, as shown in Fig. 4.18, provides a possible basis for the set-point. This is probably the simplest model that can be offered. If hypothalamic temperature falls below the set-point, then the activity of the cold sensors will increase whilst that of the warm sensors will decrease. This

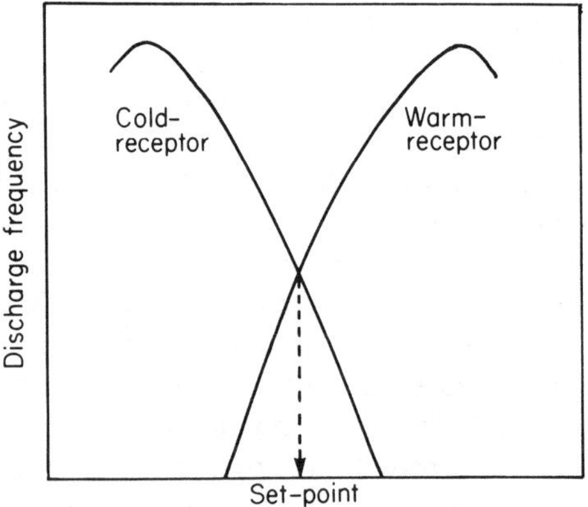

Figure 4.18 Simple model to show how 'set-point' might be determined by the interaction of the activity of cold and warm-sensitive neurons.

could result in an increase in heat production and decrease in heat loss which would drive core temperature back to the point where the activities of the warm- and cold-sensors were equal. Conversely, if core temperature rises, then the activity of the warm-sensors will be increased and that of the cold-sensors decreased, so heat-loss mechanisms will be promoted. The set-point of the controlled variable in this model resides in the reciprocal activities of the cold- and warm-sensitive neurons. It is implicit in this concept that the thermosensor neurons interface with the appropriate thermoeffector neurons. This model was originally proposed by Vendrick (1959) and could well be the basis of the temperature set-point, but only the basis, for it does not account for a number of the features displayed by mammalian and avian thermoregulatory systems (see page 126). Interestingly, this model does not incorporate a reference-signal generator, as is required in the model shown in Fig. 4.16.

The integrity of the hypothalamus is essential for the operation of thermoregulatory responses, and this implies the involvement of neurons or neural pathways in that structure in thermoregulation. Whether the hypothalamic thermoreceptors can be equated with a set-point determinant is debatable. Stitt (1983) raised the objection that there is a lack of certainty whether responses of single neurons to local temperature changes are direct effects or, rather, the result of synaptic influences of thermosensitive neurons located elsewhere. Such interpretative problems as this underlie the lack of a general consensus concerning the sites and methods by which thermal information is sensed and processed by the hypothalamus.

A further point is that although both hypothalamic and spinal-cord neurons respond to changes in temperature of 0.2–0.3°C with a change in activity, larger changes in the temperature of both the hypothalamus and spinal cord have been recorded in active animals without the activation of thermoregulatory responses (Simon, 1981). Clearly, if the thermosensitive neurons in the CNS provide a mechanism to monitor disturbances in T_{core}, converging afferent nervous pathways from thermoreceptors elsewhere in the body, and from other non-thermal sensors, must be able to modify the responsiveness of these central thermosensors.

There is, in fact, ample evidence to show that such interaction takes place between skin and central thermosensors. The data of Brück and Schwennicke (1971), shown in Fig. 4.19, demonstrate that the interaction is a multiplicative process, as witnessed by the hyperbolic relationship. Similar results are obtained for both heat-producing and heat-loss mechanisms in both birds and mammals, although some simple additive relationships have also been described.

Figure 4.19 shows the interacting skin and hypothalamic threshold temperatures at which NST can be induced. Generally NST will only occur at a relatively high T_{hyp} if T_{skin} is low, or at a high T_{skin} if T_{hyp} is relatively low.

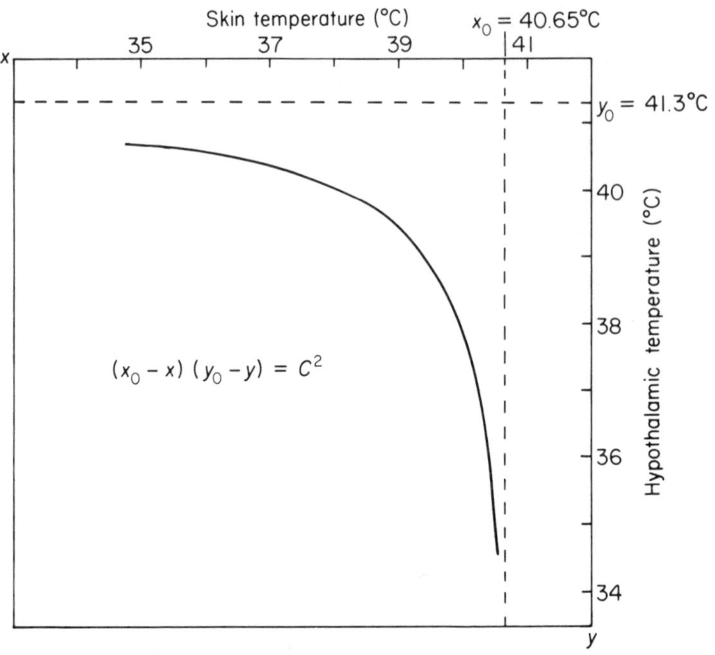

Figure 4.19 Thershold temperature for NST in the guinea pig. The curve was obtained by changing hypothalamic and skin surface temperatures independently. The equation for the best fit for the hyperbola is shown, and x_0 and y_0 are the asymptotes. These show that above $T_{skin} = 40.65$°C NST cannot be elicited no matter how low T_{hyp} is driven. Similarly, when T_{hyp} is above 41.3°C NST cannot be elicited no matter how low T_{skin} is. (After Brück and Schwennicke, 1971.)

The characteristic feature of this multiplicative processing of the two sets of information is that beyond a certain temperature (x_0 or y_0) it is not possible to evoke the compensatory effector response, no matter how far the other temperature is displaced. In the case given in Fig. 4.19, when T_{hyp} is above 41.3°C NST cannot be elicited, no matter how low T_{skin} is allowed to fall. Similarly, if T_{skin} is above 40.65°C a fall in T_{hyp} will not induce NST.

Experiments carried out on man by Benzinger (1969), in which skin and core temperatures were driven in opposite directions, have shown that the control of sweating also depends upon a complex interaction between cranial temperature and skin temperature; this is illustrated in Fig. 4.20. At skin temperatures of 33°C and above sweating does not occur if core temperature is below 36.8°C, and it increases in rate in proportion to increases in core temperature. The onset of sweating at lower skin temperatures requires a further rise in core temperature. At a constant core temperature of say 37.2°C, lowering skin temperature progressively reduced the sweating rate. So, although sweating is related to core

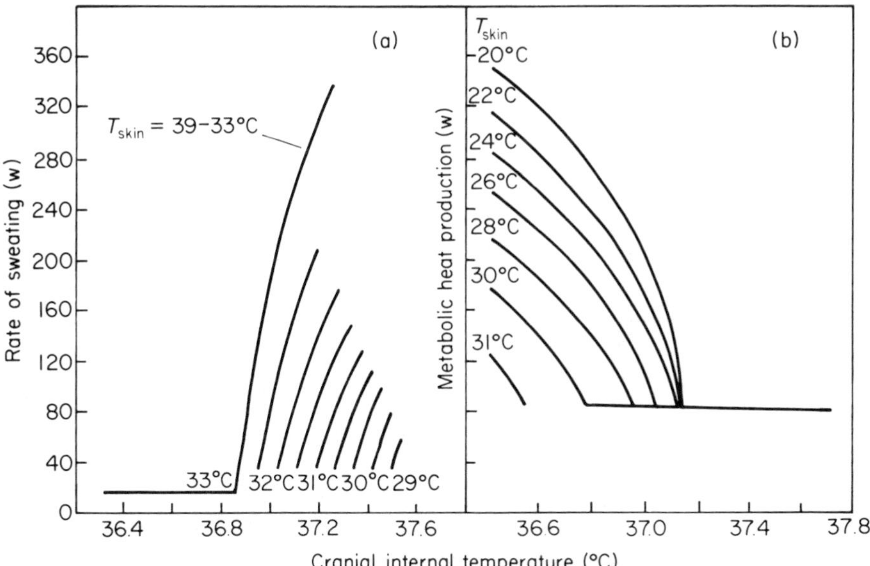

Figure 4.20 (a) Sweat rate in humans as a function of T_{skin} and tympanic temperature. At skin temperatures above 33°C sweating increases as tympanic temperature rises. However, at skin temperatures below 33°C the tympanic temperature threshold for sweating rises. (b) Heat production in humans as a function of T_{skin} and tympanic temperature. Heat production is stimulated by low skin temperatures and is inhibited by warm central temperatures. (After Benzinger, 1969.)

temperature, skin temperature determines the sensitivity of the response and also the threshold cranial temperature at which it occurs.

A very similar relationship holds between heat production by shivering and cranial and skin temperature in man (Fig. 4.20). Cranial temperature must fall below 37°C when skin temperature is above 20°C before shivering can be induced and, significantly, the higher the skin temperature is, the larger the fall in core temperature must be before shivering starts.

A further important feature of thermoregulatory responses that is demonstrated by these techniques is that there is no overlap in the different thermoregulatory effector activities of heat production and loss (Fig. 4.21). The hyperbolic shape of this curve again emphasizes the multiplicative function between peripheral and central thermoreceptors. At the threshold temperatures for shivering and panting, or between their thresholds, there is a null-point or null-range where neither panting nor shivering occurs; when core and skin temperatures both fall below their threshold value for shivering, shivering will always occur, and if both temperatures rise above the threshold for panting, then panting will always occur.

The integrative processing of thermal information must be built into, or

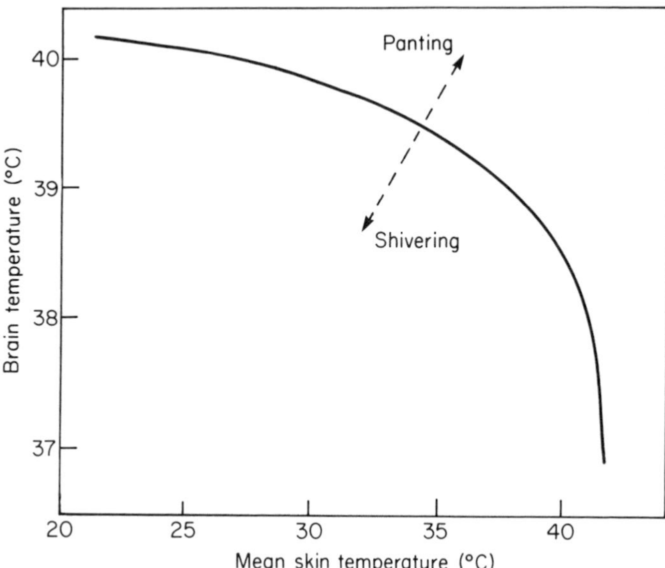

Figure 4.21 The threshold temperatures for T_{skin} and T_{brain} for eliciting either panting or shivering in the dog. It is significant that no overlap in the thermoregulatory responses occurs. (After Chatonnet, 1983.)

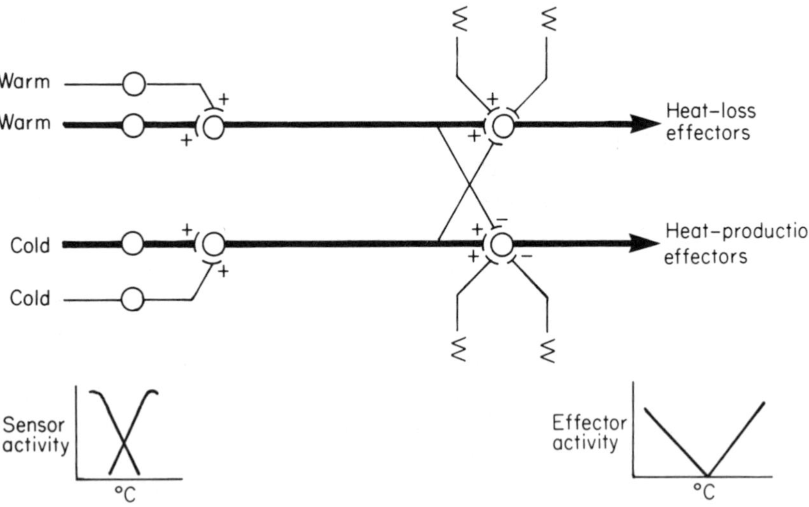

Figure 4.22 Model to show interaction of peripheral and central thermosensors, and their interface with efferent effector neurons stimulating heat loss or heat production mechanisms. $\lesseqgtr$ indicates the position of stimulatory (+) or inhibitory (−) synaptic connections from both thermosensors *and* non-thermal influences. Central pathways are considered as dominant, and these are shown in bold. Note the proposed cross-inhibitory influences. (After Bligh, 1984.)

at least taken into account, in any model that is produced to represent thermoregulatory systems. Figure 4.22 is a neuronal model, as proposed by Bligh (1973, 1984), and is a sophisticated form of a model produced earlier by Vendrick (1959). The principles of this model are that information from central and peripheral thermoreceptors converge on an interneuron which subsequently interfaces with effector neurons. This means that stimulatory and inhibitory influences can interact at specific synapses, and the resultant effect be integrated and transmitted by the interneuron receiving those diverse inputs. A further important feature is the proposal that cross-inhibitory connections are made between the pathway from warm-sensors in heat-loss effectors and that from cold-sensors to heat-production effectors: this will, of course, satisfy the observations that effector mechanisms of heat production and loss are non-overlapping. It will also fit the observations made from the Benzinger experiments that the core temperatures at which an effector mechanism is evoked can be modified by peripheral information. The synapses then are seen to act as signal mixers, which permit information from non-thermal parameters and physiological disturbances to affect the thermoregulatory device, and modify the set-point of body temperature relationships between thermosensor activities and thermoregulatory effector activity.

Some supporting evidence for the specific involvement of neuronal pathways in thermoregulatory signal processing comes from the work of Bligh and others, who administered neurotransmitters, agonists and antagonists, as well as pyrogenic agents, to specific areas of the brain. The application of these agents close to the CNS site of the interface between the thermosensors and effector neurons results in a clear and repeatable pattern of thermoregulatory responses. Whilst it is difficult to interpret this work in detail owing to the multisynaptic nature of the pathways involved, it has been possible to relate some of the putative transmitter substances used with specific responses, and so to predict the presence of specific synapses in particular pathways.

Simon (1981) has discussed the phylogeny of temperature regulation, and emphasizes the likelihood that birds and mammals have inherited the same basic integrative network for thermoregulation. From this premise the properties of thermosensory afferents in the CNS of birds and mammals are compared, and a unifying model for thermoregulation in birds and mammals is proposed.

4.10 Disturbances of the set-point

A number of small mammals and birds are able to suspend the normal control over body temperature, allowing body temperature to fall towards environmental temperature. Such a lowering of T_b is evidently an adaptive strategy for conserving energy at times when the food supply is insufficient for the maintenance of homoiothermy. This suspension of homoiothermy may be a daily or a seasonal event. The latter is referred to as hibernation in winter and aestivation in summer. This phenomenon is widespread and of polyphyletic origin; however, there are a number of features common to most species.

The major difference between daily and seasonal torpor is the length of the torpid phase. Deep hibernation may require complex physiological changes, not needed in light torpor, as it is recorded that some small rodents can still gather food when body temperature is as low as 25°C. So it is not known how the suspension of homoiothermy occurs, nor whether daily torpor and deep hibernation, even when they occur in the same species, involve the same adjustments to the processes of thermoregulation.

4.10.1 *Daily torpor*

This is a feature of some small tachymetabolic endotherms. The temperature to which T_b is allowed to fall varies between species, but approximately correlates with the ambient temperatures normally experienced. It is

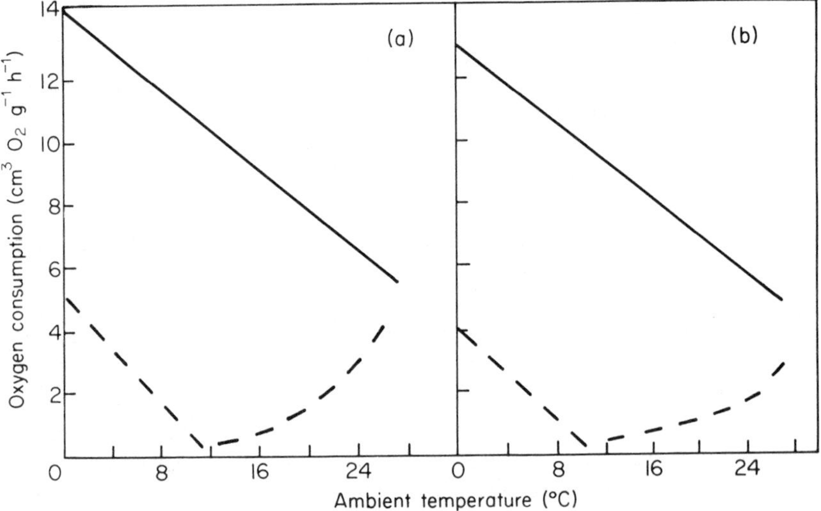

Figure 4.23 Metabolic rate in two hummingbird species as a function of ambient temperature. (a) *Panterpe insignus*; (b) *Eugenes fulgens*. Solid line is for active birds and shows the typical homoiothermic response of a rise in metabolic rate with falling T_a. The broken line shows a different pattern in torpid birds. Initially metabolic rate falls with a fall in T_a, because T_b will also fall. However, at a T_a of lower than 10°C metabolic rate rises again, showing heat production occurs to prevent further lowering of T_b. (After Wolf and Hainsworth, 1972.)

significant that animals in the torpid phase are not at the mercy of changes in T_a, since defence against a further dangerous fall in core temperature persists. In hummingbirds, as is shown in Fig. 4.23, metabolic rate rises when T_a falls below about 12°C. Note also that the MR/T_a curves are equal in slope in torpid and euthermic birds between 2°C and 12°C. Wolf and Hainsworth (1972) argue that these birds thermoregulate during torpor, but at a lower set-point.

Torpor is evidently an adaptive strategy, to save energy in animals that have a high weight-specific metabolism (see page 104) coupled with a poor capacity to store large amounts of fuel in their bodies in fat deposits. Not all small endotherms enter torpor, and whether they do so relates to the availability of a food supply. Shrews, for example, can and do feed almost continuously, and can thus fuel their maintenance of endothermy. The energy saving of torpor is considerable, since energy expended when core temperature is lowered may be only 10% of that when euthermic. Some costs seem to be associated with torpor, as hummingbirds were not found to enter torpor every night but only on those nights when their energy reserves were low. The most obvious penalty is that of increased risk of predation when torpid.

4.10.2 *Hibernation*

Small and intermediate, but not large, mammals from several different orders enter seasonal hibernation. It is probably inappropriate for large mammals to hibernate, since they are capable of storing enough fat to bridge temporary periods when food is scarce. Furthermore, their high thermal inertia, owing to their size, would make both entry into and arousal from hibernation a slow and costly process.

Hibernators use the weeks preceding hibernation to lay down food reserves for the prolonged winter period. Some, such as hedgehogs and marmots, lay down fat deposits and these species are not observed to feed during hibernation. Other species store food in caches from which they feed during periodic arousal.

It is likely that hibernation is also a strategy for energy conservation; however, its costs and benefits are difficult to evaluate over such a long period.

The principal events of interest to the thermal biologist are the preparation for and entrance into hibernation, the torpid state itself, and the periodic and terminal arousals, since each involves dramatic changes in the properties of the thermoregulatory system.

A variety of patterns of entry into hibernation have been described. The simplest state is when the balance between heat production and heat loss is changed; T_b falls abruptly and is then maintained at a new level characteristic of the torpid state. In other species, the best examples of which are the ground squirrels, entrance occurs through a series of test drops over a period of several days. Strumwasser (1960) described this as a series of falls in T_b which occurred regularly and progressively, but were interspersed with bouts when T_b was returned to normal (Fig. 4.24). In many species entrance occurs during the normal sleeping phase; indeed, hibernation is often referred to as winter sleep.

Such data emphasize that entrance is not haphazard, but a programmed and controlled event. However, little is understood of the changes that occur in the thermoregulatory mechanism to permit these changes in T_b to occur. Heller *et al.* (1977) were able to show in a ground squirrel that entrance was associated with a progressive reduction in T_{set} (Fig. 4.25). The preoptic-anterior hypothalamic (POAH) area of the hypothalamus was cooled during entry, but not to the extent that T_{hyp} fell below a level interpreted as the lowered T_{set}, then T_b continued to fall along a smooth cooling curve. However, if T_{hyp} was taken to below T_{set} metabolic rate was increased and T_b was raised, giving rise to an irregular cooling curve. This experiment shows that entrance into hibernation is accompanied by a progressive change in the hypothalamic neuronal complex associated with thermoregulation, and also that T_b is still controlled, even though the controlling mechanisms are themselves changing.

Once in the torpid state the hibernator typically has a reduced heart and respiratory rate. Heart rate and metabolic rate typically fall to be only 1/50

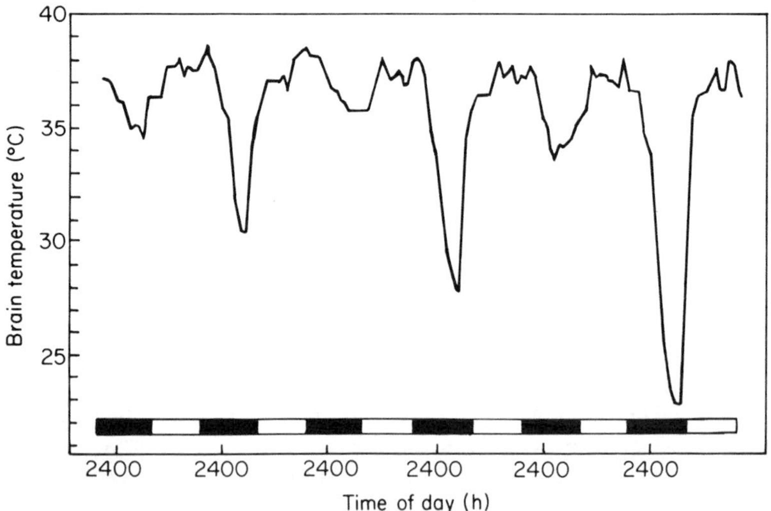

Figure 4.24 Sequential lowering of body temperature (T_{brain}) in a Californian ground squirrel over a 6-day period. Note the progressive lowering of T_{brain} occurs every other day, and that the decline in temperature begins during the night phase in each case. (After Strumwasser, 1960.)

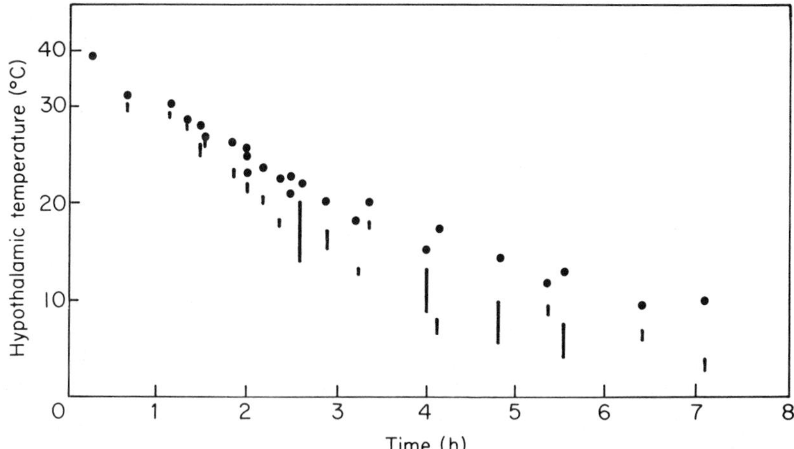

Figure 4.25 The decline in T_{set} in the ground squirrel on entrance into hibernation. The points represent actual temperature of the POAH area of the hypothalamus. The bars show the range of T_{set} at successive time intervals during entrance. This was determined by manipulation of POAH temperature. Note that actual T_b is always above T_{set}, if it were to fall below the range of T_{set}, then the animal responds by increasing heat production to raise T_b above T_{set}. This shows that, although T_{set} is changing progressively, thermoregulatory responses can still be evoked. (After Heller *et al.*, 1977.)

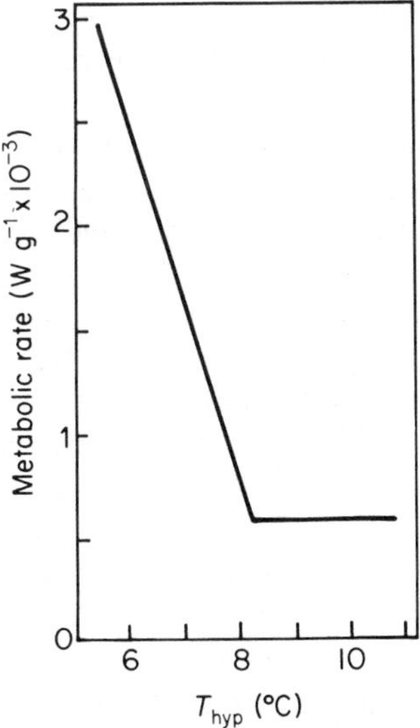

Figure 4.26 Change in metabolic rate in a hibernating ground sequirrel with lowering of the hypothalamic temperature. T_a was 3°C and when the POAH area of the hypothalamus was cooled below 8°C metabolic rate was progressively increased. In this way T_b was maintained above T_a. (After Heller and Colliver, 1974.)

of the normal rates. Cardiac output is also greatly reduced, and at about $1\ cm^3\ min^{-1}$ it is only about 1/70 of that at a T_b of 37°C. Blood is allowed to pool in some tissues whilst cardiac output is selectively supplied to others, notably brown fat. Breathing is also uneven, with long periods of apnoea separating bursts of respiration.

The actual temperature maintained in deep hibernation is usually a degree or so above the prevailing T_a. Most hibernators only hibernate when T_a is low. However, some hibernators can hibernate over a wide range of ambient conditions. In the chipmunk *Tamias strialus*, for example, this range is from 2°C to 25°C. Once in hibernation animals can increase heart rate and metabolic rate should T_a fall too low. This increased heat production provides against the threat of freezing. This is clear evidence, for example in the work of Heller and Colliver (1974) on ground squirrels, that thermoregulation is not abandoned in hibernation and that the animals are more than poikilothermic. The hypothalamus is still receptive to sensory information since, if the POAH area of a ground squirrel is cooled heat production occurs and, significantly, as is shown in Fig. 4.26, this is proportional to the hypothalamic cooling. Whether this regulation uses the same mechanisms as those which operate in the euthermic state is not known, nor is it certain that the animals defend a new T_{set}, as some animals do not respond to peripheral cooling even though they do to POAH cooling.

All deep hibernators arouse periodically from hibernation, a property not shared with the hypothermic mammal. Heat production increases dramatically and T_b is rapidly returned to normal. The periodicity maybe as short as 2–3 days, as in the hamster, or maybe longer, as in the European hedgehog.

The necessity for periodic arousal is not clear. In those hibernators that store foodstuff, rather than building up fat depots within their bodies, arousal is necessary for periodic feeding, but after feeding the animals return to the torpid state again. In hibernators that rely upon fat depots for fuel the need for periodic arousal is not obvious. It would seem to be exceedingly wasteful of energy reserves, since most of the total energy expenditure during hibernation relates to periodic arousal. It is suggested that arousal is necessary for the removal of toxic wastes, as the kidneys are reported to be inactive during hibernation. The evidence for this, however, is not strong.

The classical work of Lyman (1948) on arousal in hamsters showed that expression of an endogenous and normal periodicity, in which case hibernation could be a modification of the sleep pattern.

The classical work of Lyman (1948) on arousal in hamsters showed that at the initiation of arousal heart rate increases about 100-fold and blood pressure also rises dramatically. The heart works against a head of pressure, and this is probably due to a general vasoconstriction in the posterior region of the body. This differential vasoconstriction is essential for successful

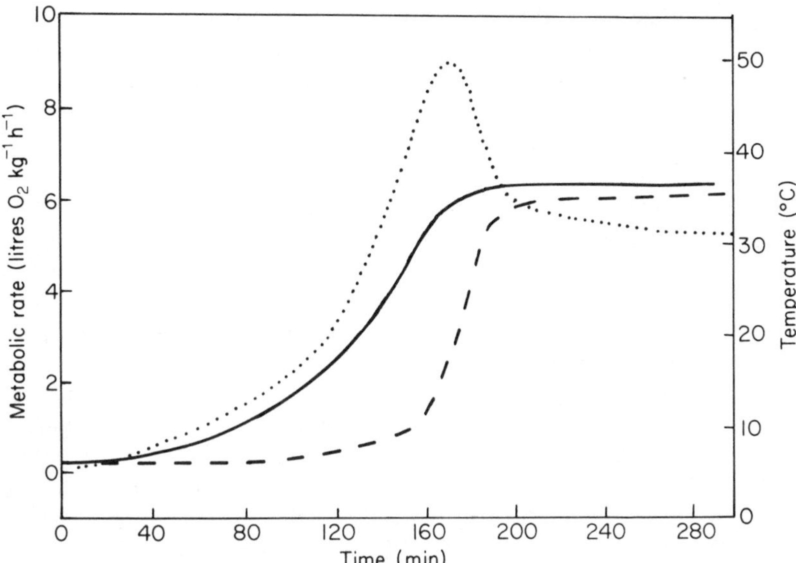

Figure 4.27 Oxygen consumption (.........), cheek-pouch temperature (——) and rectal temperature (– – –) in a hamster during arousal from hibernation. There is a dramatic rise in metabolic rate in early arousal, but the heat is not evenly distributed in the body, as cheek-pouch (head) temperature is close to normal levels before the marked rise in rectal (hindquarter) temperature occurs. (After Lyman, 1948.)

arousal, since the anterior of the hibernator warms first, whereas posterior warming is delayed (see Fig. 4.27).

The sources of heat during arousal are probably from both shivering and from NST, since shivering is often visually evident though curarized animals are able to rewarm even though shivering is prevented. The most likely source of heat from NST is the brown fat (see page 107). In hibernators this is located in the thoracic region, often close to the heart and major blood vessels. Infrared-sensitive film has shown that the temperature of the interscapular brown fat in bats is high during arousal. The warm blood is carried directly to the heart for distribution, and the restricted blood flow to the posterior ensures that the spinal cord and brain warm quickly. How arousal is initiated and controlled is not clear, but is obviously under powerful nervous control.

Hibernation and torpor are not then the abandoment of thermoregulation, and seasonal hibernation in particular requires a lengthy preparation. The hormonal events underlying the preparation and entry into hibernation are in debate but, there is a general reduction in endocrine activity that continues into the torpid phase. The status of the endocrine system during periodic arousal is also not established.

A feature of the hibernator is the wide range of body temperature over which function must be accomplished. The observed general suppression of metabolism may well be greater than that predicted from a simple Q_{10} response to a lowering of tissue temperature. Malan (1982) has suggested the observed respiratory acidosis during hibernation, which is similar to the acid–base state changes described in torpid lungfishes and turtles (see pages 182–4), is responsible for some of the reduction in metabolism.

The ability of the hibernator's heart and nervous system to function close to 0°C is dramatic, for in non-hibernating mammals function ceases at temperatures between 10°C and 20°C. The biochemical basis of this continued function at low T_b in hibernators has been the subject of much study, but it remains obscure.

Finally, the idea that hibernation is caused by the production and release into the bloodstream of a substance(s) which triggers the preparation of the tissues for torpor must be raised. Some evidence exists that hibernating, but not euthermic, ground squirrels and woodchucks have, in their blood serum, a dialysable substance that will cause hibernation when injected into other animals. At present no proven mode of action for such a putative trigger is available. Clearly, more work is required to determine whether such substances are present in all hibernating species, as well as to determine their site(s) and mode(s) of action.

4.10.3 *Fever*

Fever is associated with an elevation of the set-point, and in this respect it differs from hyperthermia, which by definition relates to body temperatures which are above the set-point when heat loss is activated. Figure 4.28 shows, diagrammatically, what seems to happen in fever. The onset of fever results from an elevation of the set-point such that normal core temperature is interpreted as being hypothermic. Thus, heat production (shivering) and heat conservation will be initiated. As a result of these processes core temperature will rise until it equals the new set-point, and normal thermoregulation will then take place about the new set-point temperature.

It is generally agreed that fever results from the release of endogenous pyrogenic substances, which are probably released from circulating white blood cells and other endothelial reticular cells, in response to an endotoxin released by invading micro-organisms. These endogenous pyrogens are proteins, and it is thought that they act on hypothalamic neurons involved in thermoregulation, but whether by direct or indirect means remains uncertain. It is suggested that the endogenous pyrogens cause the release of mediator substances that alter the sensitivity of hypothalamic neurons involved in the thermoregulatory response. There is some evidence that the mediator substances are a breakdown product of membrane phospholipid, probably a derivative of arachidonic acid, of which there are

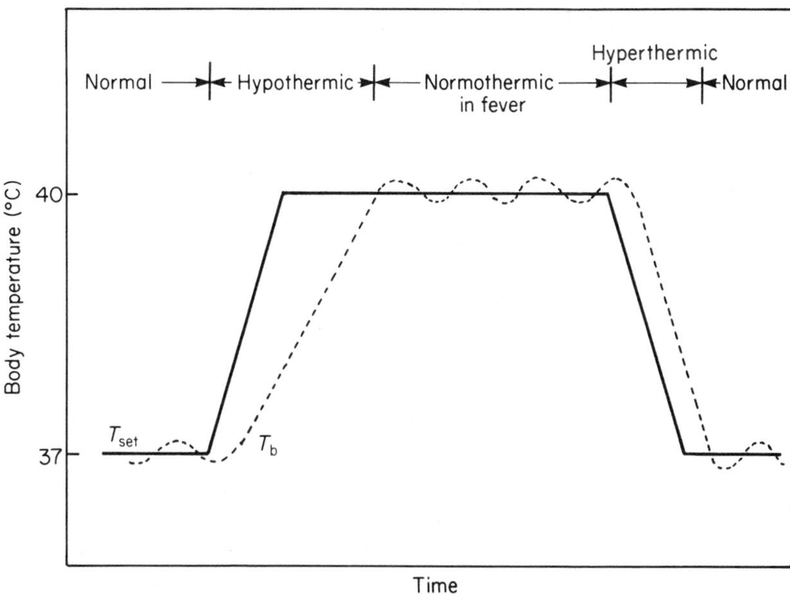

Figure 4.28 Diagram to show the possible relationships between 'set-point' temperature (——) and core temperature (– – –) in fever. It is proposed that T_{set} rises and so heat production increases and T_b rises to the new set-point and is regulated at the new level. At the end of the fever T_{set} falls to normal levels and now heat loss is effected and T_b falls to normal. (After Bligh, 1973.)

several candidates: leucotrienes, prostaglandins, prostacyclins and thromboxanes (see Hellon *et al.*, 1984).

Hyperthermia will persist as long as pyrogenic substances are being produced and released within the animal. Eventually, when the production of pyrogen ceases and its level declines, the set-point will return to the 'normothermic' level. When that occurs core temperature is then above the set-point value, so thermoregulatory heat production ceases and heat-loss mechanisms will be evoked. Sweating or panting and vasodilatation in the skin will stimulate heat loss, and core temperature will quickly fall to normal. Body temperature will then restabilize T_b at its normothermic level.

The role of fever in the body's defences during bacterial or viral infection is not clear, but it may arise from a modulation of the immunological response. It is interesting that several recent studies on fish, amphibians and reptiles have shown that a variety of ectothermal animals also display a 'behavioural' fever when injected with pyrogens, or in bacterial infection. Kluger (1978) shows that this takes the form of the behavioural selection of a higher environmental, and therefore body, temperature than that which is normally selected. Thus, fever may well be widespread in the animal kingdom, and may constitute a defensive physiological response to infection.

4.11 Adaptive and acclimatory mechanisms

In coping with changes in prevailing thermal conditions birds and mammals make a variety of responses, some physiological, some morphological and some behavioural. It is important to stress the unity of purpose of the different mechanisms employed, since they summate to form the operant response to climate changes.

4.11.1 *Homoiotherms in the cold*

Three basic genotypic responses of unchanging endothermic homoiothermic animals to low temperature are possible. All are directed to the maintenance of a constant T_b whilst minimizing energy expenditure. First, animals are adapted to minimize heat loss by possessing a more effective insulation (Figs. 4.13 and 4.29). Secondly, animals must possess a means of augmenting heat production if hypothermia threatens. The third adaptive strategy is the behavioural one of cold avoidance. This may simply be the construction of a den or nest, or the seeking of available shelter, but may also involve complex migratory behaviour to seek a more favourable, but distant, habitat.

In addition to those genotypic features which adapt a species to environmental cold, endotherms can make seasonal phenotypic adjustments to insulation and metabolism. The thermoregulatory consequences of these acclimatizational responses are shown in Fig. 4.30. Consider first that the animal changes its insulation and so its thermal conductance, without any change in standard metabolic rate (Fig. 4.30(a)). Winter acclimatization in this case reduces the slope of the MR/T_a plot, reflecting a lower requirement for heat production as T_a falls. The change in slope also predicts that the lower critical temperature (T_2) will be lowered since the lowered slope of the plot of metabolic rate against T_a still extrapolates to T_b on the x-axis. Thus, the thermoneutral zone (TNZ) is extended to lower ambient temperatures as the thermal insulation increases. The second case, (Fig. 4.30(b)), is where there is no seasonal reduction in thermal conductance (no increase in insulation) but standard metabolism is elevated. Here the slope describing the relationship between metabolic rate and T_a below T_2 is unchanged, but there is an extension of the TNZ to a lower T_2, because the raised standard metabolic rate delays the need to augment it with heat production by shivering. The actual responses made by endotherms to the cold is a mixture of those two responses.

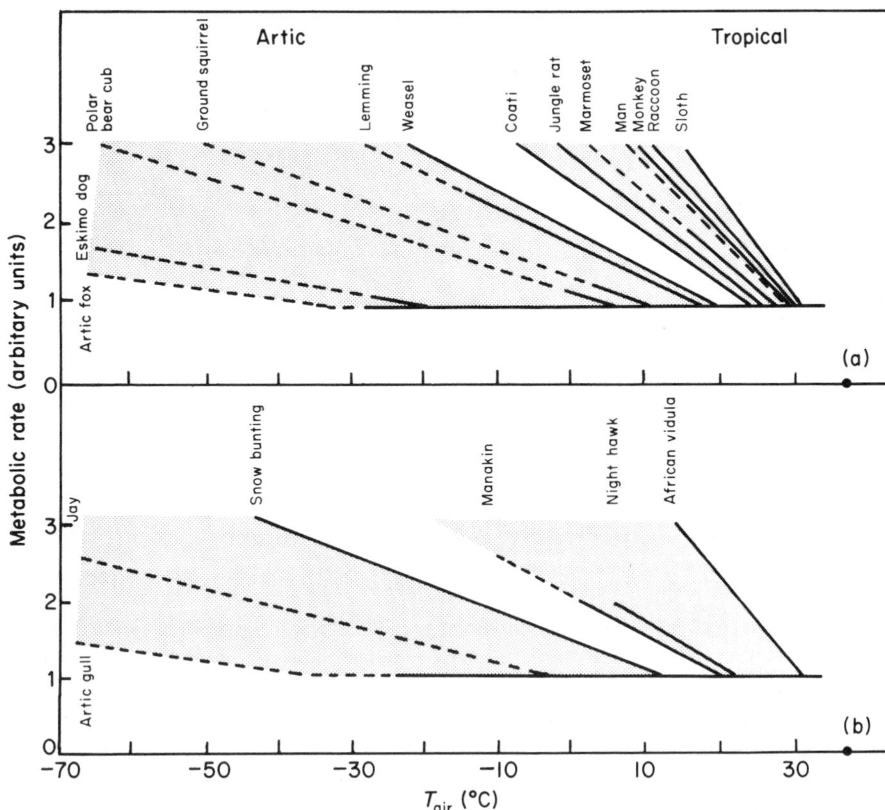

Figure 4.29 Metabolic rate ambient temperature curves for some (a) tropical and Arctic mammals and (b) birds. In each case the metabolic rate in the thermoneutral zone has been normalized to unity. In tropical endotherms the lower critical temperature is close to normal T_b and the metabolic rate–temperature curves are steep. This is because these animals are relatively poorly insulated compared with Arctic species. The broken lines are extrapolated values. (After Scholander *et al.*, 1950b.)

Larger endotherms, because of their greater capacity to increase coat and plumage quality, are likely to rely more heavily on the former strategy. For example, Fig. 4.31 shows that in red squirrels there is a much smaller increase in insulation than in the larger wolves and wolverines. The semi-aquatic polar-bear, which has to rely on its subcutaneous fat layer when in water, still improves the quality of its fur to cope with lower winter air temperatures when on land or ice.

Very small endotherms are physically unable to increase fur or plumage thickness or density to the same extent as larger endotherms. If they remain active in the winter they must, of necessity, elevate their metabolic rate. An

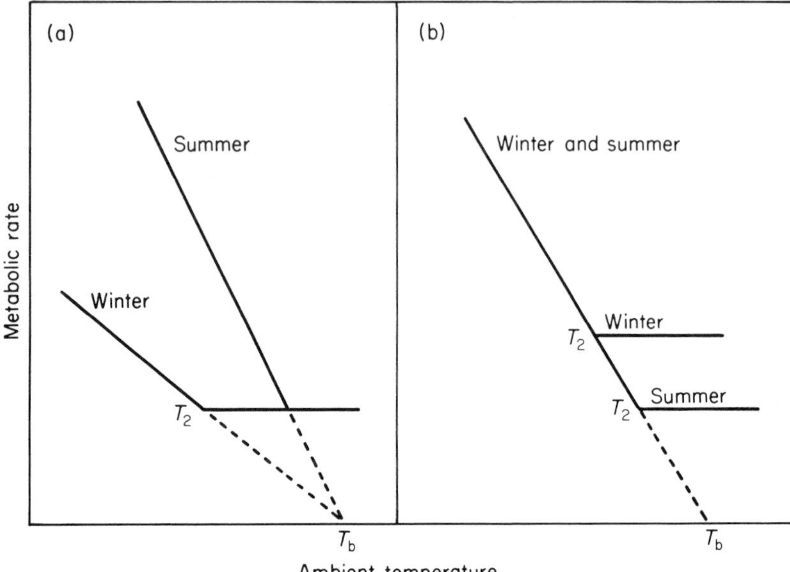

Figure 4.30 Acclimatization strategies available to homoiothermic animals. (a) Insulation increases (conductance decreases) with no change in standard metabolic rate. As a consequence, the TNZ is extended to a lower ambient temperature. (b) No change in insulation (conductance is unchanged) but there is an increase in standard metabolic rate. The TNZ is slightly extended to a lower ambient temperature.

increase in basal metabolic rate will provide an increased capacity for heat production, even if the capacity to produce heat by shivering or NST is unaltered. However, in small endotherms there is generally also an increase in heat production by NST. In cold-acclimatized rats, for example, exposure at temperatures below T_2 results in a higher metabolic heat production than in the same rats when warm-acclimatized. The second type of response of Fig. 4.30 is thus seen principally in very small mammals, such as shrews.

The metabolic basis of such cold-induced seasonal elevations in heat production and metabolic rate is species-specific, but a scheme is shown in Fig. 4.32. Both cold- and warm-acclimated rats elevate heat production in response to cold stress, but by NST in the former and shivering in the latter. This greater capacity for NST is a feature of cold-acclimatization in many mammals, and takes over from shivering as the principal method of heat production. The endocrine system, particularly the thyroid gland, has been implicated in the observed elevation of standard metabolic rate in cold-acclimatization, but the increase in NST capacity is probably a function of brown adipose tissue (Fig. 4.32), which may also hypertrophy under the influence of thyroid hormones.

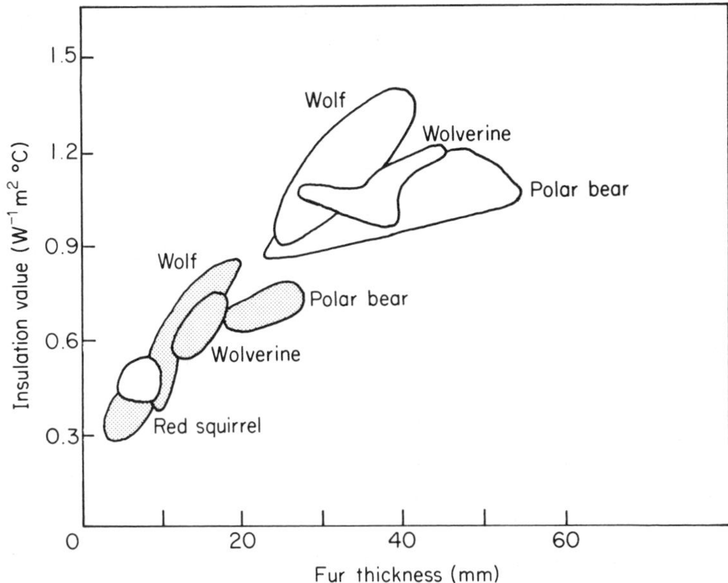

Figure 4.31 Seasonal changes in coat thickness is some mammals. Summer condition is shaded and winter condition is open. The small red squirrels make a smaller seasonal increase in fur thickness than the larger species. (After Hart, 1956.)

Cold-acclimatization has also been demonstrated in birds. Most of the responses concern an increase in plumage thickness and density, particularly of down feathers. In some larger species this so lowers the T_2 that metabolic responses to the cold are unimportant, but in smaller species, migration or shelter seeking are necessary to avoid cold. In birds NST has not been demonstrated with certainty, and shivering is the major method of increased heat production in cold stress.

4.11.2 *Homoiotherms in the heat*

Endotherms in temperate and Arctic environments need to restrict heat loss to maintain a constant high T_b. On the other hand, endotherms living in hot environments frequently experience conditions where T_a is higher than T_b, so that there is a net gain of heat. Thus, the animal is continually threatened by hyperthermia, and the only possible avenue for heat loss is by the evaporation of water. This point is emphasized in Fig. 4.3, where non-evaporative heat loss falls progressively as T_a rises towards T_b and, concomitant with this, there is a dramatic rise in evaporative heat loss. In desert environments water is usually scarce and body fluid lost in thermoregulatory evaporation is not readily replaceable. Thus, many desert

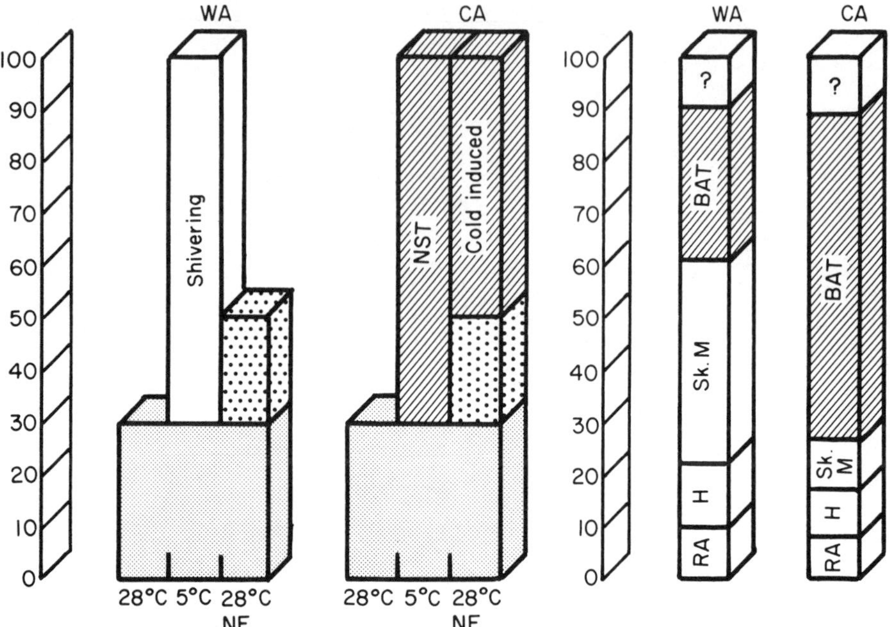

Figure 4.32 Thermal acclimation and metabolism in the rat. Basal metabolic rate is little affected by cold or warm acclimation. In warm-acclimated rats (WA) exposure to 5°C causes an increase in metabolism, as a result of shivering, over that at 28°C (the lower critical temperature). The administration of noradrenalin (NE) produces only a small rise in metabolic rate. In the cold-acclimated rats exposure to 5°C causes a large increase in metabolism due to NST. Note the large difference in NE-elicited metabolism between cold- and warm-acclimated rats. The other histograms show the approximate tissue contributions to heat production in cold- and warm-acclimated rats. Cold-acclimation causes an increase in BAT heat production and reduced contribution from skeletal muscle (Sk.M). The contribution by the heart (H) and respiratory musculature (RA) and by other tissues is changed little by acclimation. Cold-acclimation increases BAT metabolism and NST, whereas in warm-acclimation thermoregulatory heat production is from shivering. (After Cannon and Nedergaard, 1983; Foster, 1984.)

animals conserve water by producing dry faeces and urine which has a higher tonicity than sea water.

In a hot environment heat gain by an animal comes in part from its metabolism, but also from the environment. The latter component is gained across the surface area of the animal and, as metabolic heat production is also roughly proportional to surface area, the total heat load experienced is related to surface area. Thus, a small animal, with its relatively greater surface area/volume ratio, suffers a greater heat load per unit mass than does a larger animal. It is estimated that during severe heat stress a 100 g rodent would have to evaporate body water at a rate amounting to about 20% of its body weight per hour if it were to maintain a constant T_b. In contrast, a 500 kg camel would need to loose less than 10% of its body water per hour. Size, therefore, is also an important factor in the adaptive response to heat.

Since, in such environments, small animals clearly cannot replace such large volumes of water expended in thermoregulation, other strategies are needed. Many use avoidance behaviour and, when environmental heat is most intense, seek shelter in burrows or crevices. They come to the surface only at night or at periods around dawn and dusk. Some small mammals, such as the antelope ground squirrel, however, are active even in the heat of the day. Bartholomew (1964) has reported that these animals allow rise in T_b during exposure to solar radiation; however, they make repeated returns to their cool burrows, where the stored heat can be unloaded by non-evaporative routes (Fig. 4.33).

Heat storage is a strategy that is more common and of more value to larger animals living in arid tropical environments. The large size of a camel, for example, means avoidance behaviour is of limited value. However, it is able to store large amounts of heat during the day. A camel's body temperature may rise from 34°C to 41°C between dawn and sunset.

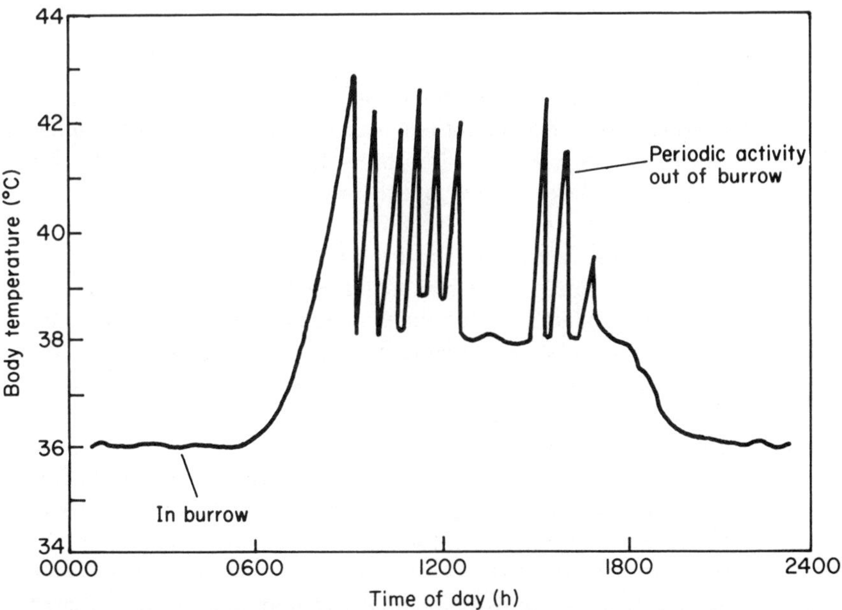

Figure 4.33 Body temperature during the day in the antelope ground squirrel. When active out of its burrow in the day T_b rises. This causes frequent returns to the cold burrow, where the stored heat can be lost. (After Bartholomew, 1964.)

Schmidt-Nielsen (1983) estimated that this saves some 5 litres of water which would be lost on evaporative cooling were the camel to be strictly homoiothermic. Indeed, when the camel has access to water to drink, it then stores less heat and its daytime rise in T_b is only about 2°C, since evaporation is effective. The heat stored by day is lost largely by radiation to the cold night sky and by convection to the cold night air. In the camel heat storage and tolerance of hyperthermia is estimated to reduce water loss by some two-thirds that of a hydrated camel. Other large desert-living mammals, notably the oryx, have been shown to tolerate large rises in T_b for long periods, contrasting with the oscillation in T_b employed by the ground squirrel (Fig. 4.33). In some cases T_b rises above T_a, so the animal can lose some heat by non-evaporative means, but still the heat gain from solar radiation will be the major factor.

As was mentioned earlier, the pelage affords protection against intense radiation loads. This thermal shield can be extremely effective, as in the case of the merino sheep (Fig. 4.12) where fleece surface temperatures can be as high as 85°C but skin temperature is still only 40°C. Much of the heat absorbed at the surface of the coat is either re-radiated or lost by convective cooling, with a steep gradient existing across the thickness of the coat. In some species the coat is sleek and glossy, as in Brahman cattle, and much radiant heat which would otherwise enter the animal is reflected and re-radiated (Fig. 4.12).

The ability to tolerate a rise in T_b during heat stress is widespread in birds. Because of their relatively small size this is not as significant in terms of heat stored, as in a large mammal, but it allows the thermal gradient to be restored. Heat loss can then occur by non-evaporative means, which conserves body water. Birds in heat stress cannot sweat but cool by evaporative water loss by panting or by gular fluttering (Fig. 4.10), which, as was explained earlier, requires the expenditure of less energy and so produces less metabolic heat than panting.

Warm acclimatization results in a reduction in basal metabolic rate, and so in heat production. This phenomenon is probably hormonal in basis, and in mammals it is associated with reduced thyroid activity. Salt and water loss are also reduced in warm-acclimatized animals, but this is probably an osmoregulatory requirement and only indirectly concerned with thermoregulation.

The rise in body temperature necessary to cause functional impairment is only some 4–6°C in most birds and mammals. The structures most at risk by such hyperthermia are the brain and the liver. Taylor (1970) reported that an interesting adaptation in some mammals (notably carnivores and artiodactyls) and birds is the ability to maintain brain temperature lower than body core temperature during heat stress. The morphological basis of the brain-cooling system lies in the vascular supply to the brain (Fig. 4.34). The brain is supplied with arterial blood which in some species is distributed

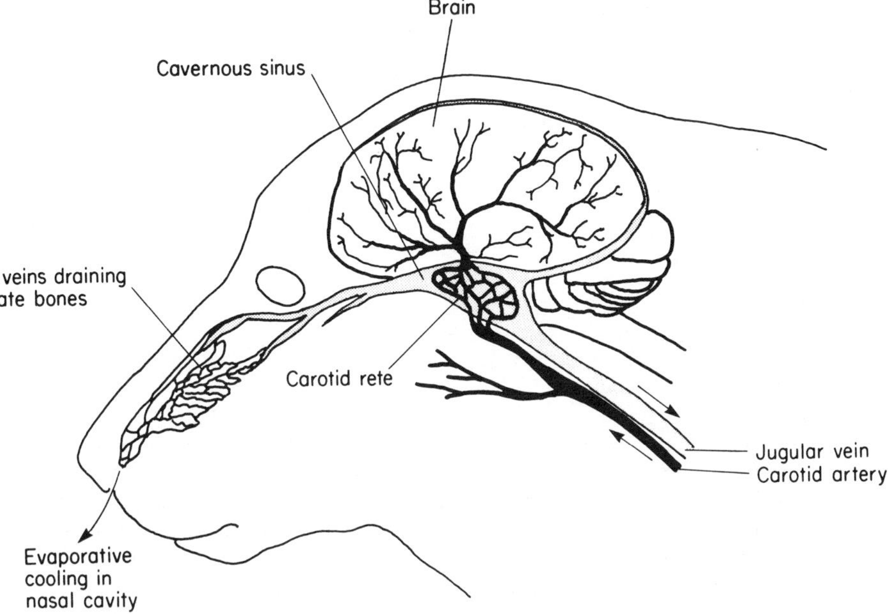

Figure 4.34 Supply of blood to brain in the dogs. The brain is supplied by the carotid artery. It breaks into a capillary rete within the cavernous sinus before entering the brain. Heat exchange occurs here between cool venous blood draining into the cavernous sinus from the nasal cavity and the warm arterial blood in the carotid rete. This prevents blood at high temperature supplying the brain. (After Baker, 1979.)

from the circle of Willis in the floor of the cranial cavity (Baker, 1979). The circle of Willis receives blood from the carotid arteries. In carnivores and artiodactyls the carotid artery enters the cavernous sinus and breaks up into a capillary network, the carotid rete, before going on to feed the circle of Willis. Blood enters the cavernous sinus from the veins draining the nasal cavity, ophthalmic region, face and mouth. Whilst no mixing of the venous and arterial blood occurs, the cooled venous blood draining from the nasal mucosa cools the arterial blood passing through the rete to the circle of Willis.

In sheep, cat and dogs brain temperature has been found to fall when exposed to heat stress. Deep-body temperature on the other hand rises. Brain temperatures can be maintained 1–1.5°C below the carotid arterial blood temperature only as long as the animal is allowed to pant. However, there is evidence that even in species lacking a carotid rete, blood draining the mucosa and facial surfaces can cool the brain, albeit in a less effective manner.

However, it is during exercise that the mechanism in carnivores and artiodactyls is most effective at brain cooling. Figure 4.35 shows that in an

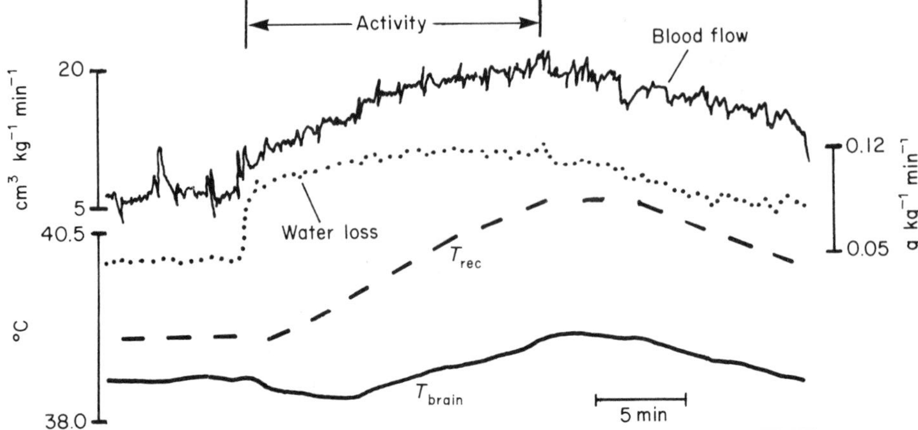

Figure 4.35 The effect of exercise in carotid blood flow, evaporative water loss and T_b in dogs. The onset of exercise causes a rapid rise in water loss, associated with evaporative loss from the nasal region. Body temperature rises steadily during activity; initially brain temperature falls slightly, but shows only a small increase compared with body temperature (T_{rec}). (After Baker, 1982.)

exercising dog the cranial blood supply increases and there is a rapid onset of evaporative water loss. Brain temperature initially falls slightly, but at the peak of exercise has risen by about 0.5°C when rectal temperature has risen by almost 3°C. In long-term exercise, dogs can maintain brain temperatures 1–2°C below carotid artery temperature. Similar results have been obtained with running gazelles.

Many birds also maintain a brain temperature lower than their body core temperature. This seems to be a feature of the ophthalmic rete, in which a network of veins carrying cool blood from the front of the head, the eyes and upper respiratory tract run close to arteries carrying blood to the brain. This provides a similar morphological heat exchanger as the carotid rete in mammals.

An interesting corollary to this story is the recent suggestion by Wheeler (1978) that some dinosaurs cooled their brains in a similar fashion. A number of dinosaurs possessed large venous sinuses that may have cooled the brain directly. In others, the hadrosaurs for example, elaborate head and nasal structures may have served thermoregulatory brain-cooling functions. If this were so the existence of these structures might indicate that these animals were liable to hyperthermia.

4.12 Ontogenetic aspects of thermoregulation

There is a marked variation between species in the capacity for thermoregulation in the young. In some species, e.g. guinea pigs and domestic fowl, the young are relatively large at birth and have a well-developed pelage of hair or down, and within the first few hours of birth attain adult levels of thermoregulation. These species also display a precocious ability for locomotion that makes behavioural selection of ambient temperature possible. In other species, such as small rodents and passerine birds, young are small, naked and helpless. In these thermoregulation is poorly developed and they are essentially poikilothermic in early life, but they develop adult levels of thermoregulation over a period of days or weeks.

The relatively small size of newborn mammals and newly hatched birds means that they have a large surface area/volume ratio. As a consequence they will be liable to a relatively large heat loss in the cold and be susceptible to hyperthermia in the warm. They will therefore have a narrower range of tolerated T_a than the adult. Thus, size is an important factor that must be considered in the ontogenesis of the thermoregulatory system.

In species with very small young it would be metabolically expensive to maintain homoiothermy; examination of Fig. 4.4 shows these young would be on the steepest part of the metabolism curve. It is therefore not surprising that thermoregulation is poorly developed at first, and nest-building, huddling and parental behaviour are essential features ensuring the T_b of the young is maintained within viable limits. Poikilothermy in these young is an important strategy that allows development at a minimal energy cost. Food energy is channelled into growth rather than fuelling thermoregulation. In the vesper sparrow Dawson and Evans (1960) showed that a 2 g hatchling grew to 18 g in 9 days, the young were essentially poikilothermic for the first 4 days, growing at about 40% body weight day^{-1}. Geiser *et al.* (1986) report a very protracted development of endothermy in a small marsupial (*Dasyuroides byrnei*). The young remain sheltered in the mother's pouch for about 30 days, after which time progressive exposure of the young occurs and at about 55 days they may be left protected only by the nest. They are essentially ectothermic up to this period. Endothermy slowly develops over a subsequent 4–5 week period, again associated with the development of fur and increased body weight. Full endothermy is only attained after about 90 days of age. Such slow development of thermoregulation may be adaptive, allowing nutrients to be utilized for growth rather than for thermoregulation. An additional interesting feature of this is that the achievement of endothermy is accompanied by the ability to enter torpor in winter. During torpor T_b falls to 15–24°C and the periods last for up to 10 h.

In these species there is also the gradual attainment of hair or feathers, and so a progressive increase in the insulation value of the pelage. This parallels a dramatic fall in surface area/volume ratios, so heat loss becomes a progressively smaller problem. Concomitant with this is the development of an increased thermogenic capacity of the young, through either NST or shivering. The attainment of the capacity for heat production and the reduction in thermal conductance is accompanied by the functional development of the hormonal and neural mechanisms of thermoregulation. On this latter aspect, that is the development and refinement of the neural mechanisms, little information is available.

It must not be supposed that in all of those species which are adapted to tolerate periods of hypothermia the young cannot increase heat production in the cold. In the rat, even in the newly born, a thermoregulatory increase in metabolism occurs when T_a falls. The size of this response rises with age until at 10 days the adult response is approached. In the ground squirrel (Gelineo and Sokic, 1953), however a reduction in T_a results in a fall in metabolism until the young are about 3 weeks old. In species with large neonates and in precocious species such as the guinea pig (Brück and Wünnenberg, 1965) metabolism is at the level predicted for an adult of the same body weight within hours of birth. In the domestic chick the capacity to defend a T_b of 40°C when put in an environment at 10°C improves rapidly within the first few days after hatching.

There are clear advantages from this pattern of development found in precocious species, since the young develop adult capacities for thermoregulation very quickly and can be active and alert within hours of birth or hatching.

4.13 Conclusions

Each group of tachymetabolic animals appears to maintain a characteristic body temperature. For eutherian mammals this is about 38 ± 1°C, it is somewhat higher for birds at 40 ± 1°C and lower for marsupials (34 ± 1°C) and monotremes (30 ± 1°C). However, for any one species the temperature of the core (T_b) is accurately maintained. No obvious relationship between the T_b maintained and size or climatic zone has been found.

What is understood by body temperature is not simple, as the whole body is not at the same temperature. The shell (periphery) is usually several degrees Celsius lower than the core, moreover the shell is also variable in size. In many species core temperature, although accurately maintained, does fluctuate in a regular daily fashion, with the lowest T_b being recorded during periods of rest.

This constant core temperature is a result of a balance between heat loss and heat production. Clearly size is an important feature in homoiotherms, since if different sized mammals maintain the same T_b the smaller animal will have the highest weight-specific metabolism, because of its relatively higher surface area/mass ratio. This argument can be extended to provide an explanation for the apparent 2–3 g lower limit for the size of birds and mammals. Such small animals have a very high surface area/mass ratio, and the resulting high conductance would require very high metabolic rates to be sustained. It has been suggested that these high metabolic rates could not be continuously fuelled, or may not be biochemically feasible.

A critical examination of this argument would suggest metabolic rate should scale with $mass^{0.67}$ because surface area scales with $mass^{0.67}$. In fact, metabolism scales to $mass^{0.75}$, an exponent significantly different from 0.67. It must be concluded that factors other than surface area must influence metabolism in homoiotherms, and so cause the limit on body size. This criticism is supported by the fact that the metabolic rate of bradymetabolic ectotherms also scales with mass by the same exponent of 0.75, so the underlying cause of this relationship must be more fundamental than thermoregulation.

However, size and scaling are clearly important factors in homoiotherm adaptive strategies. McMahon (1973) has shown that the power output of muscle scales with $mass^{0.75}$, and it is likely that the same elastic-force arguments can be applied more generally. So the physiological functions of the cardiovascular and respiratory systems would also scale with $mass^{0.75}$. This provides a plausible theoretical basis for the lower size limit in birds and mammals, as very small animals would require high levels of oxygen supply to fuel their high metabolism. These demands for oxygen may well be beyond the design limits of the vertebrate cardiovascular system. Similar arguments can be applied to account for the limit in range of body form in birds and mammals. The tendency is for a compact body with limbs that are either well insulated or show marked heterothermy. This minimizes the area over which heat is lost – an important design factor in animals with high energy costs associated with thermoregulation. This contrasts markedly with the frugal energy strategy of ectotherms, which display a wide variety of body forms; some, like the cylindrical plan of worms and snakes, would not be adaptative in birds or mammals.

In a number of species of birds and mammals the facilities for the periodic relaxation of homoiothermy has evolved. This can occur daily (torpor) or seasonally (hibernation), but both are thought to effect savings in energy when food is in short supply.

The transition from bradymetabolic ectothermy to tachymetabolic endothermy of birds and mammals required a complex army of physiological, anatomical and biochemical changes to have occurred. These changes can be resolved into mechanisms associated or involved with the control of heat exchange with the environment and an increase in weight-specific heat production (metabolism). What is clear in this complex event is that the birds and mammals had a separate origin from the reptiles, and so it is natural to look to that group for clues to the origins of homoiothermy. Although Bartholomew (1982) pointed out that it may be misleading to look for these origins in modern reptiles, a group adapted to ectothermy, nevertheless they display a sophisticated behavioural thermoregulation. This is a feature also found in lower vertebrates, and it is significant that in fish, amphibians and reptiles it has been shown that experimental changes in hypothalamic temperature influence thermoregulatory behaviour. It is therefore reasonable to predict that the basic central neural networks and thermosensors required in both avian and mammalian thermoregulation pre-existed. This basic integrative network was inherited, and qualitatively and quantitatively shaped further, as the mechanisms for heat-production and heat-loss mechanisms evolved, to produce a more sophisticated and precise system of control.

The metabolic level of modern reptiles is some 20 times less than that of birds and mammals of the same size (Nagy, 1983). Modern reptiles also show no obvious transition or shift towards the higher basal metabolic level of birds and mammals. It is considered this lower metabolic level is characteristic of the reptilian stock. If this is the case, then tachymetabolism evolved relatively late, and independently, in birds and mammals, owing to their distinct origins from the reptiles. The higher metabolic level of mammals has been shown to be a function of the larger mass of mammalian tissues which possess a greater volume of mitochondria per unit mass. This gives mammalian tissues a three- to fourfold higher mitochondrial surface area than reptilian tissues (Else and Hubert, 1985). However Jansky (1962) showed that in modern mammals tonic muscle provides about 30% of heat production at basal levels of metabolism. It is of interest that the therapsid reptiles, from which modern mammals evolved, had also adopted the upright stance of mammals, rather than the sprawling posture. This new posture would be expected to have required a greater energy expenditure, and so it is likely that the cardiovascular and respiratory systems would have also required remodelling. Clearly a remodelling of the CNS control of posture and locomotion would have been necessary, but whether any remodelling of the thermoregulatory networks occurred at this time can only be speculation. It is reasonable to suppose that the requirement to supply higher levels of oxygen to the muscles made increased oxygen available to other tissues too. This would support an increase in the volume of mitochondria in these tissues, perhaps heralding the transition to tachymetabolism.

The reptile stocks from which birds and mammals evolved had been separated for some 200 million years. Thus, the similarities in thermoregulation shown in the two modern groups must represent parallel evolution. It is, however, probably based on their common inheritance of an even longer-established neural network. The similarity in the level at which body temperature is regulated in birds and mammals suggests that the level was established at that regulated behaviourally in the ectothermic phase. Later, and independently, tachymetabolism evolved in the two groups.

5 Rate compensations and capacity adaptations

5.1 Introduction

Variations in body temperature have profound effects upon virtually all aspects of animal performance. Yet this is a situation that most ectotherms and some endotherms experience either on a daily or on a seasonal basis. Other species have also experienced shifts in temperature over evolutionary time as they invade new geographical or climatic zones. There is a large body of evidence to show that animals are not totally passive in the face of varying body temperatures. Instead they display an impressive ability to adjust their physiology in ways which overcome the direct effects of temperature variations.

Precht (1958) has drawn a useful distinction between two different forms of adaptive response to temperature, namely those which modify the ability of animals to withstand the lethal effects of extreme temperatures (*resistance adaptations*, see Chapter 6) and those which modify the physiological performance of an individual or the rate of its vital processes over the normal range of temperatures (*capacity adaptations*). In most cases these latter adjustments serve to compensate for, or offset, the direct effects of temperature variations (a phenomenon called *physiological compensation*) thereby providing these animals with a degree of functional independence from variations in their body temperature. However, this is not in conflict with the widely held belief that temperature is a very real and important environmental factor which influences activity, reproductive success and even geographical distribution, since even in those cases where such compensations occur they are rarely complete, and even then are complete only over a restricted range of temperatures. There is no clear or easily understood relationship between capacity and resistance adaptations. For example, there are reports of animals which display resistance adaptations but not capacity adaptations and vice versa, so that it is more convenient to treat them separately, even though they may be different manifestations of the same basic adaptive mechanism. In this chapter we consider the phenomenon of capacity adaptation and its physiological basis.

Adaptation to a given situation usually consists of a genetic and a non-genetic component. Genetic adaptation is the basis for evolution. Individuals with favourable characteristics in a given environment survive and reproduce better than less-fit individuals, and if these favourable characteristics have a genetic basis they gradually become fixed within a population over a large number of generations. Documented cases of genetic adaptation to temperature are usually based upon comparisons of populations from different latitudes or microclimatic zones, or of species from different thermal habitats. Often these comparisons are difficult to evaluate because temperature is rarely the only environmental factor to consider and the observed characteristics may make sense only when the wider ecological context of the species is considered.

Non-genetic adaptations, which are commonly termed *acclimatizations*, involve changes in the performance of an individual during its lifetime. Usually these adjustments are compensatory in nature but other types of response exist that are no less adaptive within their particular ecological context. So marked are these adaptive abilities that it is generally recognized that the thermal experience of animals must be carefully controlled before the experiment if the results are to have validity. Indeed, information on the lethal limits of animals, or upon their physiological performance, is of little value unless accompanied by a description of their previous thermal history. Acclimatization involves changes in the phenotype, the limits of which are set by the genotype. These adaptations are induced by environmental stimuli, are usually reversible but, in contrast with genotypic adaptations, are not passed on to succeeding generations. Rather, it is the ability to acclimatize which is inherited. Genetic and non-genetic adaptations may be distinguished by long-term breeding programmes, by transplantation experiments or by laboratory-conditioning experiments.

Acclimatization in a seasonally-varying environment occurs in response to changes in several interacting factors such as temperature, photoperiod, oxygen tension, salinity, and food and water availability. This complexity makes experimental study rather difficult, and it is usual to vary one factor only in an otherwise controlled laboratory environment. Although experimentally convenient, such procedures may give rise to rather different responses than might occur naturally with the progression of the seasons. The term *acclimation*, though etymologically indistinguishable from *acclimatization*, has been reserved for laboratory-induced adaptations (Prosser, 1973). This distinction is important, since there are often major differences between an individual acclimated to low temperatures in the laboratory and one acclimatized to winter conditions in the field. The detection of seasonal clues other than temperature also creates the possibility of anticipatory adjustments which may pre-adapt the individual to rapidly deteriorating conditions. Laboratory acclimation will certainly not mimic either the sequence or the intensity of these stimuli.

According to our definitions, acclimatization and acclimation are physiological responses with specific adaptive benefits. Unfortunately, these words are more commonly used to designate the treatment that animals have undergone before experimentation, so that animals maintained under defined conditions may be described as acclimatized or acclimated, even though no adaptive change may have been demonstrated. Since this ambiguity has not led to obvious confusion, we shall not restrict the meaning to one or other of these common usages. The term physiological compensation, however, does imply adaptive changes in physiology following a change in environmental conditions.

5.2 Patterns of compensation

Theoretical considerations suggest that both evolutionary and seasonal adaptations to temperature would lead to quantitative changes in the various physiological processes; that is, a change in rate. This is because acute or sudden variations in temperature over the normal range of temperatures have a mainly quantitative effect upon those processes, and it is this 'rate' effect which must be offset to preserve the *status quo ante*. However, within the context of seasonal changes in temperature, reproductive condition and food availability, etc., in which these adaptive mechanisms operate, this simple expectation may be frustrated and other responses may occur. One obvious example occurs during times of extreme cold and food scarcity, when a compensatory strategy is not possible, and animals may adopt a quiescent state. It is essential that responses during acclimation be carefully interpreted within ecological and seasonal context of the species concerned. Ideally, acclimations observed after laboratory conditioning should be compared with the responses of animals in their natural environment.

Several patterns of acclimation have been observed with respect to temperature. These responses are usually classified according to schemes devised either by Precht (1958) Precht *et al.* (1973) or by Prosser (1973). Precht's scheme is based upon the comparison of the rate of any particular process after direct transfer to a new temperature with that measured at the new temperature after a period of acclimation. For example, for animals acclimated at temperature T_1 will yield rate measurements at T_1 and T_2 and a normal rate–temperature curve (Fig. 5.1). If after a period at T_2 the rate is unchanged, then clearly no adaptation has occurred. However if the rate is altered, then some adjustment has occurred which is termed *ideal* (or *complete*) if the rate at T_2 is identical to the original rate at T_1, or *partial* if the rate lies between the ideal case and the unaltered case (points A and B on Fig. 5.1). Rarely, the rate is depressed below point B to give a *supraoptimal* response, or elevated above point A to give an *inverse* or *paradoxical* response.

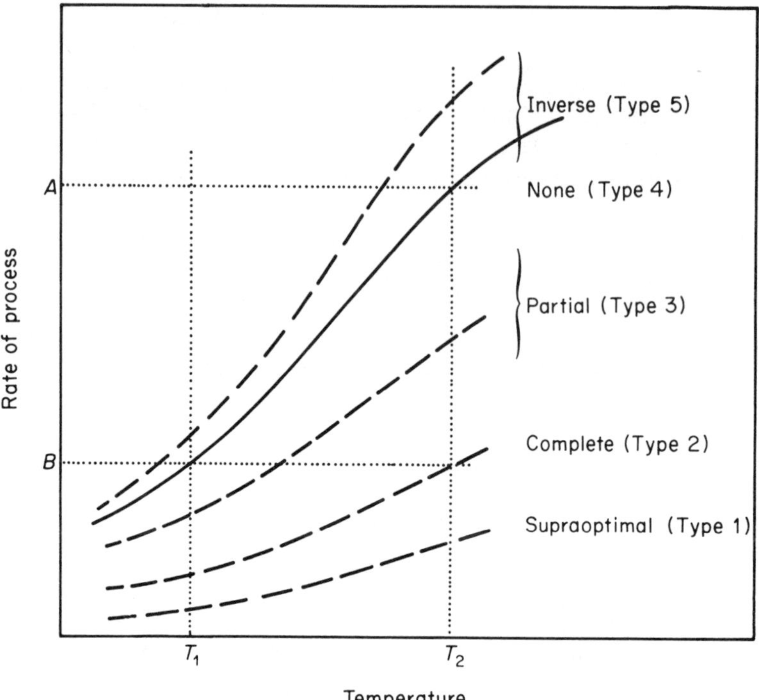

Figure 5.1 Schematic diagram to illustrate the terminology of Precht (1958, 1973 concerning the compensation of rate processes following transfer of animals from temperature T_1 to T_2. The solid line represents the increase in rate upon transfer to the higher temperature. After a period of days or weeks at T_2 the rate may become altered to give new rate–temperature curves which lie above or below the original depending upon the type of compensation. The broken lines represent the rate–temperature curves after acclimation for each of the categories. (After Precht, 1958.)

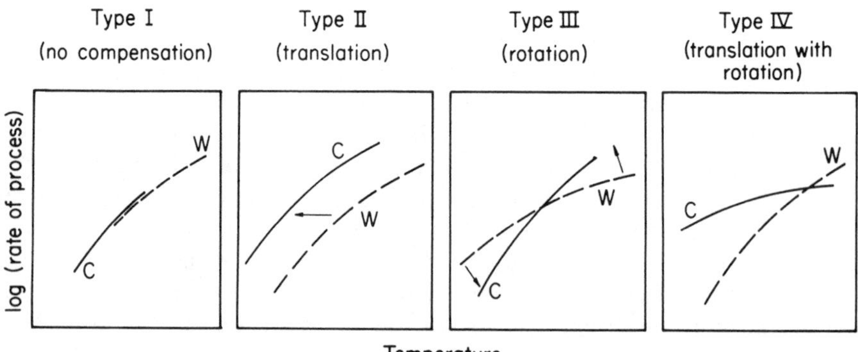

Figure 5.2 The scheme devised by Prosser to classify the compensation of rate processes following the temperature acclimation of animals. The rate–temperature curves for cold- and warm-acclimated animals are shown (C and W, respectively). (After Prosser, 1976.)

Whilst Precht's scheme has the convenience of measurements at only two temperatures, it does not describe the complete response. The classification of Prosser (1973) relies upon a comparison of the complete rate–temperature curves for differently acclimated animals. Figure 5.2 shows the basic patterns of response. If the temperature dependence of a process is identical but the rate–temperature curves are shifted as a result of acclimation, then the response is termed a 'translation'. If the curves simply have different Q_{10}s the response is termed a 'rotation' because one curve is rotated with respect to the other. Most often, however, there is a combination of rotation and translation, with one effect predominating over one range of temperatures and the other over a different range.

5.3 Temperature acclimation

5.3.1 Energetics

Many different aspects of animal performance and physiology are subject to change during acclimation. Certainly the most intensively studied is the rate of oxygen consumption, mainly because this was commonly seen as the most general indicator of overall energy usage and metabolic status of the animal. Some illustrative examples of partial and complete compensations of oxygen consumption are shown in Fig. 5.3(a)–(d). In the beetle *Melasoma* acclimation is associated with a simple translation of the rate–temperature curves. The net effect of this was that at 15°C the standard rate for the 12°C-acclimated beetles was substantially greater than that of the 25°C-acclimated beetles. This increase in standard rate during cold-acclimation was approximately 50–60% of that required to bring it to the level found in 25°C-acclimated beetles at 25°C. Partial compensations of this type are typical for the majority of studies for both vertebrates and invertebrates.

Figure 5.3(b) shows a near perfect compensation (Prosser type II, Precht type 2) for the water flea, *Daphnia* sp., the respiratory rates being almost identical at the respective acclimation temperatures. Figure 5.3(c) shows a more extensive study of the salamander *Plethodon dorsalis*. The rate–temperature curves for 5, 10 and 15°C-acclimated salamanders were almost superimposable so over this range of temperatures no compensation has occurred. However, above 15°C the rate–temperature curves were displaced or translated along the temperature axis (Prosser type II). Relative to 20°C-acclimated animals the 25°C-acclimated animals showed a perfect compensation (Precht type 2), but between 15°C and 20°C there was overcompensation (Precht type 1). Thus different types of compensation occur over different parts of the temperature range for this species. By connecting the measurements at each acclimation temperature an

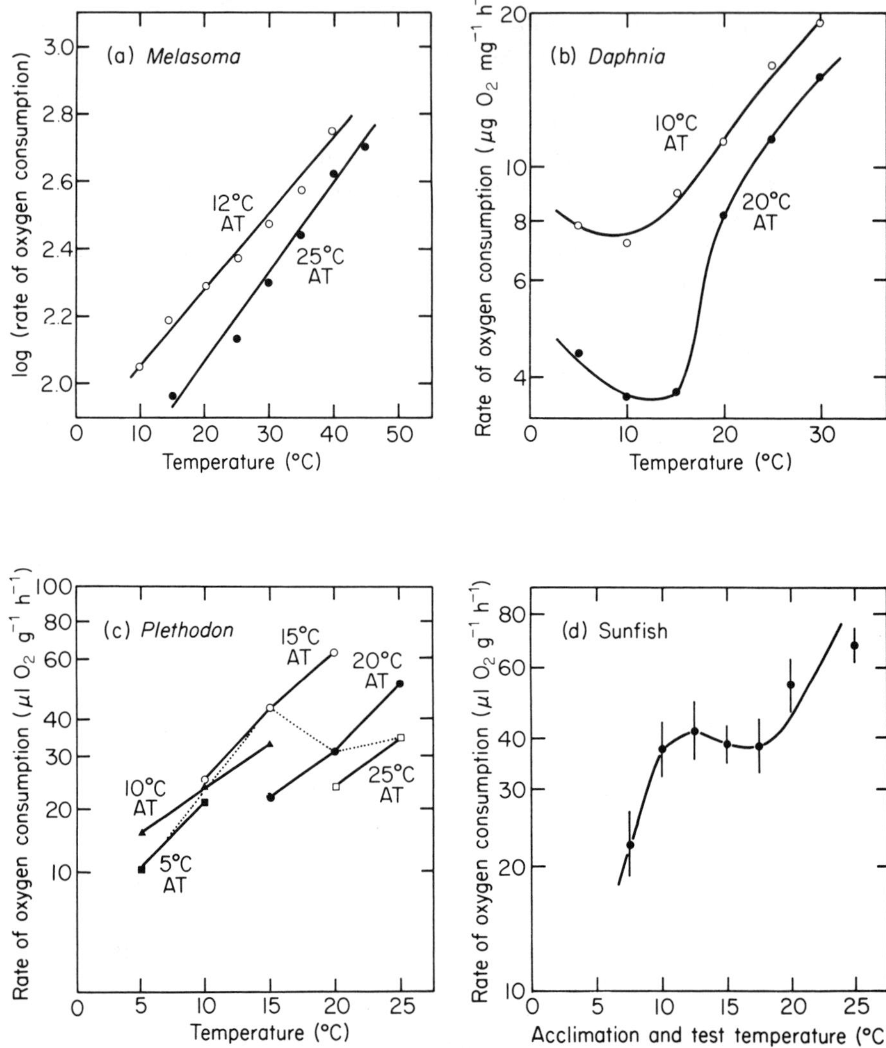

Figure 5.3 Some examples of compensations of oxygen consumption as a result of temperature acclimation. AT – acclimation temperature. (a) The beetle, *Melasoma* sp. (After Marzusch, 1952.) (b) The water-flea, *Daphnia ambigua*. (After Armitage and Lei, 1979.) (c) The Ozark salamander, *Plethodon dorsalis*. (After Brown and Fitzpatrick, 1981.) The dotted line connects the values recorded at each acclimation temperature. (d) The pumpkinseed sunfish, *Lepomis gibbosus*. (After Roberts, 1964.) Values were measured at each acclimation temperature. Bars represent ± 2 standard errors.

'acclimated' rate–temperature curve (in contrast with the acutely-determined curve) has been described which displays the ecologically pertinent relationship between rate and temperature, and the range of temperatures over which a seasonal temperature independence may be achieved.

Figure 5.3(d) shows similar observations from the classical study of Roberts (1964) on the pumpkinseed sunfish, *Lepomis gibbosus*. The most striking feature of this acclimated rate–temperature curve is the plateau extending from 10°C to 17.5°C, which represents the temperature range over which compensation was perfect (Precht type 2). At lower temperatures the rate of oxygen consumption declined more than would be expected from Q_{10} of 2–3, and Roberts suggested that metabolism is adaptively depressed in the cold by an inverse compensation as an energy-conservation measure. Above 17.5°C the acclimated rate–temperature curve increases with temperature, there being no obvious compensation over this upper range.

The number of studies in which partial or complete compensations have been recorded is considerable (see Table 5.1), and compensation is a feature of all phylogenetic groups. Compensation may occur to a greater degree in some species than in others. It is commonly thought that the salmonid fish display smaller compensatory responses of oxygen uptake than do fish species from more seasonally varying or continental climates (Roberts, 1967). Animals which remain active during the winter months would naturally display larger compensations than those which become winter-torpid (see, for example, Ragland *et al.*, 1981). Similarly, in those animals whose life-history and thermal habitats permit the behavioural regulation of body temperature, metabolic compensation is of less importance. Thus, insects and reptiles are commonly thought to have small compensatory responses. Nevertheless, there are many documented examples of compensation by both of these groups, but in some they are evident at temperatures below which behavioural thermoregulation is effective.

Table 5.1 A list partial or complete compensations of oxygen consumption.

Arthropods	
Cockroach	Dehnel and Segal (1956)
Spiders	Anderson (1970), Moeur and Eriksen (1972)
Daphnia sp.	Armitage and Lei (1979)
Millipede	Dwarakaneth (1971)
Crabs	Robert and Gray (1972)
	Roberts (1957)
	Vernberg (1969)

(*Contd.*)

Table 5.1 (*Contd*)

Molluscs	
Limax sp.	Segal (1961)
Mya sp.	Anderson (1978)
Heliosoma sp.	Wood (1978)
Mytilus edulis	Widdows and Bayne (1971)
Crepidula sp.	Newell and Kofoed (1977)
Philomyscus sp.	Rising and Armitage (1969)
Corbicula sp.	McMahon (1979)
Other invertebrates	
Aurelia sp.	Mangum *et al.* (1972)
Chryosoaro sp.	Mangum *et al.* (1972)
Lumbriculus sp.	Kirberger (1953)
Strongylocentrus sp.	Percy (1974)
Fish	
Atlantic salmon	Peterson and Anderson (1969)
Rainbow trout	Evans *et al.* (1962)
Goldfish	Kanungo and Prosser (1959a)
Carp	Meuvis and Heuts (1957)
	Suhrmann (1955)
Blenny	Campbell and Davies (1975)
Eel	Precht (1951)
Toadfish	Haschemeyer (1969b)
Yellow bullhead	Morris (1965)
Sunfish	Roberts (1964)
Bitterling	Kruger (1969)
Lungfish	Grigg (1965)
Cichlid	Morris (1962)
Amphibia	
Salamander	Vernberg (1952)
	Feder (1978)
	Fitzpatrick *et al.* (1971, 1972)
	Brown and Fitzpatrick (1981)
Frog	Precht (1958)
Reptiles	
Tortoise	Wood *et al.* (1978)
Lizards	Murrish and Vance (1968)
	Ragland *et al.* (1981)
	Patterson and Davies (1978a)
	Dawson and Bartholomew (1956)
	Dutton and Fitzpatrick (1975)

Table 5.2 A list of studies demonstrating inverse compensations to temperature.

Nematode	Dusenbery *et al.* (1978)
Sea urchin	Ulbricht (1973)
Sunfish	Roberts (1964)
Frog	Holzman and McManus (1973)
	Packard (1972)
	Dunlap (1972)
Salamander	Fitzpatrick *et al.* (1971, 1972)
	Fitzpatrick and Brown (1975)
Snakes	Jacobson and Whitford (1970)
Lizard	Patterson and Davies (1978b)
	Aleksiuk (1971, 1976)
Horned lizard	Mayhew (1965)
Desert iguana	Moberly (1963)

Table 5.2 shows a number of instances of inverse compensation. In virtually all cases they are manifest as profound depressions of oxygen consumption in the cold, below that found for warm-acclimated animals at cold temperatures. Most of the documented cases occur in the amphibia and reptiles, animals which by common observation do not maintain normal levels of activity in the cold.

The obvious advantage of Precht type 2 or 3 compensations is that they promote the relative constancy of various rate processes despite seasonal fluctuations in temperature. If one assumes that an individual exists in an optimal state over a narrow range of temperatures, then responses of this type will tend to preserve this optimal state. Thus, compensation in the cold is seen principally as a means of maintaining metabolic performance and an adequate rate of energy production, whereas compensation in the warm reduces an energy expenditure which is unnecessarily high for maintenance purposes and which could be used more profitably for growth and reproduction.

The depression of activity and metabolism during cold-acclimation below that predicted from the acute rate–temperature curve of a warm-acclimated animal (i.e. inverse compensation) represents an adaptation to reduce energy expenditure in times of inclement conditions or reduced food-availability. It is a strategy that is widely adopted by both poikilotherms and hibernating mammals in the cold and is distinct from the simple cold-induced torpor of animals when cooled. The energetic savings may be great. For example, metabolic depression in the lizard *Lacerta vivipara* meant that only 5% of its total annual energy budget was expended over the winter months, which occupy approximately 44% of the year (Patterson and Davies, 1978b) and even greater energetic savings have been recorded in ground squirrels during hibernation (Wang, 1979). Inverse compensation is usually associated with specific patterns of behaviour. Thus, some fish species, such as the eel (Walsh *et al.*, 1983), the bullhead and the striped bass (Crawshaw, 1984), cease feeding altogether, burrow into the mud and adopt a sleep-like state for a winter period which may last up to 6 months. A second important feature of inverse compensations is that they are usually coupled with high temperature coefficient which permits an abrupt transition from torpidity to activity when temperature rises in the spring.

An objection that could be made to the supposed adaptive benefits of metabolic compensation in the cold is that it involves an increase in the energetic costs of standard metabolism above that required for maintenance purposes, which might be used more profitably for growth or reproduction. In other words, why do organisms not take advantage of reduced maintenance costs in the cold to support activities which contribute directly to the fitness of the individual? This view implies that maintenance metabolism is simply the energetic cost of preserving the animal as an organized entity. An alternative view is that maintenance metabolism directly contributes to the ability of an animal to perform work. In other words, maintenance metabolism must be kept at a minimal 'tick-over' level to enable the animal to gear-up to maximal levels of activity. Recent work suggests that there is an obligatory linkage between resting and maximal levels of oxygen consumption in the vertebrates (Bennett and Ruben, 1979; Taigen, 1983), the so-called 'factorial aerobic scope for activity'. Thus, if maintenance metabolism was depressed in the cold, then the scope for activity would also be reduced and the ability of the animal to escape predation, etc., would suffer. An example of the adaptive value of acclimation has been provided by Newell and Pye (1971), who showed that the aerobic scope for activity of the winkle *Littorina littorea*, and hence the capacity for work, was kept maximal at each of the acclimation temperatures by means of perfect translations of the rate–temperature curves for standard and active rates of oxygen consumption.

A crucial and generally unrecognized problem in understanding the significance and mechanisms underlying compensatory responses of oxygen consumption, lies in separating the possible adjustments of maintenance metabolism and capacity of the cellular apparatus for work from changes in the level of routine physical activity. Compensations of oxygen consumption for whole animals are usually interpreted as indicating adaptive changes in the cellular machinery. Yet we have seen that even small changes in activity may result in large increases in oxygen consumption. Indeed, Fry (1971) maintains that a fish resting quietly may consume oxygen at a rate varying from its standard rate (i.e. its minimum) to well over half its maximal rate. Thus, almost all measurements of oxygen consumption, even on quiescent post-absorptive animals includes a quantitatively unpredictable combination of maintenance metabolism and activity metabolism. An example of the effect of behavioural drive on the rate–temperature curve is the demonstration by Schmeing-Engberding (1953) that a plateau in the acute rate–temperature curve for oxygen consumption in the carp disappeared in anaesthetized animals. Thus, uncontrolled and spontaneous activity may alter the position and the shape of rate–temperature curves with no effects upon maintenance metabolism, and so studies of oxygen consumption are only as good as the methods used to control for physical activity.

When an understanding of animal energetics at different times of the year is sought this routine activity is not of concern, since it presumably forms part of the normal behaviour and energy expenditure of the species. However, in attempting to understand the physiological basis of compensation these considerations are of paramount importance, because it is difficult to separate adaptations of standard or maintenance metabolism from ill-defined changes in spontaneous activity, muscular tone and irritability as controlled by the central nervous system. This is a more severe problem in studies of continuously active animals which display an elaborate behaviour and a heightened irritability, than in sessile, filter-feeding animals whose routine rate is equivalent to their standard rate. It is unfortunately a problem that has no satisfactory solution and because the activity levels are not clearly defined or controlled, the number of whole-animal studies noted in Table 5.1 which satisfactorily demonstrate compensations of maintenance metabolism is small.

Despite these interpretative difficulties with whole-animal respirometry, some persuasive evidence for a compensation of standard metabolism and of the capacity of the underlying cellular processes for work comes from *in vitro* studies of tissue oxygen consumption (Hazel and Prosser, 1974). In these experiments all considerations of directed or spontaneous activity can be set aside and the *in vitro* conditions can be maintained constant for tissues of both cold- and warm-acclimated animals. The magnitude of compensatory increases in oxygen consumption may be quite impressive. Thus, Jones and Sidell (1982) found that the rate for skeletal muscle in cold-acclimated striped bass was greater than that of warm-acclimated bass even when measured at their respective acclimation temperatures. Similar responses have been observed in goldfish brain (Freeman, 1955), goldfish liver (Kanungo and Prosser, 1959b) and isolated thoraces of the blowfly, *Calliphora* (Tribe and Bowler, 1968).

A complication is that not all tissues reflect the compensatory response shown by the whole animal, nor, in some cases, do all tissues display the same pattern of compensation as each other. For example, in the rainbow trout the brain showed complete compensation, the liver showed overcompensation and the gill showed no compensation (Evans *et al.*, 1962). Similarly, in the American oyster thermal acclimation resulted in perfect compensation for gill respiration, a partial compensation of mantle tissue respiration and no compensation of muscle tissue respiration (Bass, 1977).

It seems that the apparently straightforward energetic compensation of the whole animal is, in reality, the result of a complex reorganization of tissue metabolism.

5.3.2 *Spontaneous activity and locomotory performance*

Perhaps the most obvious manifestation of physiological compensation is the improved locomotory performance and similar levels of spontaneous activity of poikilotherms at different acclimation temperatures. It is a matter of common observation that cold-acclimated fish show more normal routine activity and show better neuromuscular control in the cold than warm-acclimated fish. Peterson and Anderson (1969) have quantified the spontaneous activity of the Atlantic salmon in the 'stabilized phase' following an abrupt change in temperature; that is, after the initial excitement caused by the shift in temperature had subsided. As might be

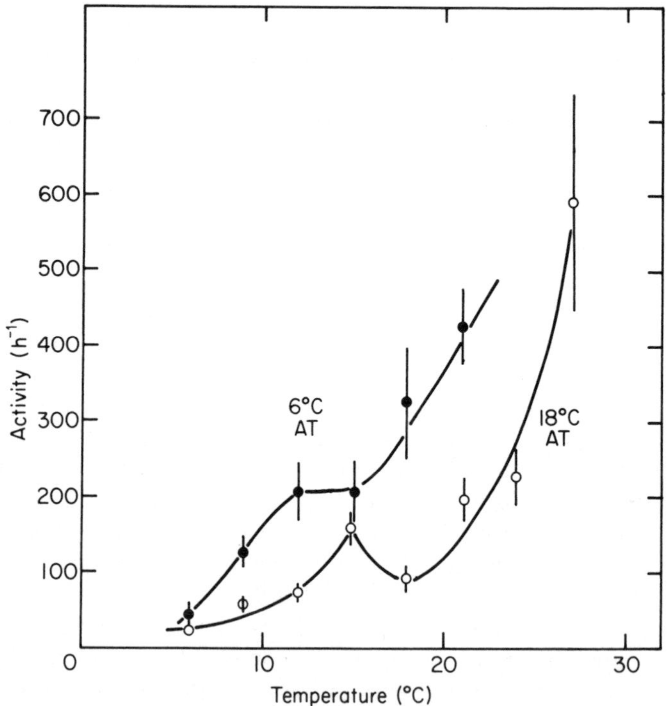

Figure 5.4 The effects of thermal acclimation at 6°C and 18°C upon the spontaneous activity of the Atlantic salmon, *Salmo salar*. Each estimate of activity was obtained during the 'stabilized' phase after transfer to a new temperature and is the average ± s.e.m. for nine experiments. AT – acclimation temperature. (After Peterson and Anderson, 1969.)

expected, the activity of the 6°C-acclimated fish was greater than that of 18°C-acclimated fish at all test temperatures (Fig. 5.4). However, an important factor in controlling the level of spontaneous activity was the size of the temperature shift rather than the temperature *per se*, since even a temperature decrease produced an elevation in activity in the 18°C-acclimated fish. At 6°C, the spontaneous activity in both groups was greatly reduced and almost identical to each other, suggesting that low temperature is acting to depress activity in the 18°C-acclimated fish.

There is a clear and obvious advantage to animals having an optimal locomotory performance at the temperature to which they have been acclimated. Such is certainly the case for the maximal sustained cruising speed in the goldfish (Fig. 5.5(a)) and in the juvenile coho salmon (Fig. 5.5(b)). The latter study was presented as a response surface relating speed to both acclimation temperature and test temperature. If transects are taken parallel to the acclimation temperature axis, then the maximal performance occurs at the acclimation temperature. Similar observations have been made in a variety of fish (green sunfish, *Lepomis cyanellus*, Roots an Prosser, 1962; the salmon, *Salmo salar*, and in the minnow, *Notropsis* sp.; McCrimmon in Fry, 1967); and in the American lobster (McLeese and Wilder, 1958).

In seeking to understand these whole-animal responses it is useful to distinguish between cruising, which is a sustainable activity and is dependent upon the aerobic red muscles, and burst swimming, which is non-sustainable and requires activation of the anaerobic white muscles. As swim speed increases above the maximal cruising speed there is a progressive recruitment of white fibres. One possible mechanism of adaptation is to vary the pattern of muscle-fibre recruitment or the number of active muscle fibres as temperature changes. Indeed, Rome *et al.* (1984) have recently shown that white-fibre recruitment occurs at lower swim speeds when temperature is reduced. Thus, to maintain a given swim speed in the cold more white fibres must be active. Other studies have shown that the proportion of red fibres increases on cold acclimation (goldfish, Johnston and Lucking, 1978, Sidell 1980; striped bass, Jones and Sidell, 1982) which may increase the maximal cruising speed and thereby offset the effects of cold. However, this is achieved at the expense of the white-fibre population such that burst performance is reduced (Sidell and Johnston, 1985). The relative importance of maintaining cruising or burst performance probably depends upon the lifestyle of the species in question. For a 'sit-and-wait' predator, such as the pike, burst performance may be more critical than in continually foraging species such as the carp or striped bass.

A second adaptive mechanism is to alter the contractile properties of individual muscles fibres by compensations of either energy metabolism or of the myofibrillar apparatus itself. The latter response has been elegantly

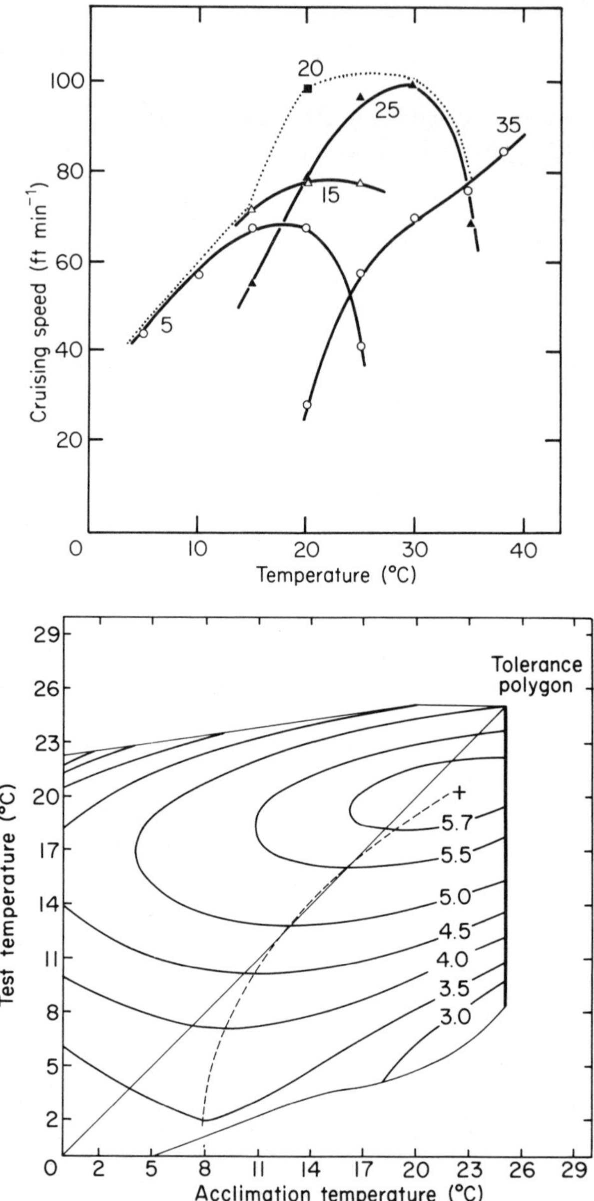

Figure 5.5 Thermal compensation of locomotory performance. (a) The effects of temperature upon the cruising speed of goldfish acclimated to different temperatures is shown. Note that the maximal cruising speed was usually observed at the acclimation temperature. (After Fry and Hart, 1948.) (b) Isopleths of maximal cruising speed (body lengths. s^{-1}) of juvenile coho salmon, *Onchorhynchus kisutch*, as function of test temperature and acclimation temperature (see Chapter 6 for a description of response surfaces). The lines were calculated using response surface methodology (see Chapter 6 and Alderdice, 1972) and were plotted within the tolerance polygon already established for this species. Note that the maximal cruising speed is obtained at test temperatures that encompass each acclimation temperature as shown by the broken line. (After Griffiths and Alderdice, 1972.)

demonstrated in carp by Johnston *et al.* (1985). They used a technique in which the sarcolemma of single isolated muscle fibres was chemically removed or 'skinned' and the mechanical properties during contraction recorded as the exposed myofibrils were flooded with calcium. Figure 5.6 compares the force–velocity curves for fibres from 8°C- and 23°C-acclimated carp. Contraction velocity was about 100% higher and maximal force was about 70% higher in fibres of cold-acclimated fish. Accompanying these changes in muscle performance were substantial changes in the capillary bed of skeletal muscle (Johnston, 1982).

Adaptations of the myofibrillar apparatus can be achieved in at least two ways. First, by varying the number of myofibrils per cell the force-generating properties of the fibre are directly affected. However, because so much of the cell is already composed of myofibrils it is difficult to see how this could be substantially increased in the cold. A second mechanism is to vary the catalytic properties of the myofibrillar ATPase, the enzyme

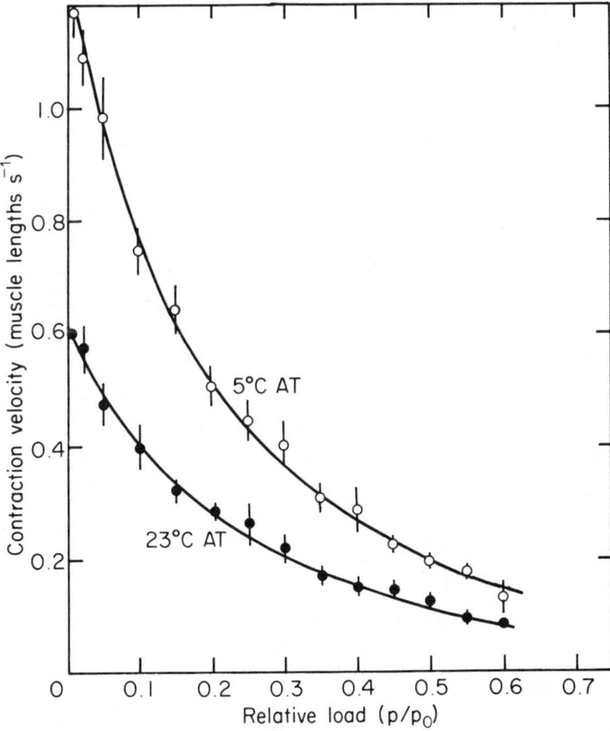

Figure 5.6 The effect of thermal acclimation of carp upon the contractile properties of skeletal muscle. The force–velocity curves for isolated, 'skinned' muscle fibres from 5°C- and 23°C-acclimated carp are compared. Note that fibres from cold-acclimated carp display a two-times greater contraction velocity whether loaded or unloaded. AT – acclimation temperature. (After Johnston *et al.*, 1985.)

responsible for force generation between the actin and myosin filaments. There is good evidence of significant increases in myofibrillar ATPase activity in cold-acclimated goldfish (Johnston, 1979) which is linked to changes in the protein composition of this multienzyme complex (Johnston *et al.*, 1985). However, this mechanism may be of only limited significance, since compensations of ATPase activity are absent in several other non-cyprinid fish species.

5.3.3 *The nervous system*

The central nervous system of higher multicellular animals is of great significance to temperature acclimation. We have seen that the CNS plays a major role in controlling spontaneous activity and muscle recruitment as temperature changes, and the neuroendocrine system may be involved in controlling adaptive responses in other tissues. There is also, some evidence that the electrical properties of central and peripheral neurones, their axons and synapses, may be altered in a compensatory manner (Lagerspetz, 1974; Prosser and Nelson, 1981).

The giant nerve fibres of the earthworm have proved a useful preparation for acclimation studies. The fibres of cold-acclimated worms show greater conduction velocity, faster rising and falling phases of the action potential, shorter absolute refractory periods and, thus, greater maximal impulse frequencies than the fibres of warm-acclimated worms (Lagerspetz and Talo, 1967; Talo and Lagerspetz, 1967). Thus, virtually all aspects of giant-fibre performance were altered in a compensatory manner by cold-acclimation. These adaptations seem to be responsible for a significant increase in the speed of the escape reflex of the earthworm during cold-acclimation. However, because giant fibres are interrupted at intervals by septate junctions, it is difficult to judge whether the adaptive change is brought about by alterations in the conduction velocity *per se* (i.e. cable resistance, voltage-dependence and resting ionic permeabilities, etc.) or by changes in synaptic transmission. Dierolf and MacDonald (1969) claim that thermal acclimation brings about changes in the time-course of ionic conductance changes during excitation.

Thermal acclimation has also been found to produce perfect compensatory changes in the spontaneous electrical activity of an identified neurone in the snail *Helix aspersa* (Zecevic and Levitan, 1980). The normal bursting activity in 20°C-acclimated snails was blocked by cooling to 5°C. After a 2-week period at 5°C the levels of spontaneous activity in the neurone were comparable with those of warm-acclimated snails at 20°C. Although the duration of the action potential and the electrogenic Na^+–K^+ pump were as temperature sensitive, they were unaffected by acclimation.

In vertebrate animals thermal acclimation also causes changes in the degree of central nervous activity, the speed of central processing of neuronal inputs and in the thermal limits for normal function (see Chapter 6). In the rainbow trout Konishi and Hickman (1964) found that the speed of electrical responses in the midbrain following electrical stimulation of the retina was altered as a result of thermal acclimation. Transfer of 10°C-acclimated fish to 4°C caused a 40% protraction of the entire midbrain electrical response. After 21 days at 4°C the protraction was reduced to less than 20%. In a similar study of the brown bullhead Bass (1971) found complete compensation of carp between 10°C and 20°C in the duration of slow negative potentials produced in the facial lobes of the brain following electrical stimulation of the barbels. These slow potentials were thought to represent simultaneous post-synaptic events in a large number of central neurones. Because the conduction velocity of peripheral nerves was not affected by acclimation, it seems that the compensation in duration of these evoked potentials is due to alterations in the process of synaptic transmission.

The sensitivity of sensory receptors may also be adaptively altered by acclimation. Spray (1975) has studied cold receptors in frog skin by recording from the dorsal cutaneous nerve as the skin was progressively cooled. The cold-acclimated preparation showed high levels of continuing activity at all temperatures compared with the warm-acclimated preparation particularly at temperatures below 20°C. In addition, the sensitivity to changes in temperature was greater in cold-acclimated frogs.

These studies show that the functional properties of single neurones and their axons may be adaptively altered during thermal acclimation. Although there may be changes in the conduction properties of the axons, it is probably in the synapses and neuromuscular junctions that the most important changes occur. However, there is also the possibility of the recruitment, at different acclimation temperatures, of neuronal populations with different and potentially adaptive characteristics.

5.3.4 *Digestion and intestinal absorption*

The restoration of energy demand following the compensation of metabolism and locomotory performance demands an equivalent change in the rate of energy intake. Unfortunately, the rates of digestion and assimilation are difficult to quantify in any precise way because of the difficulty of maintaining adequate control over meal size, appetite, gut distension, etc. An additional difficulty is that the measurement of digestion rate at a changed temperature requires an extended period during which additional adaptive changes may occur.

Nevertheless, there are some good indications in fish, at least, that thermal acclimation is accompanied by perfect compensations in the specific activity of gut enzymes (Owen and Wiggs, 1971; Hofer, 1979) and the secretion of gastric HCl (Smit, 1967) as well as partial compensations in the rate of absorption of digested molecules and inorganic ions (Smith, 1966).

5.3.5 *Epithelial transport of ions and metabolites*

Acclimation produces clear and impressive changes in the functional properties of some secretory epithelia (Smith, 1976). The intestinal epithelium of the goldfish is a site where complete or partial compensations of absorptive function might be expected. In this system Na^+ is thought to enter the epithelial cells from the lumen by a facilitated diffusion mechanism, and is then actively transported across the base of the cell to the serosal surface by a Na^+ pump. In moving Na^+ the epithelium develops a transepithelial potential of up to 2–3 mV which is roughly equivalent to the net transport of Na^+ to the serosal surface (Smith, 1966).

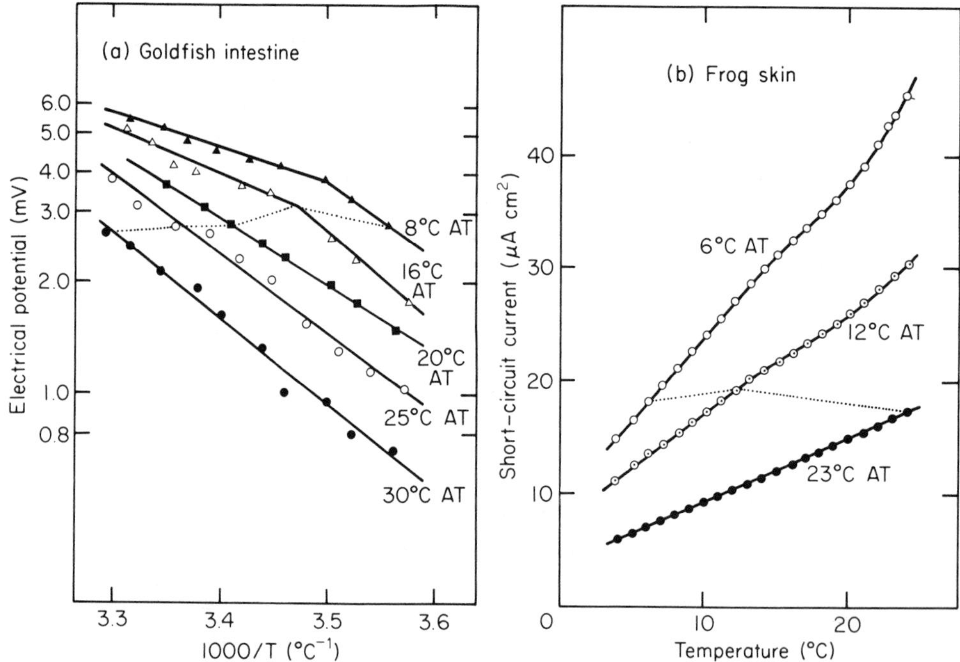

Figure 5.7 The effects of temperature acclimation upon the transport properties of epithelial tissues. (a) A near-perfect compensation for the steady-state electrical potential across the gut of temperature-acclimated goldfish. (After Smith, 1966.) (b) A perfect compensation of the short-circuit current, I_{sc}, across the isolated frog skin. I_{sc} is a good indicator of the rate of Na^+ transport across this epithelium. (After Lagerspetz and Skytta, 1979.) In both parts the dotted line connects the values measured at the respective acclimation temperatures. AT – acclimation temperature.

Figure 5.7(a) shows the stimulatory effect of acute increases of temperature upon the transmural potential of the intestine of goldfish acclimated to different temperatures. However, the leftward shift of each graph with increasing acclimation temperature tends to compensate this effect and the potential measured at each acclimation temperature (as indicated by the dotted line in Fig. 5.7(a)) hardly varies. A similar situation occurs in frog skin where the short-circuit current is perfectly compensated during acclimation to temperatures between 6°C and 25°C (Fig. 5.7(b); Lagerspetz and Skytta, 1979). The short-circuit current reflects the net sodium transport across the epithelium in the absence of a transepithelial potential. It is clear that the osmoregulatory performance of frog skin is not apparently affected over this wide range of acclimation temperatures.

These compensations in transport capacity of these tissues may be accompanied by substantial structural modifications. In the carp the mucosal surface area of the intestine increases by almost 100% on cold-acclimation (Lee and Cossins, 1987). This is brought about by an increase in the length of the villi and by a smaller increase in the diameter of the gut itself. The increase in mucosal surface area in cold-acclimated animals may be a result of a different temperature dependency of proliferation of the absorptive cells which line the intestine (enterocytes) and their loss from the epithelium. For example, a reduction in gastrointestinal mobility caused by cooling may reduce the abrasion of the mucosa by the luminal contents and this reduces the rate of cell sloughing at the tips of the villi. A reduction in the rate of cell death, if not compensated by a reduction in cell proliferation, would lead to an increase in the standing population of enterocytes. In any case the increase in absorptive area acts additively with compensations of transport capacity (Gibson *et al.*, 1985).

5.4 Seasonal effects upon acclimation

There are factors other than temperature which influence the physiological status of animals under natural conditions. For example, reproduction and growth occur at specific times during the annual cycle, and the physiological requirements for these processes may override or modify the normal compensatory patterns discussed previously in acclimated animals.

Dunlap (1980) caught stocks of the frog *Pseudacris* sp. in the spring and summer, and subjected them to identical acclimation procedures at 5°C or 25°C for 5–7 days. The routine rates of oxygen consumption of the spring and summer frogs showed substantial differences despite similar acclimation treatment. Thus spring frogs had increased rates relative to summer frogs when both were acclimated to 5°C, but not when acclimated to 25°C. Dunlap noted conspicuous plateaux in the rate–temperature curves for both spring and summer frogs which corresponded with the temperatures experienced in the field; thus spring animals showed a temperature-insensitive zone which extended 5–10°C lower than summer frogs (Fig. 5.8).

An environmental factor which provides a more accurate trigger for seasonal events than temperature, is photoperiod. Roberts (1964) has elegantly demonstrated an effect of day length in his classical studies upon the sunfish, *Lepomis* sp., by comparing the acclimated rate–temperature curves for routine oxygen consumption of animals conditioned under long- or short-day photoperiods (Fig. 5.9). The curves for each photoperiod were quite different, yet temperature is a crucial factor in the expression of this effect. Interestingly, Roberts did not ascribe a metabolic basis to this effect, with all of the possible implications regarding a controlling role for the endocrine system, but suggested that the higher rates of oxygen consumption under short day length is probably due to increased routine activity and muscle tone.

Subsequently, Burns (1975) measured the routine rates for sunfish taken at monthly intervals from an otherwise undisturbed pond. In the laboratory each batch of fish was brought to a measurement temperature of 17.5°C over 20–30 h. The rate at 17.5°C showed an inverse relationship with the

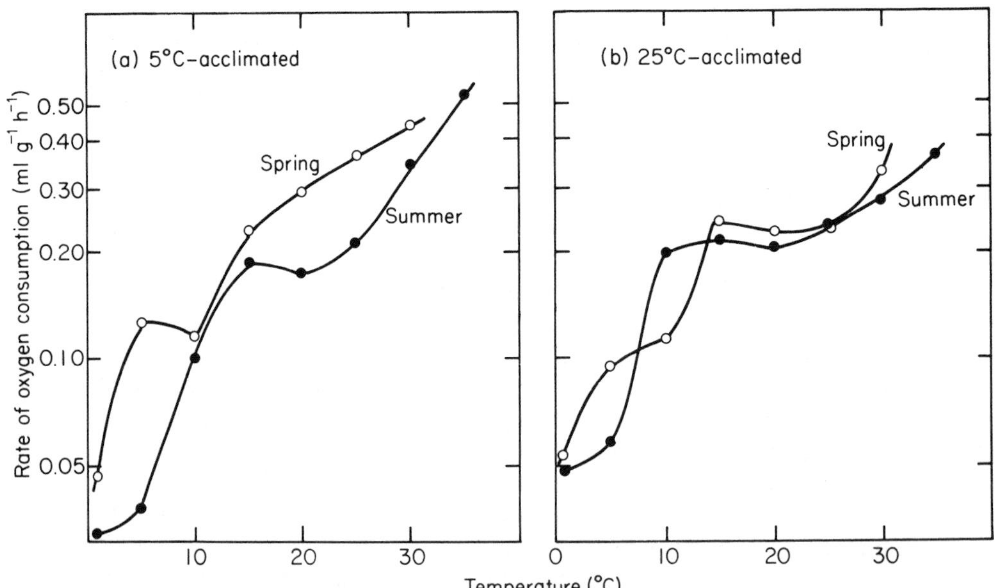

Figure 5.8 The effect of season upon the acute rate–temperature curves for routine oxygen consumption of the Chorus frog, *Pseudacris triseriata*. Note that season-dependent differences were especially pronounced in cold-acclimated frogs where spring frogs displayed high levels of activity at 5°C relative to summer animals. (After Dunlap, 1980.)

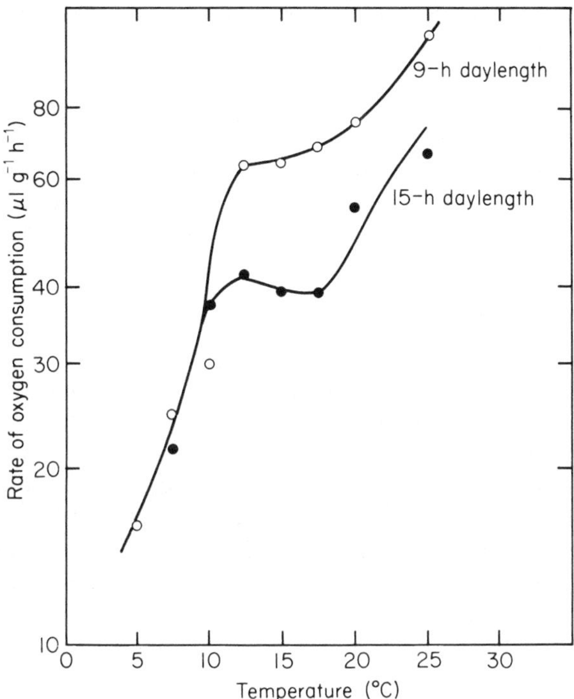

Figure 5.9 The acclimated rate–temperature curves for the rate of routine oxygen consumption in the sunfish, *Lepomis gibbosus*, acclimated under a short (9-h) or a long (15-h) day length. (After Roberts, 1964.)

water (and acclimatization) temperature at the time of capture, as would be expected if compensation had occurred, except for the months of May and June when the fish were nesting and spawning, and when day length increased from 13 to 15 h and temperature from 8°C to 17°C. This refractory period was thus related to reproductive events which were probably triggered by a photoperiod and/or temperature cue. Courtship and other behavioural activities elicit higher levels of spontaneous activity, and hence routine oxygen consumption, despite rapidly rising temperatures.

5.5 Cellular mechanisms of compensation

The compensatory changes in whole-animal and tissue function described in the preceding sections are based upon changes in the properties of their constituent cells, subcellular organelles and molecules. The study of cellular and molecular acclimatizations has proved a fertile area of investigation, and much of fundamental significance has been learned as a result. Despite the wide variety of species and tissues under consideration there are, at the cellular level of organization, common patterns of adaptive response. These are described in some detail by Hazel and Prosser (1974), Cossins and Sheterline (1983) and by Hochachka and Somero (1984). Because temperature not only has a 'rate' effect but also disturbs the normal functioning of the cellular apparatus in a qualitative way, we should expect adaptations which minimize the temperature-induced malfunctions as well as those which lead to compensated activities.

5.5.1 *Metabolic reorganization*

The compensatory changes in tissue oxygen consumption certainly indicate that the intensity and perhaps the pattern of cellular metabolism is subject to regulation. The simplest approach to this is to compare in tissue extracts from cold- and warm-acclimated animals the *in vitro* activity of enzymes which indicate the relative capacities of the various metabolic pathways (see Hazel and Prosser, 1974). Often the different enzymes of a given metabolic pathway, such as glycolysis, show parallel changes within a particular tissue. However, enzymes in different pathways may display changes in opposite directions. Those enzymes which show Precht type 2 or 3 responses (complete or partial) are mainly associated with the pathways of energy production. Thus, enzymes of glycolysis, gluconeogenesis, the hexose monophosphate shunt, the citric acid cycle, the electron-transport chain and digestive enzymes all show an increase of between one- and two-fold in specific activities during cold-acclimation.

Inverse compensations of enzymatic activity have been consistently observed with enzymes associated with lysosomes, peroxisomes, urea and nitrogen metabolism. All of these enzymes are involved in the elimination of toxic waste products of metabolism or with the breakdown of metabolic intermediates (Hazel and Prosser, 1970, 1974). Bearing in mind that metabolic compensation to temperature is frequently partial and that metabolic activity is greater at higher acclimation temperatures then the increased activity of these enzymes in warm-acclimated animals may enable the animal to cope with an increased production of metabolic waste products.

The increased importance of the pentose shunt pathway during cold-acclimation has been confirmed by metabolic studies. An ingenious technique in this respect is to monitor the appearance of radiolabelled CO_2 when either [1-^{14}C]glucose or [6-^{14}C]glucose is used as substrate, since the pentose cycle specifically releases radiocarbon from the 1-position while glycolysis leads to release of both 1-position and 6-position radiocarbon at equal rates. Any differences between the rate of $^{14}CO_2$ evolution with the two substrates therefore indicates the relative importance of the pentose shunt. Both Hochachka and Hayes (1962) and Stone and Sidell (1981) have found that the C_1/C_6 ratio was significantly greater in the cold. The significance of this is the greater capacity for the production of the NADPH that is required for enhanced lipid biosynthesis in the cold.

Various fuels are available for oxidation by individual tissues (e.g. triglycerides, glucose, glycogen and amino acids) and these may vary with changes in acclimation temperature. Figure 5.10 provides a simplified metabolic map to illustrate the two most important storage forms: triglyceride and glycogen. It is clear from this that a crucial control point occurs at acetyl-CoA where both fuels compete for access to the Krebs cycle. Consequently there is frequently an antagonistic relationship between lipid oxidation and glucose catabolism. A second important branch point is at glucose-6-phosphate, since this can be used as a substrate for glycolysis, for the hexose monophosphate shunt or for glycogenesis.

Switching between different energy sources can be demonstrated by measuring the relative contents of glycogen and triglycerides. For example, Stone and Sidell (1981) have found greater concentrations of triglyceride and lower concentrations of glycogen in the liver of cold-acclimated striped

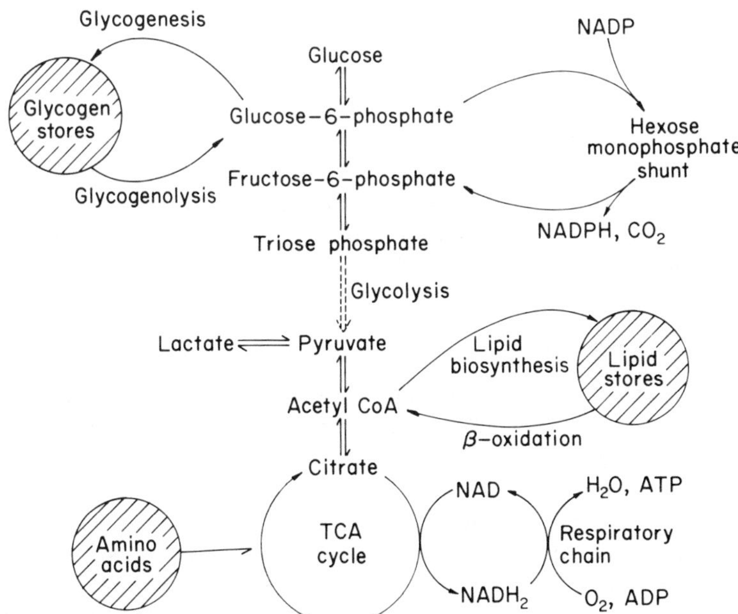

Figure 5.10 Highly simplified and schematized metabolic map to illustrate the relationship of the principal energy-storage forms (triglyceride, glycogen and amino acid) with each other and with the pathways of energy metabolism. A more complete description of these pathways and their adaptive regulation can be found in Hochachka and Somero (1984).

bass. That this is related to changes in the predominant substrates for cellular energy production is suggested by three additional observations. First, the respiratory quotients for liver slices were approximately 1.0 and 0.8 for cold- and warm-acclimated fish, respectively. This is consistent with the preferential use of glycogen in the cold and fats in the warm. Secondly, iodoacetate, an inhibitor of glycolysis, produced a greater percentage inhibition of tissue oxygen consumption in warm-acclimated fish, suggesting that an increasing proportion of glycolytic substrate was used in the warm. Thirdly, the rate of oxidation of radiolabelled glucose was compensated in 5°C-acclimated fish whereas the rate of oxidation of radiolabelled fatty acids was not. Thus, it seems that the liver of the striped bass stores fat in the cold and preferentially oxidizes carbohydrate for oxidative phosphorylation. In warm-acclimated fish this is reversed.

This pattern of carbon flow does not necessarily apply to all fish species. Trout liver (Hazel, 1979) and hepatocytes of the killifish, *Fundulus heteroclitus* (Moerland and Sidell, 1981) possess greater triglyceride levels and lower glycogen levels in warm-acclimated animals compared with cold-acclimated fish. A further complication is that different tissues within the same animal may display quite different temperature-specific patterns of carbon flow. Thus, in contrast with the liver, the muscle of striped bass predominantly utilizes fatty acids in the cold and glucose in the warm (Jones and Sidell, 1982). These complex modifications of metabolic pathways flow are clearly rather tissue- and animal-specific, and are probably related to factors other than temperature compensation *per se.* One obvious factor is the relative dietary intake of lipid and carbohydrate, while another is the seasonal occurrence of gametogenesis, a process which requires the production of large quantities of storage lipids. A third factor is the seasonal preparation for food scarcity.

5.5.2 *Enzymatic adaptations*

It is clear that at the cellular level the process of temperature acclimation is the result of a complex reorganization of cell metabolism and an increased capacity for energy metabolism. The changes ultimately depend upon the regulation of the activity of key regulatory enzymes. Hochachka and Somero (1984) describe how enzymatic performance can be affected by temperature variations in more subtle ways than are suggested by the 'rate' effect. In particular, the binding affinity of the enzyme for substrate, cofactors and allosteric modulators is often perturbed by changes in temperature, sometimes dramatically. Thus, temperature acclimatization requires mechanisms to nullify or ameliorate these effects, as well as to produce compensations in capacity. Hochachka and Somero (1973, 1984) have described three principal mechanisms by which these regulations can be achieved:

(a) A change in the cellular concentration of enzyme (the so-called 'quantitative strategy');
(b) A change in the types of enzymes isoforms present (the 'qualitative strategy'); and
(c) The modification of kinetics of pre-existing enzymes (the 'modulative strategy').

5.5.3 *Quantitative changes in enzyme concentration*

The changes in enzymatic activity of tissue extracts described earlier does not provide an unequivocal measure of the cellular enzyme concentration. The unequivocal demonstration of changes in enzyme concentration has only been achieved in a few studies. First, Wilson (1973) raised antibodies in rabbits to the cytochrome oxidase of goldfish muscle. Incubation of rabbit serum with tissue extracts from goldfish caused a precipitation of the antibody–cytochrome oxidase complex. The degree of precipitation thus gives a relative measure of enzyme concentration. Wilson found that muscle extracts from 5°C-acclimated fish gave approximately 66% greater precipitation than extracts from 25°C-acclimated goldfish.

Secondly, Sidell (1977) developed methods for the quantitative extraction and purification of cytochrome c from the epaxial muscle of thermally-acclimated green sunfish, *Lepomis cyanellus*. The cytochrome c content of the extract was then quantified from the difference spectrum of oxidized minus reduced cytochrome c. Sidell found that 5°C-acclimated fish possessed 54% more cytochrome c than 25°C-acclimated fish, a value that was close to that found by Wilson for cytochrome oxidase.

The cellular concentration of an enzyme is a function of the rates of both synthesis and degradation. It is important to appreciate this dynamic balance, because it means that changes in enzyme concentration can be brought about by changes in the relative rates of these two processes even though the absolute rates of both may be significantly reduced at the lower temperature. This is exactly the case with cytochrome c in green sunfish muscle. Sidell (1977) measured the absolute rate constant of degradation of cytochrome c by following the rate of change in enzyme concentration after transfer from one acclimation temperature to another and by determining the rate of loss of radioactivity from ^{14}C-labelled cytochrome c at each acclimation temperature. Knowing the cellular concentration of the enzyme (E) and the degradative rate constant (K_d), Sidell was able to calculate the rate of synthesis (S) from the steady-state relationship

$$S = K_d E \tag{5.1}$$

Even though the absolute rates of synthesis and the degradative rate constant were both affected by temperature, Sidell found that the latter was significantly more depressed by a reduction in temperature. As a result synthesis was greater than degradation, leading to an obligatory increase in enzyme concentration in the cold until the equilibrium described by Equation 5.1 was satisfied. Furthermore the values of S and K_d, isothermally at 25°C, were not statistically different to those following transfer from 5°C to 25°C. This suggests that after transfer of animals from one temperature to another there is an instantaneous or rather rapid readjustment of S and K_d. Thus, this response may be brought about by a direct effect of temperature at the cellular level upon the synthetic and degradative processes, and may not require a more sophisticated control mechanism.

A second possible mechanism of increasing enzyme concentration is simply to increase the rate of protein synthesis by compensatory increases in the activity of the enzymes responsible for protein synthesis. Indeed, Haschemeyer (1969a) has shown that the rate of polypeptide elongation *in vivo* was increased on cold-acclimation. More recently, Nielsen *et al.* (1977) found that the activity of the aminoacyltransferase is increased in the liver of cold-acclimated toadfish. Since this enzyme plays a key role in directing the binding of amino acid–tRNA complexes to the codon-recognition site of the ribosome, its activity may be rate-limiting to protein synthesis as a whole. Compensations in the rate of protein synthesis have been widely observed (carp muscle, Loughna and Goldspink, 1985; catfish liver, Koban, 1986; green sunfish liver, Kent and Prosser, 1980) and this certainly seems to be a phenomenon of general significance. Moreover, since both synthesis and degradation form part of the normal maintenance activities of the cell, the compensation of these processes reinforces the conclusion made previously regarding the nature of maintenance metabolism. It is certainly set at levels which are higher than that needed solely to sustain life. Indeed, we can now appreciate that cellular responsiveness to stress requires the turnover of its component parts; the more rapid the turnover is, the faster the response.

5.5.4 *Enzyme isoforms*

The realization that in some cases the catalytic properties of enzymes are severely disturbed by temperature changes suggests that simply increasing the quantity of all enzymes would serve no useful purpose. Alternatively, changes in the types of enzyme isoforms present within a cell provides a means whereby the functional characteristics of the enzyme may be more closely matched to the prevailing environmental conditions. This may be achieved either by on–off synthesis of different enzyme isoforms over particular ranges of temperature, or by maintaining a complex set of isoforms and varying the relative proportions of each.

A clear example of the first type of response has been provided by Baldwin and Hochachka (1970). They found that 2°C-acclimated and 18°C-acclimated trout possessed electrophoretically distinct forms of acetylcholinesterase (Figure 5.11(a)). The adaptive significance of the two forms is related to the perturbations of catalytic properties by temperature. Figure 5.11(b) shows that the enzyme–substrate affinity was greatly affected by temperature especially at low temperatures where K_m was so high that the catalytic rate was greatly reduced. However, each isoform maintained K_m at minimal levels (i.e. maximal enzyme–substrate affinity) over a different range of temperatures which approximately corresponded to the acclimation temperatures at which it was observed. Trout acclimated at an intermediate temperature possessed both isoforms, and presumably benefited from the simultaneous activity of both. The second type of response, the modification to the proportions of a complex set of isoforms or

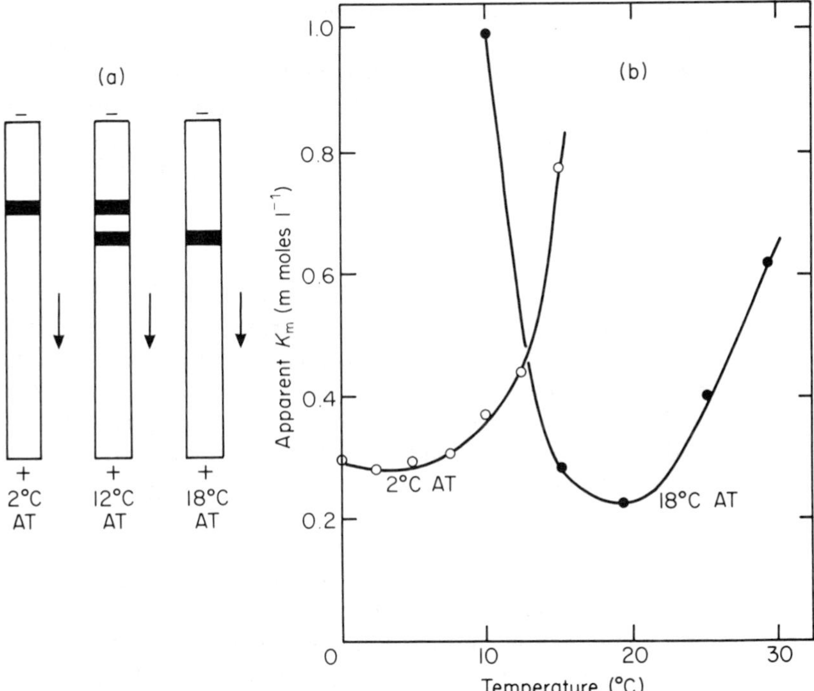

Figure 5.11 Temperature-induced synthesis of acetylcholinesterase isoforms in trout brain. (a) Electrophoretically distinct isoforms were produced in cold- and warm-acclimated trout, whereas at intermediate temperatures both forms were produced. (b) Comparison of the effect of temperature *in vitro* upon the Michaclis constant (K_m) for acetylcholine. Note that K_m was maintained low and relatively constant over overlapping temperature ranges. Thus despite the disruptive effects of temperature variations upon substrate-binding of each isoform, effective enzymatic function was ensured by swapping one isoform for the other. (After Baldwin and Hochachka, 1970; Hochachka and Somero, 1973.)

subunits, has been observed in isocitrate dehydrogenase of trout (Moon and Hochachka, 1971) and in a few other cases.

Virtually all of the demonstrations of temperature-specific isoform synthesis have been performed on trout or goldfish. Some doubt exists about just how widespread this response is, because no changes have been found in some thorough studies of other eurythermal fish. For example, Shaklee *et al.* (1977) examined 12 enzymes from six tissues of cold- and warm-acclimated green sunfish. They found changes only in the non-specific esterases of the eye and liver. A correlation between isoform composition and acclimation temperature was noted in preliminary studies with some enzymes, but when larger numbers of samples were analysed this relationship was obscured. This highlights a very real problem in studies of this sort; that is, enzymic polymorphism exists in many loci in many animals, and it is neccessary to demonstrate in a statistically valid way that this polymorphism is related to temperature and is not random or related to other factors. Hochachka and Somero (1984) suggest that because trout and goldfish are tetraploid and possess more than the normal complement of genetic information, this redundancy enables environment-specific genes to be maintained. This may not be a viable strategy for diploid organisms.

5.5.5 *Enzyme modulation*

Many enzymes are exquisitely sensitive to the microenvironmental conditions that exist within cells. It is certainly possible that alterations in the cellular concentrations of H^+, cations and various allosteric modulators could be instrumental in producing compensatory adaptations without the need for changes in the enzymes themselves. This strategy has a response time which is potentially much faster than the quantitative and qualitative mechanisms outlined previously, since synthesis of macromolecules is not required.

One important modulative effect on enzymes is exerted by the effect of intracellular pH. We have noted how variations in temperature may lead to changes in the ligand-binding properties of enzymes when measured *in vitro* under conditions of constant pH. These changes are frequently caused by changes in the protonation of titratable groups on the enzyme, the most abundant of which is the α-imidazole group of histidine. The p*K* for α-imidazole has a temperature-dependence of -0.0017 pH units $°C^{-1}$, so that cooling at constant pH causes a progressive protonation and warming a deprotonization (Fig. 5.12; Reeves, 1977). Since interactions between these titratable groups are crucial to the tertiary and quaternary structure of proteins, the advantage of maintaining a constant fractional dissociation of the α-imidazole groups is obvious. This can be achieved by altering pH as temperature changes by the same amount as the p*K* for α-imidazole varies. A regulated change in the pH of blood and cytosolic fluids of this magnitude

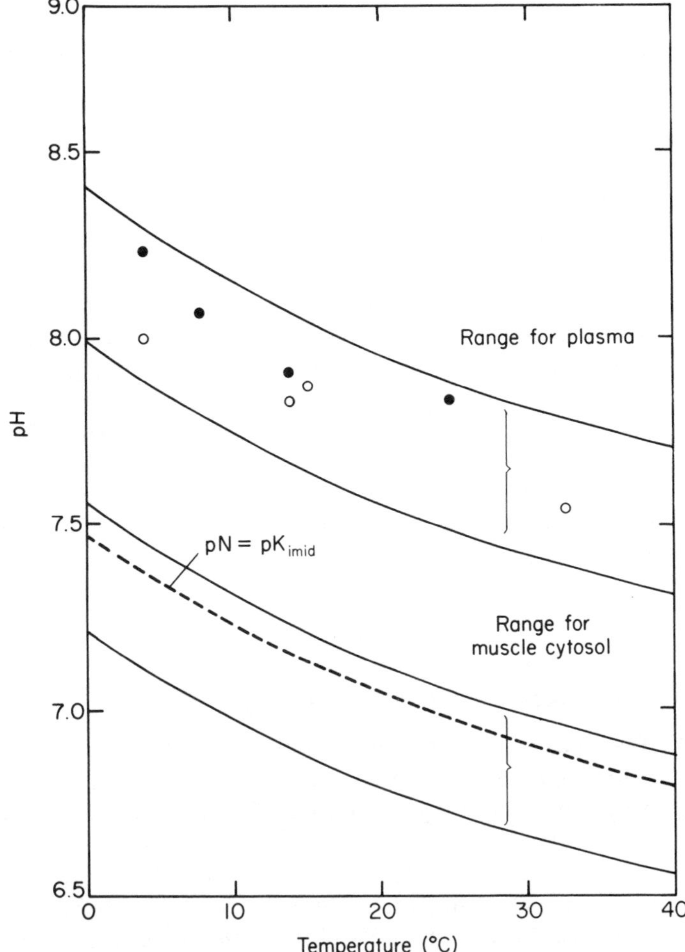

Figure 5.12 The relationship between acclimation or body temperature and the pH of blood or muscle cytosol in different animal species. The effect of temperature upon the pK of α-imidazole and on the neutral pH of water (i.e. when $[H^+] = [OH^-]$) are also shown as the broken line. According to the alphastat hypothesis the variation of blood or cytosolic pH with temperature maintains a constant difference between body fluid pH and the neutral pH of water (pN). This maintains a constant $[H^+]/[OH^-]$ ratio even though $[H^+]$ varies with temperature, and is thought to preserve the net charge on proteins. The variation of pH with temperature is achieved by active regulation as shown by data for the rainbow trout (●) and the common carp (○), summarized by Cameron 1978. (After Somero, 1981.)

has, in fact, been widely observed as animals have become acclimated to an altered temperature or when comparing species which live in different thermal environments (Reeves, 1977, but also see Heisler, 1979). Thus, the regulation of the dissociation state of the α-imidazole groups is of more importance in conserving enzymic function than in maintaining a constant pH.

Variations in cellular pH may also elicit changes in enzymic function which may be responsible for the compensation of cellular processes, for metabolic reorganization or even for the metabolic depression by torpid or hibernating animals (Malan *et al.*, 1985; Walsh *et al.*, 1983). Direct evidence for this during temperature acclimation is lacking, though because intracellular pH is now widely regarded as an important regulator of cell function (Boron, 1986; Nuccitelli and Deamer, 1982) it would be surprising if this mechanism were not involved in temperature acclimation. One potential candidate is phosphofructokinase, which is the first important regulatory enzyme in glycolysis. This enzyme is allosterically regulated by a variety of modulators, as befits a key regulatory enzyme, including pH. In goldfish muscle the activity of this enzyme was increased from 20% to 80% of its maximal rate by a change in pH of only 0.1 (Freed, 1971). In a hibernating mammal, changes in the oligomeric state of the enzyme caused by the non-alphastat regulation of cytosolic pH may lead to enzyme inactivation (Hand and Somero, 1983).

5.5.6 *Homeoviscous adaptation*

A second and more widely recognized modulative effect upon membrane-bound enzymes is exerted by membrane lipids. It is a long-standing observation that the unsaturation of membrane lipids (i.e. the number of unsaturated bonds in the fatty-acid chain) in animals, plants and micro-organisms depends upon the ambient temperature. In animal cells the fatty acids are mainly straight-chain molecules with 16–22 carbon atoms and with up to six unsaturated bonds. Because there may be up to 30 species of fatty acid which may be combined in different pairings on the phospholipid, and because there are often changes in other lipid components, the details of the compositional differences between cold- and warm-acclimated animals are extremely complex. However, the most common pattern is that cold-acclimation leads to an increased proportion of fatty acids containing unsaturated bonds at the expense of saturated homologues.

It has also been known for many years that the physical properties of fatty materials is dependent upon their unsaturation. Early work using monolayers of phospholipids showed that the introduction of the kinked unsaturated bond into a saturated fatty acid leads to an expansion of the monolayer by interfering with the otherwise close packing arrangements of the parallel-aligned hydrocarbon chains (Quinn, 1981). This naturally leads to an increased rotational mobility and flexing of the chains about the carbon–carbon bonds, which can offset the ordering influence of cooling (Hazel and Prosser, 1974). Several biophysical techniques have recently become available to measure the magnitude and rate of the motion of the membrane constituents directly. Each technique is sensitive to a different aspect of the molecular motion, and each gives a somewhat different measure of 'membrane fluidity'. Most work relevant to this discussion has

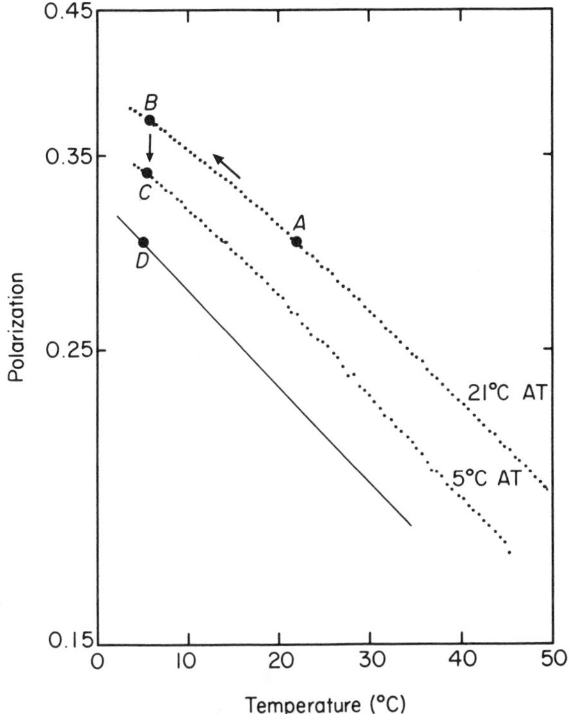

Figure 5.13 Seasonal homeoviscous adaptation of brain membrane preparations isolated from trout acclimated to 5°C and 21°C. Fluidity was estimated by measuring the polarization of fluorescence from a 'probe' molecule (1,6-diphenyl-1,3,5-hexatriene, DPH) positioned within the membrane bilayer when illuminated with vertically polarized light. A low value of polarization means that the probe has rotated through a large angle before it fluoresces and is therefore relatively free to rotate. A high value of polarization means that probe motion is more restricted, and hence fluidity is lower. Sudden transfer of fish from 25°C to 5°C results in a substantial decrease in fluidity (point *A* to point *B*) as indicated by the increased polarization. During the following weeks at 5°C the fluidity is adjusted downwards (*B* to *C*). The solid line indicates the position of the curve required for complete compensation during cold-acclimation. Note that the adjustment observed in cold-acclimated animals is not sufficient to completely compensate for the effects of cooling. AT – acclimation temperature. (Data of M.K. Behan and A.R. Cossins.)

measured the rotational properties of a rod-shaped fluorescent probe dissolved in the membrane bilayer in order to compare the fluidity of membranes from differently-acclimated animals (Cossins, 1977, 1981). In most cases the membranes of cold-acclimated animals were, indeed, found to be more fluid than the corresponding membranes of warm-acclimated fish (Fig. 5.13). Because this response restores fluidity towards its pre-existing value it is termed 'homeoviscous adaptation', though in temperature acclimation it is usually sufficient to overcome only 20–50% of the temperature-induced changes in membrane properties (Cossins, 1983). The time-course of change during acclimation varies from 2 to 30 days, depending upon the turnover rate of the membrane phospholipids and the direction of temperature change.

The changes in lipid composition and membrane fluidity are caused by complex changes in lipid biosynthesis (Hazel, 1984) though how this is regulated to produce the homeoviscous response is not known. Figure 5.14 shows that cold-acclimation of carp leads to a rapid increase of 30–40-fold in the specific activity of the Δ^9 desaturase of liver microsomes, the enzyme responsible for inserting the first unsaturated bond into an otherwise saturated fatty acid. This increased activity results from a greater number of enzyme molecules in the cold, which implicates the protein synthetic system and perhaps gene expression in the control mechanism. Note that a

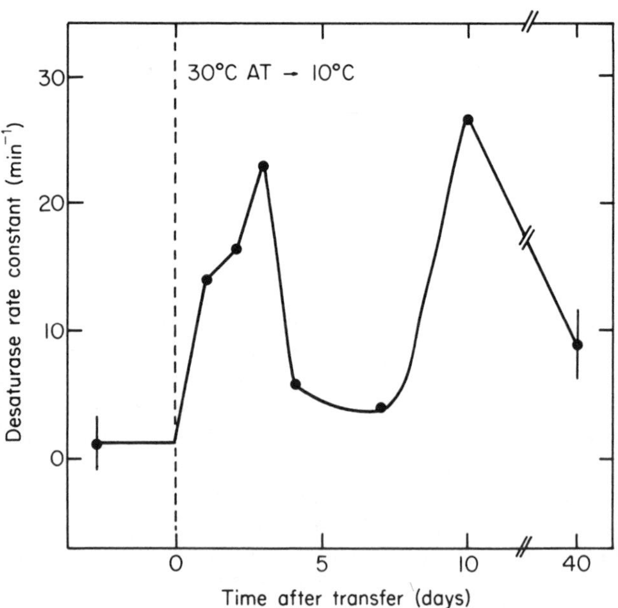

Figure 5.14 The time-couse of Δ^9- and Δ^6-desaturase activity in liver microsomes following transfer of 30°C-acclimated carp to 10°C. The vertical bars represent ± s.d.m. (After Schüenke and Wodtke, 1983.)

second peak of activity occurs 7–40 days after transfer, though this is not associated with any observable change in fluidity.

The adaptive significance of this partial preservation of membrane fluidity is due to the many and varied effects of membrane fluidity upon membrane properties and functions. The fluid-mosaic model of membrane ultrastructure (Singer and Nicholson, 1972) suggests that the phospholipid bilayer forms the effective solvent environment for many of the functional components of membranes (i.e. proteins). It is clear that membrane-bound enzymes, which undergo conformational transitions or flexing movements during catalysis, are greatly influenced by the fluidity of the surrounding bilayer; the more fluid the bilayer is, the less constrained the protein and the greater its conformational flexibility. The dependence of membrane-bound enzyme function upon fluidity can be demonstrated in several ways. For example, the removal of phospholipids by detergent extraction or by phospholipase activity leads to a loss of enzymatic activity which can be restored by reconstitution with purified phospholipids. The effect of membrane fluidity upon enzymatic activity has been most elegantly demonstrated in a clutured cell line which was deficient in cholesterol biosynthesis. Cholesterol has an important ordering effect upon phospholipid membranes. Supplementation of cholesterol into the culture medium produced a concentration-dependent increase in membrane order, and a correlated decrease in activity of the Na^+K^+-ATPase (Sinensky *et al.*, 1979).

That the changes in membrane fluidity during thermal acclimation are responsible for changes in the activity of membrane-bound enzymes is suggested by increases in the turnover number for the Na^+K^+-ATPase of goldfish intestinal mucosa as a result of cold-acclimation; in other words, the specific activity was increased without any change in the number of active enzyme molecules (Smith and Ellory, 1971). Similarly, changes in the respiratory activity of muscle mitochondria of carp during acclimation were not accompanied by changes in the total content of cytochrome (Wodtke, 1981). A more direct demonstration of this effect was provided by Hazel (1972) who found that reconstitution of succinic dehydrogenase activity of lipid-free extracts of goldfish muscle mitochondria was greatest with lipid extracts of cold-acclimated goldfish muscle. In other words, the lipids of cold-acclimated fish provided a less restrictive environment which allowed greater enzymatic activity than is found with lipids of warm-acclimated fish. The regulation of membrane fluidity is thus a highly potent means of regulating many aspects of membrane function.

Other important properties of membranes are their permeability properties. Studies with artificial membranes have shown that the introduction of unsaturated hydrocarbon chains into phospholipids produces a substantial increase in the permeability of electrolytes and non-electrolytes. Indeed, the permeability of artificial membranes and mitochondria prepared from cold-acclimated animals has been found to be much greater than those prepared from warm-acclimated animals (Cossins, 1983).

5.5.7 *Cellular restructuring*

A quite separate adaptive strategy that is gaining widespread recognition is through ultrastructural alterations to the cell and tissue morphology. For example, cold-acclimation of goldfish produces an increase in the total surface area of sarcoplasmic reticulum in skeletal muscle, together with a more intimate relationship between the reticulum and the myofibrils (Penny and Goldspink, 1980). In the crucian carp, cold-acclimation was associated with an increase in proportion of cell volume occupied by mitochondria (i.e. mitochondrial volume density) while myofibrils showed a decrease (Johnston and Maitland, 1980). Moreover, the proportion of mitochondria between the myofibrils increases during cold-acclimation at the expense of those at the periphery of the cell (Sidell, 1983).

Ultrastructural adjustments such as these become intelligible when the role of diffusion as a limiting factor in cellular systems is appreciated. Because of the stepwise nature of the metabolic pathways, the products of one enzyme are the substrates of the next. Metabolic intermediates, cofactors and ions must therefore diffuse across finite distances which may be rate-limiting to the pathway as a whole. The creation of multi-enzyme pathways is one means of reducing diffusional distances, and the arrangement of subcellular organelles with respect to their function is probably another. The increase in mitochondrial volume density clearly serves to increase the capacity for ATP production, particularly in the region of the myofibrils, while the elaboration of the sarcoplasmic reticulum creates a more intimate connection between the stores of calcium and the myofibrillar ATPase. The important point here is the reduction of diffusion distances for metabolites and ions such as ATP and Ca^{2+}. Tyler and Sidell (1984) have calculated that ultrastructural alterations in mitochondrial distribution in fish red muscle during thermal acclimation are sufficient to compensate for a $Q_{10} = 1.8$ for diffusion rates, which is the same as the Q_{10} for lactate diffusivity through fish muscle cytoplasm.

5.6 Genotypic adaptation to temperature

Genotypic or evolutionary adaptations to temperature can be demonstrated by comparing the physiology of species which inhabit different thermal environments, or latitudinally- or altitudinally-separated populations of the same species. As with temperature acclimation, compensation should produce translations of the acute rate–temperature curves along the temperature axis such that the compared populations or species possess equivalent rates at their respective environmental temperatures as well as adaptations to extend their tolerance limits.

In all such cases it is vital to exclude phenotypic adaptations (i.e. acclimatizations) from consideration by carefully controlling the previous thermal experience of the experimental subjects. Another complicating phenomenon is *canalization*, in which the physiological status or properties of an organism is influenced by the thermal experience of earlier developmental stages, irrespective of its recent thermal experience. The importance of this process is largely unexplored except in a few cases, but it does point to the ultimate necessity to breed and rear animals under controlled conditions.

5.6.1 *Intraspecific latitudinal adaptation of metabolic rate*

These difficulties have been taken into account by the classical studies of Vernberg and his colleagues on the fiddler crabs of the genus *Uca*. These crabs have a wide geographical distribution in both temperate and tropical

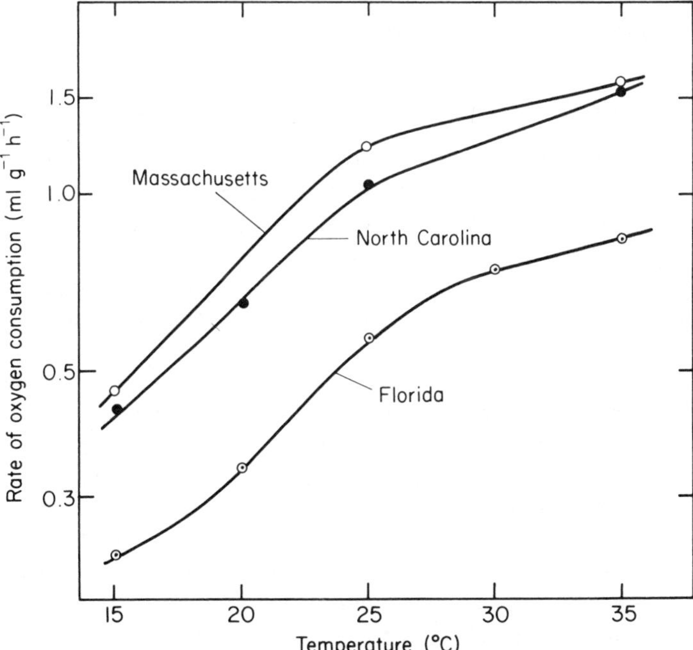

Figure 5.15 A comparison of the rates of oxygen consumption for fiddler crabs, *Uca pugilator*, obtained from populations in Massachusetts, North Carolina and Florida. All animals were reared in the laboratory under the same conditions, so differences represent genotypic variation between each population. (After Vernberg and Costlow, 1966.)

waters, and one species, *Uca pugilator*, is found along almost the entire eastern seaboard of North America. They found significant differences in the rates of oxygen consumption of laboratory-reared *U. pugilator* from populations in Massachusetts, North Carolina and Florida (Fig. 5.15). Their experiments were complicated by a variable relationship between oxygen consumption and size which depended on the stage of the life-cycle as well as the experimental temperatures. Since individuals from each population varied in size it was necessary to discover the nature of this size effect and compare their weight-corrected rates of oxygen consumption. Crabs originally collected at the more northerly latitudes displayed higher rates of oxygen consumption than those from lower latitudes, despite constant rearing and acclimation temperatures. On the basis of this index of physiological activity each population should be considered as a distinct race, each of which displayed a level of energy expenditure commensurate with its respective microclimatic conditions. Whether this reflects different levels of activity or of standard metabolism is unknown, though these differences were not observed in early larval stages. Vernberg and Costlow (1966) suggest that this is consistent with the similar sea temperatures experienced at each latitude at the times of maximal larval abundance. The more northerly populations delay reproduction until the warmer months.

5.6.2 *Interspecific adaptation of metabolic rate*

Dunlap (1980) has noted interspecific differences in the rates of routine metabolism in two species of frog from South Dakota, which is correlated with their respective lifestyles. When acclimated to 15°C or 25°C the routine rates for *Pseudacris* was approximately twice the routine rate of *Acris* (Fig. 5.16). *Acris* is active throughout the day and night, and frequently basks in the sun. By contrast, *Pseudacris* is mainly nocturnal, tends to breed earlier than *Acris*, and has a lower tolerance to high lethal temperatures. These differences imply that *Pseudacris* is exposed and physiologically adapted to colder conditions than *Acris*, presumably by way of elevated metabolic rate. Davies and Bennett (1981) have compared the rates of resting oxygen consumption for two closely related species of juvenile Natricine snakes from different latitudes. The cool temperate *Natrix natrix* (Britain) showed greater rates than the warm temperate *N. maura* (Spain) at all measurement temperatures (Fig. 5.17), but when measured at their respective preferred body temperatures the rates were more similar.

The comparison of metabolic rates of different species is generally less easy than the comparison of those of populations of the same species, because species may differ in so many ways not directly connected with environmental temperature. It is essential in interspecific comparisons to compare animals which have nearly identical life-histories and lifestyles. It is obviously not very instructive to compare a continually active species

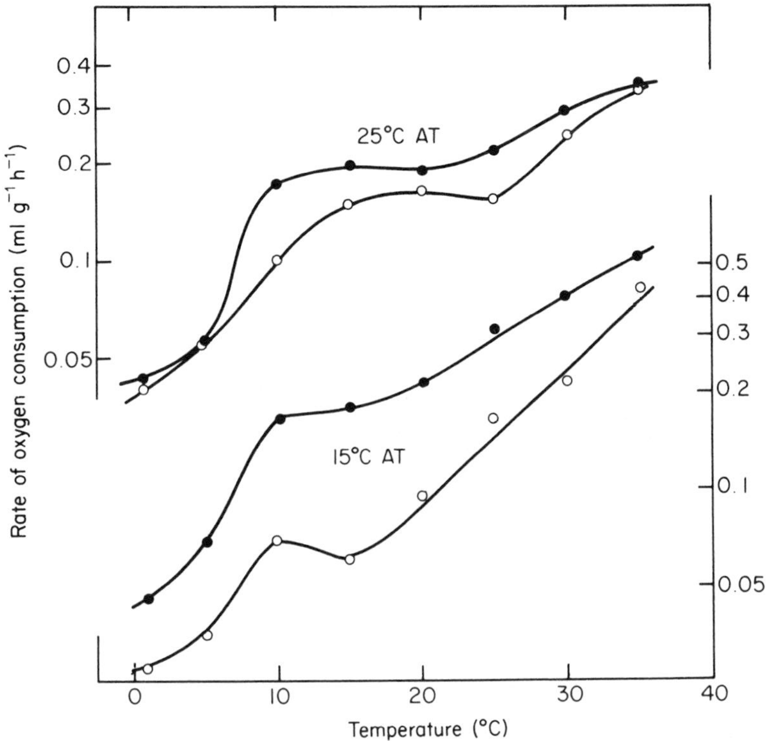

Figure 5.16 A comparison of acute rate–temperature curves for two closely related species of frog from South Dakota, USA, (●) *Pseudacris triseriata*, (○) *Acris crepitans*. The higher metabolic rate in *Pseudacris* corresponds to its more nocturnal habits. (After Dunlap, 1980.)

with a sluggish species, even if they are phylogenetically related. It is also important to understand the responses of the different species, both to acclimation regime and to the experimental system employed, so that artefactual differences are excluded. A final complication occurs when the normal temperature ranges of the organisms to be compared do not overlap, since it becomes necessary to extrapolate the rate–temperature curve of one organism to the normal temperatures of the other. The method of extrapolation is crucial to the comparison.

These problems are highlighted by studies on the genotypic adaptation of animals to extreme temperatures. Polar seas, for example, are characterized not only by extreme cold, but also by the virtual absence of seasonal temperature variations. The abundance and diversity of the Antarctic fauna are as high as in more-temperate waters, and the low temperatures clearly present no insurmountable barrier to life. The question is, therefore,

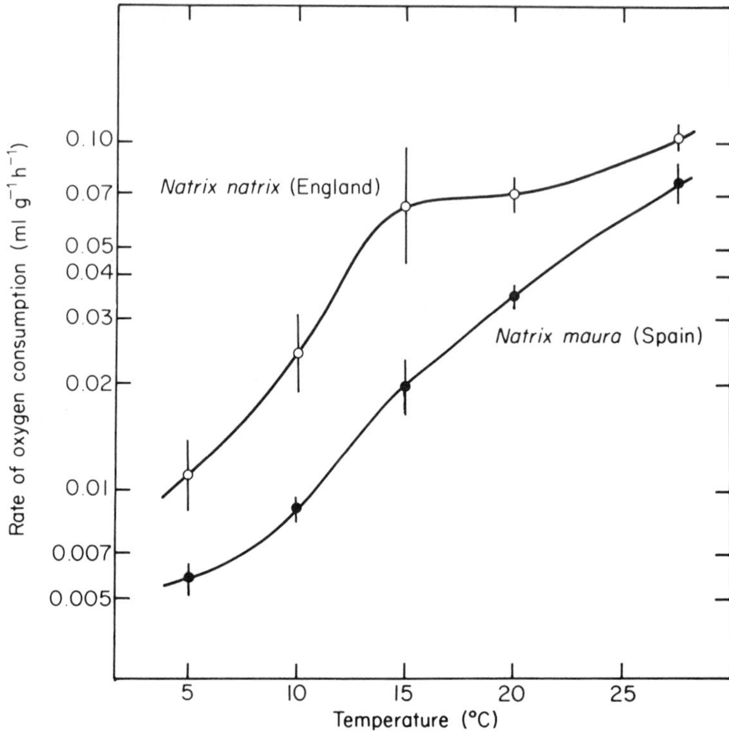

Figure 5.17 A comparison of the acute rate–temperature curves for oxygen consumption in two species of Natricine snake with similar ecological characteristics but from different climatic zones. Juvenile snakes from each species were hatched and reared under identical laboratory conditions. The vertical bars represent 95% confidence intervals. (After Davies and Bennett, 1981.)

what physiological adaptations have accompanied the invasion of this habitat?

As early as 1916, Krogh predicted that polar poikilotherms would exhibit elevated metabolic rate compared with temperate animals at 0°C, and it was a matter of common observation that species of high latitudes were as active as similar species from warmer climates. Many early studies showed that the respiratory rates of invertebrates from Greenland, the North Sea and the Mediterranean were, indeed, similar at their respective environmental temperatures (Bullock, 1955). Subsequently, in what has become a cornerstone of the hypothesis of genotypic temperature compensation, Wohlschlag (1960, 1964) showed that the rates of oxygen consumption of polar fish, as determined by closed-vessel respirometry, was between five and ten times greater than that expected by the extrapolation of Krogh's normal curve for the goldfish to polar temperatures (Fig. 5.18). As with seasonal compensation, the adaptive value of this elevation of metabolism

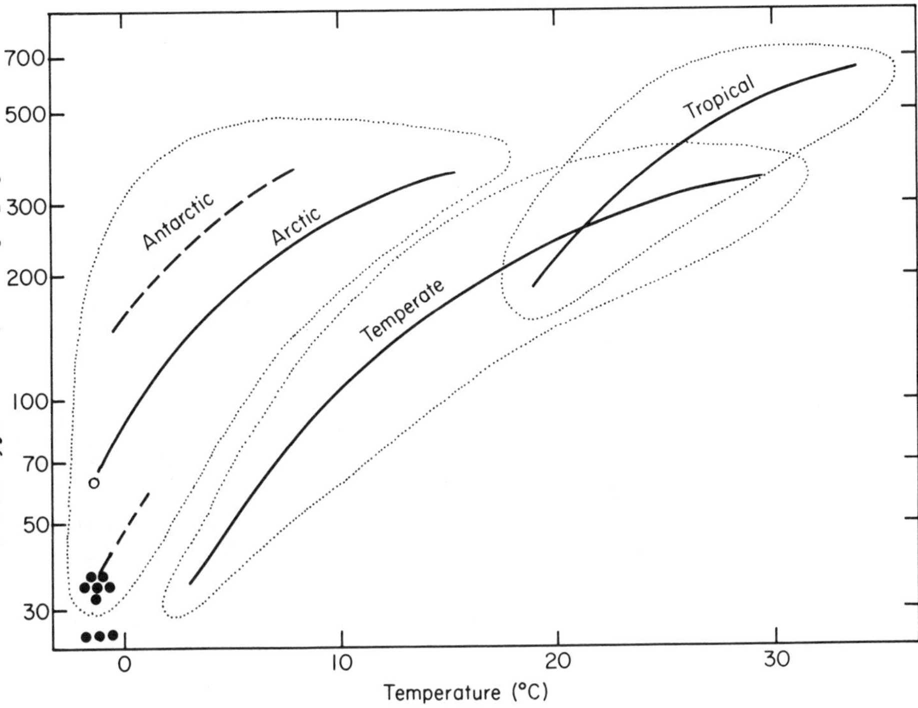

Figure 5.18 A summary of the rate–temperature curves obtained for polar, temperate and tropical fish according to the hypothesis of metabolic cold compensation. The dotted lines represent the general range of variability within each zone of latitude. The individual data points obtained by Holeton (1974) for the arctic cod, *Boreogadus saida* (○) and several other arctic species (●) are shown. (After Brett and Groves, 1979.)

in the cold was thought to be the maintenance of standard metabolism at a level sufficient for maintenance purposes. Tropical species, by contrast, displayed only a limited compensation compared with temperate species.

More recently, Holeton (1974) has determined the routine rates of Arctic fish species in greater detail using open-circuit respirometry. He noted that polar fish were much more sensitive to capture and handling stress than had been previously realized and he found it necessary to leave fish in the respirometer for 24–48 h before a reasonably stable rate was attained. It is therefore clear that the high values of oxygen consumption observed by Wohlschlag were due, at least in part, to high levels of spontaneous activity, rather than indicating some specialized metabolic adaptation. The values obtained for most species by Holeton were generally well below the levels previously considered to be cold-adapted. The value for the Arctic cod, *Boreogadus saida*, was more similar to the values published previously for Arctic species, though this species was the one most likely to display spontaneous activity in the respirometer (Holeton, 1974), and probably does not give reliable values for the standard rate.

A second criticism made by Holeton (1974) concerns the method by which the routine rates for temperate species was extrapolated to polar temperatures to form a reference with which to compare polar species. Wohlschlag used the experimental data of Ege and Krogh (1914), which were obtained using single fish after abrupt shifts in temperature, a technique which on the basis of present knowledge is not satisfactory. When more recent data, which incorporated corrections for activity and weight-specific effects, were used and extrapolated using $Q_{10} = 2$ then the difference between polar and temperate fish was reduced from seven- to two-fold. Bearing in mind that there is a considerable spread in routine rates between lethargic fish, such as the zoarcids, and the more active Arctic cod *Boreogadus* (see Fig. 5.18), this difference is not impressive. It therefore appears that Wohlschlag's curve for Antarctic fish is more representative of routine rates rather than standard rates.

Cold-compensation of metabolic rate for invertebrates has been observed in some studies but not in others. Block and Young (1978) found that the routine rate of some Antarctic mites was two to four times higher than that of temperate mites extrapolated to 0°C, bringing their routine rates to values exhibited by the temperate species in their normal temperature range. However, these studies are subject to similar criticisms as the studies on polar fish, namely that activity level was uncontrolled and it is unclear to what extent the differences stem from an elevation of resting metabolism or an increased level of activity in Antarctic mites. On the other hand, some other studies have not found elevated respiratory rates in terrestrial and sublittoral invertebrates of polar regions (Scholander *et al.*, 1953; Lee and Baust, 1982a, b; Maxwell, 1977). Houlihan and Allan (1982) have compared the respiratory rates of Antarctic and British gastropods. They found that the rates for British species measured at 0°C were, indeed, well below those of Antarctic species to give an apparent compensation. However, when animals were measured over their normal range of environmental temperatures all species fell on the same graph of rate of respiration versus temperature. Thus, in this case Antarctic species did not show an evolutionary compensation of respiratory rate, but were less susceptible to extreme cold and have extended their normal temperature range downwards.

In summary, it seems that the expectation of an elevation of metabolism of all animals in polar environments is oversimplistic. Some species may display elevated metabolic rates relative to temperate species, whereas others do not. The prediction of elevated rates in cold-adapted species depends upon the compared species possessing broadly comparable life-styles, life-histories and energetic strategies. Clarke (1979) has pointed out that the Antarctic fauna is characterized by rather low annual growth rates compared with temperate species, and, despite the constant temperature, by an intensely seasonal pattern of growth. This may be related to the seasonal pattern of primary productivity, and hence the food availability in polar seas. It is therefore clear that Antarctic species are not simply temperate species shifted with metabolic compensation to a lower range of temperatures, but rather have evolved a particular suite of adaptive characteristics which uniquely match them to their peculiar habitats. This includes not only a relatively slow annual rate of growth, but also reduced reproductive output, deferred maturity, a low rate of standard metabolism and a lack of metabolic cold-adaptation (Clarke, 1979). This general reduction in individual energy requirements allows a greatly increased standing biomass for a given energy input without the high rate of turnover observed in warmer climates. It seems that for most species it is this, rather than a straightforward metabolic compensation, which is the successful strategy in polar seas. As described presently, there are profound adaptations of the physiology in polar species which allow successful function in the extreme cold, so that metabolic compensation is possible, at least in principle. This implies that the constraints producing non-compensated standard metabolic rates, growth rates, etc., are not physicochemical but ecological (Clarke, 1980). One possibility is that the energetic input into the polar ecosystem is insufficient to support compensated rates of standard metabolism, and the animals reduce their energy requirements by adopting a quiescent state for large parts of the year. This limitation of food supply may be more of a problem for animals at the beginning of a food chain than for those in higher trophic levels.

5.6.3 *Physiological and biochemical adaptations*

The abandonment of energetic cold-adaptation as a general feature of polar organisms does not mean that these animals are not physiologically suited for continued function in the extreme cold, nor that compensations of some physiological properties do not occur. Again, most work has focused upon Antarctic species, since these represent a particularly extreme and clear-cut example of adaptive physiology. The fact of survival and continued life for these animals implies that their physiological systems are different from temperate and tropical species, and because these latter species suffer from an extreme form of the rate effect when placed in extreme cold these differences must, at least in part, be compensatory in nature.

The thermodynamic analysis presented in Chapter 2 showed that the rate of a chemical reaction was a function of the free energy of activation, $\Delta G^{\neq}$. At biological temperatures only a small fraction of molecules possess sufficient energy to overcome this barrier, and the reaction rate is accordingly rather slow. Enzymes achieve a remarkable increase in reaction rate by reducing the height of this energy barrier, and thereby

Table 5.3 Thermodynamic parameters for activation of the myofibrillar Ca^{2+}, Mg^{2+}-ATPase of fish living at different environmental temperatures (1 cal $\hat{=}$ 4.2 J). (After Johnston and Walesby, 1977.)

Species	Environment and temperature	$\Delta G^{\neq}$ (cals. mol^{-1})	$\Delta H^{\neq}$ (cals. mol^{-1})	$\Delta S^{\neq}$ (entropy units)	Relative enzymatic activity at 0°C	Inactivation halftime at 37°C (min)
Champsocephalus gunneri	Antarctica (−1 to +2°C)	15 870	7 400	−31.0	1.0	1
Notothenia neglecta	Antarctica (0–3°C)	16 130	11 300	−17.8	0.62	1
Cottus bubalis	North Sea (3–12°C)	16 290	13 550	−9.9	0.45	12
Dascyllus carneus	Indo-Pacific (18–26°C)	17 540	25 350	+25.1	0.05	60
Pomatocentrus uniocellatus	Indo-Pacific (18–26°C)	17 620	26 500	+32.47	0.05	80

increasing the proportion of molecules with the required energy of activation. The $\Delta G^{\neq}$ of homologous enzymes from different species varies (Low *et al.*, 1973; Low and Somero, 1974, reviewed in Hochachka and Somero, 1984) and Table 5.3 illustrates this with data for the myofibrillar ATPase of skeletal muscle. Cold-adapted species have somewhat smaller values of $\Delta G^{\neq}$ than warm-adapted species or compared with mammalian enzymes. As $\Delta G^{\neq}$ has an exponential relationship with catalytic rate, these differences mean that cold-adapted enzymes have a higher catalytic rate than warm-adapted enzymes when measured under the same conditions. This observation applies to all enzymes so far examined (Hochachka and Somero, 1984) and includes enzymes from several different phylogenetic groups, which suggests that cell temperature is of prime importance in defining the characteristics of enzymes.

A detailed and lucid account of how these interspecific differences are brought about is given by Hochachka and Somero (1984). Of significance here is the general interpretation that cold-adapted enzymes achieve their high enzymatic rates in the cold by possessing a less constrained and rigid secondary and tertiary structure. In simple terms it seems that because of this greater conformational freedom and flexibility the conformational transition that occurs during enzyme activation requires less thermal energy in cold-adapted enzymes than in their warm-adapted homologues, possibly because fewer weak bonds are broken in the process. Note how in Table 5.3 the activation enthalpy, $\Delta H^{\neq}$, of cold-adapted enzymes is substantially less than in warm-adapted enzymes and also that the activation entropy ($\Delta S^{\neq}$) is generally less positive or even negative compared with warm-adapted forms.

This increased catalytic efficiency of cold-adapted enzymes is, however, achieved at some cost to their thermal stability. Table 5.3 compares the half-time for thermal inactivation of myofibrillar ATPases of different species. The enzyme from polar fish was inactivated rapidly at 37°C, and clearly it would not be an effective enzyme over the upper biological range of temperatures. This increased susceptibility to heat is a consequence of the more flexible tertiary structure of cold-adapted enzymes. Thus, evolution has provided for equivalent enzymatic performance at different temperatures, though the linkage between adapted enzyme function and altered thermal stability provides absolute thermal limits for function, and explains why mammalian enzymes are less efficient than they could be.

Adaptation at the molecular level is certainly not restricted to enzymes, since there is now good evidence that the cellular membranes of cold-adapted species possess a compensated fluidity compared with warm-adapted animals (Cossins and Prosser, 1978). A major problem in interspecific comparisons is to ensure that the necessarily impure membrane fractions obtained from each species are composed of similar collections of membrane vesicles with similar functional properties. Thus, it is preferable

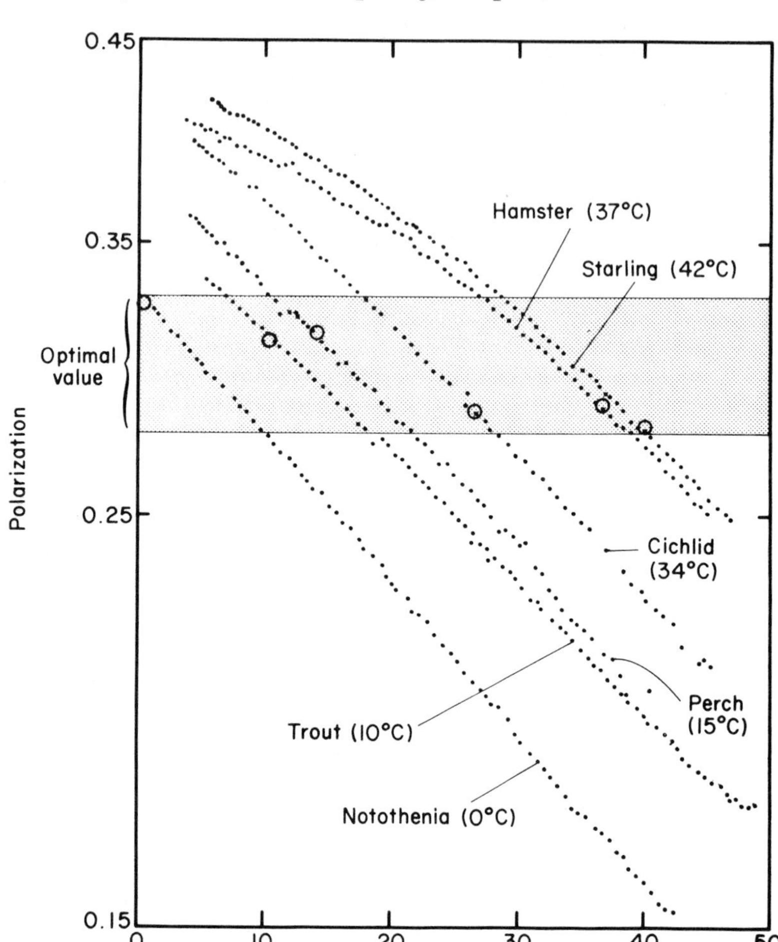

Figure 5.19 Evolutionary homeoviscous adaptation for brain membranes of different fish, bird and mammal species. Polarization of DPH fluorescence (see Fig. 5.13 for details) is plotted as a function of the measurement temperature. The values at the respective cell/body temperatures are indicated with a large circle. Note that values of polarization at the respective body temperatures are maintained within a small range (shaded zone) compared with the direct effects of temperature. (A.R. Cossins, M.K. Behan, G. Jones and K. Bowler, unpublished observations.)

to compare tissues with broadly similar cellular morphologies and membrane compositions in the different species. Figure 5.19 compares the fluidity of a purified membrane fraction isolated from the brains of various fish species, a mammal and a bird. The curves all fall on different positions on the temperature axis in an orderly sequence depending upon cell temperature, such that at any measurement of temperature the fluidity increases with reduced cell temperature.

Thus, evolutionary adaptation to a specific cell temperature involves similar compensatory responses of brain membranes as observed during thermal acclimation of fish. However, the magnitude of these interspecific differences was much larger than between thermally-acclimated fish. Figure 5.19 shows that the index of fluidity was almost identical in all of the fish, mammal and bird species when measured at their respective cell temperatures. The more impressive compensation over evolutionary time is presumably because each species specializes its membranes for a particular temperature regime rather than maintaining a competence at a variety of temperatures. Thus, we again see that cell temperature rather than phylogenetic status is a major determinant of cellular structure.

Interspecific homeoviscous adaptations are correlated with large inter-

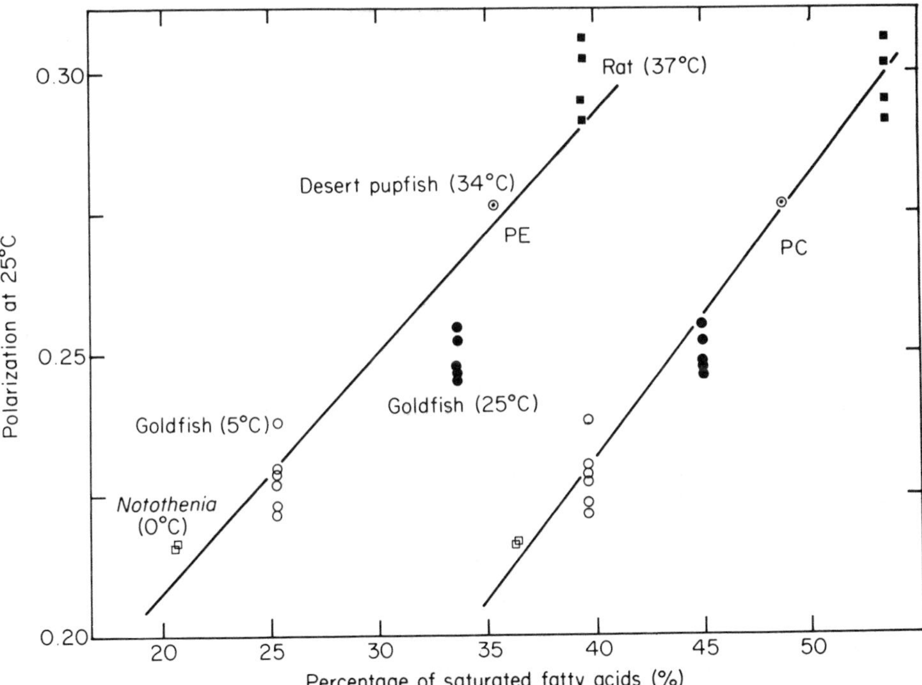

Figure 5.20 Interspecific differences in membrane fluidity are related to differences in lipid saturation. An index of fluidity (i.e. DPH polarization at 20°C, see Fig. 5.13) for brain membranes from different fish species and rat is plotted against the percentage of fatty acids which are saturated in the major phospholipid classes, phosphatidylcholine (PC) and phosphatidylethanolamine (PE). The numbers in parentheses refer to the body/acclimation temperature of each species. (After Cossins and Prosser, 1978).

specific differences in membrane lipid composition (Cossins and Prosser, 1978). Figure 5.20 shows how an index of membrane fluidity varies in relation to changes in the fatty-acid composition of the brain membranes; increased fluidity in cold-adapted species is associated with a progressive decrease in the proportion of saturated fatty acids in membrane phospholipids. The arguments for this increasing the molecular motion and disorder of the hydrocarbon chains are exactly as described for homeoviscous adaptation during temperature acclimation. Similarly, the functional significance of interspecific differences in fluidity is related to the effects of bilayer fluidity upon membrane-bound enzymes. It is interesting to speculate that because the hydrocarbon environments offered by the brain membranes of *Notothenia* at 0°C and rat at 37°C are roughly equivalent, the homologous membrane-bound enzymes may be rather similar in secondary and tertiary structure. In contrast with soluble enzymes, the required adaptations in catalytic properties and thermal stability of membrane-bound enzymes may simply be produced by adaptations of solvent properties.

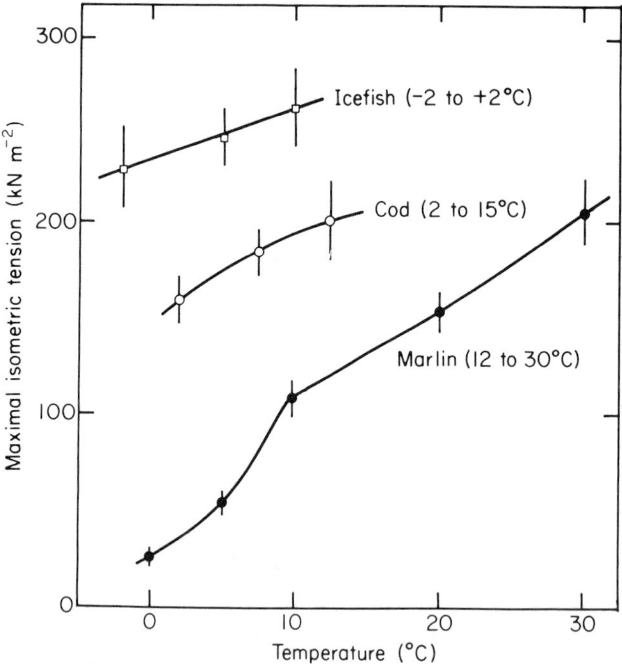

Figure 5.21 The maximal isometric tension developed by skinned muscle fibres from the white fast muscle of the icefish (*Chaenocephalus aceratus*), the North Sea cod (*Gadus morhua*) and the Pacific blue marlin (*Makira nigricans*). The contractile performance of icefish muscle is clearly enhanced in the cod compared with the marlin. (After Johnston and Altringham, 1985.)

These molecular-level adaptations of soluble enzymes and of membrane fluidity are likely to underlie the compensation of cell and tissue performance. This applies especially to studies of the neuromuscular apparatus where Johnston and Altringham (1985) have recently found some remarkable similarities in tension generation by muscle fibres in different fish species from different latitudes when measured at their normal temperatures. Using skinned fibre preparations from the marlin, an Antarctic species and the cod, they found that at 0°C maximal tension was seven times higher in icefish than in marlin fibres, though at their respective body temperatures the maximal isometric tensions were not dissimilar (Fig. 5.21). The contraction velocities of unloaded fibres from the different species were similar, though during more realistic conditions of loaded shortening the cold-adapted fish were faster. These adaptations may be related to rather subtle changes in the myofilament apparatus. Thus, muscle tension is related to the number of cross-bridges which are made between the actin and myosin filaments. These cross-bridges are continually being made and broken, so tension is a function of the lifetime of the attached cross-bridge. Normally, the average lifetime is short compared with the overall cycle time, so a small increase in the attached lifetime can increase the number of attached cross-bridges at any instant. This idea of adapted cross-bridge dynamics is supported by an extreme sensitivity of normal muscle contraction and relaxation in Antarctic fish to moderately high temperatures, which parallels the sensitivity of Antarctic fish ATPases. Muscle fibres of warm-water fish species, by contrast, show full control over contraction and relaxation at temperatures up to 35–40°C. Thus, compensated muscle performance at temperatures near 0°C have been achieved at the expense of an ability to operate above 8–10°C.

Neuronal and neuromuscular properties are also adapted in polar fish. For example, the cold-block temperatures were not only shifted to lower temperatures, but the velocity of axonal conduction showed a partial (Precht type 3) compensation relative to temperate fish species (Macdonald, 1981). Similarly, there is a compensation in synaptic properties of polar fish, since Macdonald and Balnave (1984) have shown that the rate of decay of miniature end-plate potentials at a neuromuscular junction was appreciably shorter in polar species than in temperate species. The rate of decay is thought to be limited by the conformational freedom of channels in the post-synaptic membrane, and this may be sensitive to the fluidity of the post-synaptic lipids. These various adaptations seem to underly the improved neuromuscular performance of polar species in the extreme cold (Montgomery *et al.*, 1983; Montgomery and Macdonald, 1984).

5.7 Conclusions

The capacity adaptation of animals to temperature is a highly complex syndrome, with physiological adjustments occurring at all levels of organization. Although most responses are clearly compensatory, others are not. These non-compensatory responses can usually be rationalized in terms of seasonally-driven rhythms of, for example, food availability, reproduction and moulting. Thus, the proper interpretation of acclimatization physiology requires a knowledge of the ecological constraints that operate on members of a species and the energetic strategies that they adopt. In this respect the terminology associated with Precht's classification is unfortunate, since it imposes limits and objectives to the process which may be inappropriate. 'Inverse' or 'paradoxical' energetic compensations and seasonal metabolic reorganization must be viewed within a seasonal and energetic context when their adaptive nature becomes obvious. This means that the results of laboratory acclimation experiments must be viewed with some caution, since animals will be treated in the laboratory to conditions outside their normal seasonal context.

Phenotypic adaptations clearly improve the match between an animal and its environment. As a result the animal is better able to cope with the stress of environmental variability and is 'more fit' in a Darwinian sense than individuals without this ability. The matching of acclimatization responses and environment is more clearly illustrated by comparing individuals from localities with different climatic conditions. The physiology of each population is usually uniquely suited to the prevailing conditions, such that they represent different physiological subspecies or races. This physiological plasticity illustrates an important point; that genotypic adaptation allows a more specialized fit between the individual and its environment.

Finally, the demonstrations of phenotypic adaptations have a more general and fundamental significance that is immediately apparent. The fact that the cells and tissues of ectotherms respond in an adaptive manner to environmental perturbations points to the existence of control mechanisms which dynamically regulate physiology. The concept of a dynamically-regulated and plastic physiology is frequently overlooked by those who study animals under constant conditions and who take for granted the adaptedness of their experimental subjects. Adaptive responses are not specific to temperature but may be observed in response to other types of disturbance or by altered demands. They should be regarded as just one expression of a particularly important characteristic of living systems in general, rather than a peculiar property of a restricted group of animals. The study of temperature adaptation at the molecular and cellular levels of organization have undoubtedly contributed much of fundamental significance to biology. However, despite all these advances there is no clear understanding of how compensatory responses are regulated and controlled. There are two obvious possibilities: regulation by intracellular control systems or by centrally-mediated (i.e. neuroendocrine) systems. Although hormones are involved in basal metabolic rates in mammals, no definitive role has been established in temperature acclimation of ectotherms (Umminger, 1978). The modifying effects of photoperiod in thermal acclimation indicates that hormones are to some extent involved. Intracellular control mechanisms may be revealed by studying *in vitro* acclimation responses using cell cultures. Koban (1986) has recently demonstrated some attenuated *in vitro* compensations in hepatocytes from the catfish, though it is likely that some plasma factors are necessary to mediate the full response.

6 Thermal injury, thermal death and resistance adaptation

6.1 Introduction

We have seen how temperature affects the rate of almost all biological processes. Temperature has an equally dramatic effect upon those processes outside the 'normal' temperature range, simply by interfering with the ability of living systems to perform their normal functions. Ultimately, such profound disturbances lead to the death of the individual, though a variety of damaging sub-lethal effects may occur during shorter exposure to lethal temperatures or during longer exposures to sub-lethal temperatures.

Just what constitutes the 'normal' range of temperatures varies from animal to animal, depending upon the evolutionary thermal experience of the species. Thus, polar species are cold-tolerant and warm-sensitive whilst tropical species are cold-sensitive and warm-tolerant. A second important, though commonly underestimated, influence is the previous thermal experience of an individual during its lifetime. Tolerance to high and low temperature is rarely fixed in a species and is subject to change as a result of conditioning, a process termed 'resistance acclimation'.

This chapter is concerned first with thermal tolerance and the way in which temperature may limit animal survival and, secondly, with the various mechanisms which have evolved to ensure survival in a variable and potentially hostile thermal environment. The importance of thermal death and injury in the geographical distribution of animals and their speciation will also be considered.

6.2 Methods for determining lethal limits

The destructive effects of high body temperatures are time dependent, and it is important to realize that it is not possible to quote a single lethal temperature for an animal without stipulating the exposure period. A high temperature that is tolerated for a few minutes may become lethal over a longer period. It is also important to recognize that in any sample of animals some are more resistant to lethal temperatures than others, so it is necessary to resort to statistical techniques to characterize the resistance of a population to lethal conditions.

The commonest method of defining the lethal conditions of temperature and time for a group of animals is to determine the combination of temperature and exposure time which kills a given percentage, say 50%, of the sample. This can be done in two ways; by exposing the animals to a single lethal temperature and monitoring the mortality of animals subjected for increasing periods of time (time mortality, see Fig. 6.1(a)), or to monitor the mortality of animals exposed to different lethal temperatures for a given

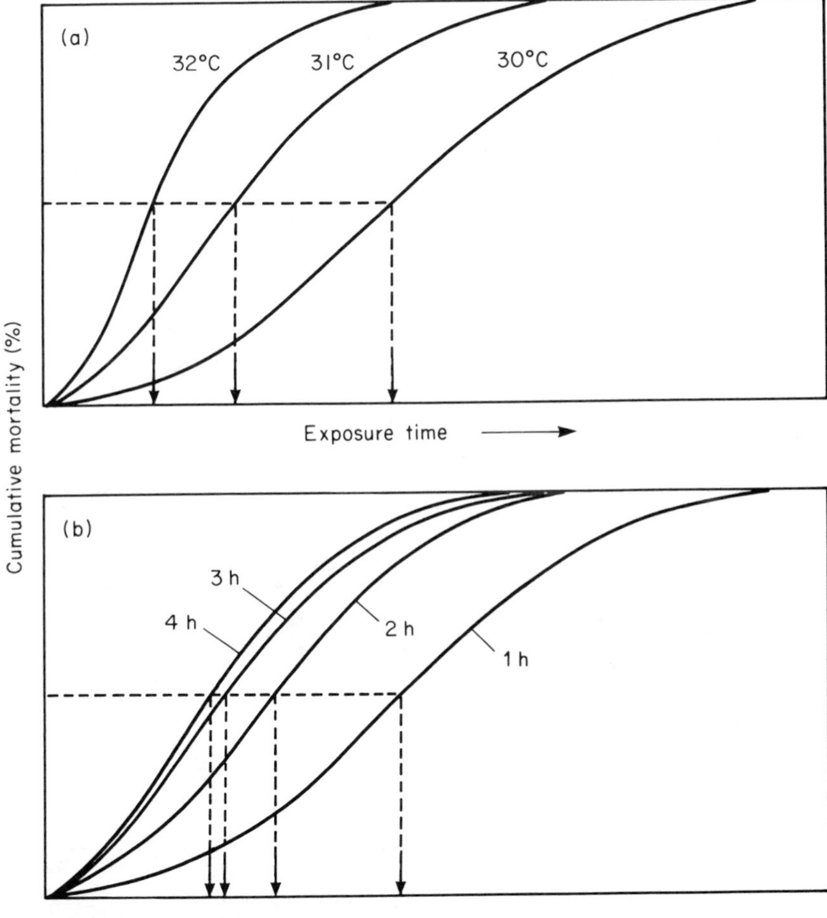

Figure 6.1 Hypothetical graphs illustrating the measurement of thermotolerance by (a) time mortality and (b) dosage mortality. The drop lines represent (a) the lethal times and (b) the lethal temperatures for 50% mortality.

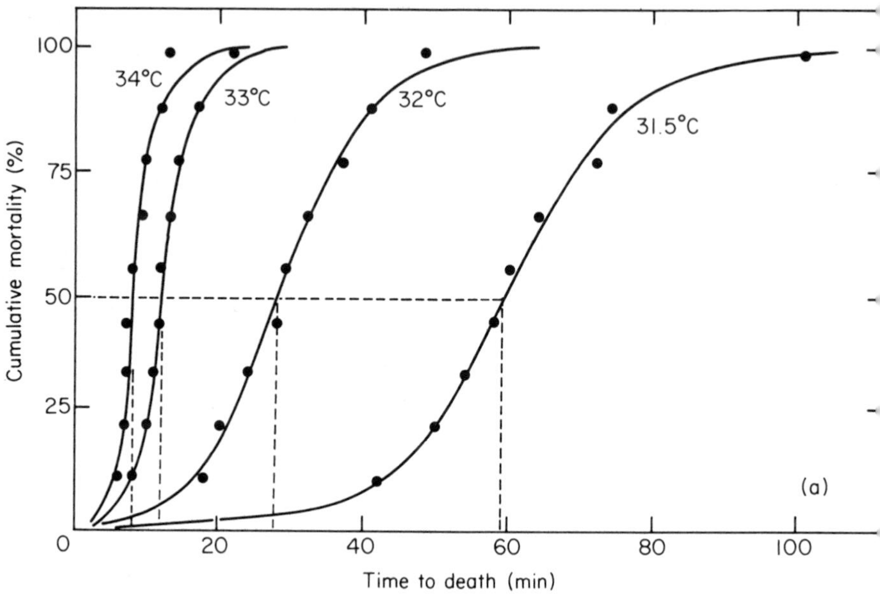

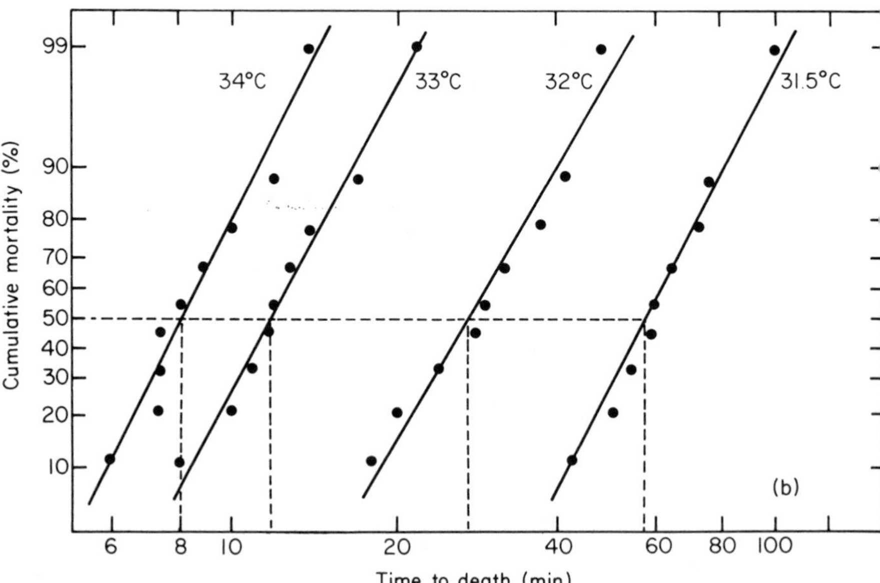

Figure 6.2(a) Time mortality curves for the freshwater crayfish, *Austropotamobius pallipes*, acclimated to 10°C. (b) The mortality plotted in probits against the logarithm of time to give a series of parallel lines. The drop-lines in both graphs represent the lethal times for 50% mortality. (Data from Bowler, 1963.)

exposure period (dosage mortality, Fig. 6.1(b)). In each case the percentage mortality is plotted as a function of time or temperature, respectively, and the time or temperature for 50% mortality (the median lethal dose or LD_{50}) is estimated graphically.

In the case of time mortality, sigmoidal graphs are usually obtained which illustrates the statistical nature of thermal mortality, i.e. most of the animals die over a fairly restricted range of lethal conditions, but some are more susceptible and some are less susceptible. If the distribution of survival times in the population conforms to a Normal distribution then plotting the mortality in probits against the logarithm of time should give a straight line as, for example, in Fig. 6.2(b). Here the probit lines are linear, parallel and regularly spaced, all of which indicates a statistical homogeneity in the population with respect to the causes and process of heat death.

A discontinuity in a probit plot is usually taken to indicate either that the sample of animals is not statistically homogeneous or that different causes of thermal distress occur at different temperatures. Two such cases are illustrated in Fig. 6.3(a) and (b). Heat death in the minnow, *Chrosomus*, at 28°C produces two intersecting probit plots with about 30% of the sample dying more rapidly than would be expected from a linear probit plot. This is probably due to a shock effect upon sudden transfer of the fish to the lethal temperature, since equilibration of fish to the lethal temperature over 15–30 min largely removed the early mortality on the section of the sample. The more-resistant individuals apparently died from a different cause. Figure 6.3(b) shows complex probit plots for cold death in the sockeye salmon. Deaths at 1°C were from a single cause, as indicated by the linear probit plot. Exposure at 2°C and 3°C, however, split the sample of animals into two groups, one which died rapidly and one in which death was considerably delayed. Exposure to 5°C resulted in only delayed deaths. The causes of cold-death in the two groups are evidently different, since they occurred over different temperature ranges.

Several experimental difficulties are inherent in these types of experiment. First, there is the possibility of causing death from the initial shock of the rapid temperature jump at the start of the exposure period. This may be avoided by gradually increasing the exposure temperature from the acclimation temperature to the exposure temperature. Indeed, this is the basis of an alternative approach proposed by Cowles and Bogert (1944), in which the temperature is gradually raised or lowered until the animal becomes so affected that its ability to escape the lethal conditions is impaired. These temperatures are termed the 'critical thermal maximum' (CT_{max}) and the 'critical thermal minimum' (CT_{min}), respectively. This procedure has been redefined by Hutchison (1961) and recently re-evaluated by Paladino *et al.* (1980). Whilst this approach does give a

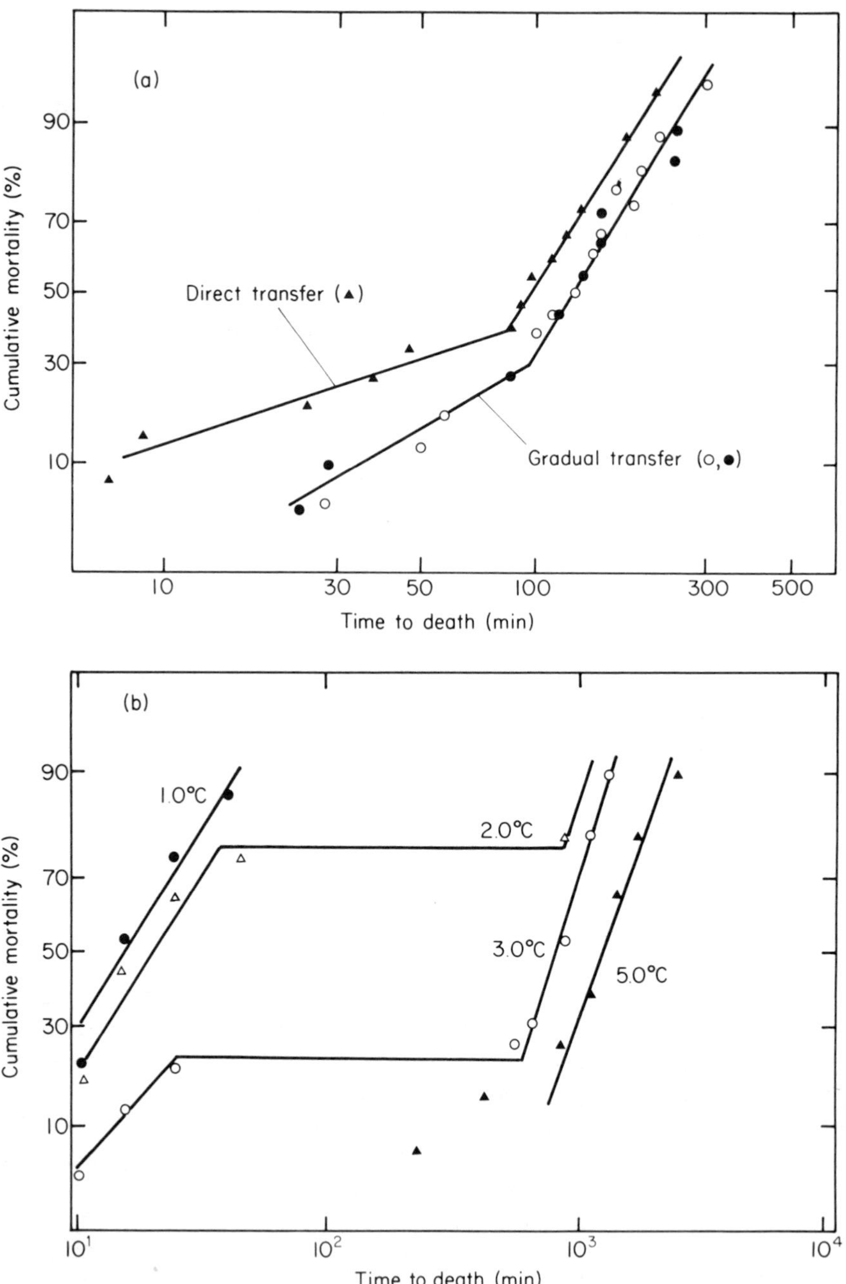

Figure 6.3 Complex probit plots of cumulative percentage mortality. (a) The minnow, *Chrosomus eos*, was exposed to the lethal temperature (28°C) either by direct transfer or following a more gradual increase in temperature over 15–30 min. (Modified after Tyler, 1966.) (b) Young sockeye salmon were exposed to temperatures between 1°C and 5°C. Mortality occurred either rapidly or after a considerable delay. The causes of death in each case were different. (After Brett, 1952.)

measure of thermal tolerance which may have more relevance to the experience of animals in their normal environment than LD_{50}, it does have the disadvantage that the change in temperature is so slow (say 0.5–1.0°C h^{-1}) that the resistance of an animal may actually change during the experiment.

A second problem with the LD_{50} type of experiment is the establishment of a suitable criterion of death. Clearly, the most unequivocal procedure is to expose samples of animals to a lethal temperature for a specific period and then to return them to their holding temperature so that the proportion killed can be estimated by counting the number that fail to recover. This has the disadvantage of requiring large numbers of animals as each temperature–time combination requires a separate sample of animals. However, it is unavoidable when the lethal conditions of eggs or other immobile stages of a life cycle are being determined.

Alternatively, it is possible to use a symptom of thermal death, such as the loss of righting response or the cessation of respiratory movements, and to determine the time taken for that point to be reached. It is necessary to establish the relationship between the chosen criterion of thermal death and true death (i.e. the inability to recover) in separate experiments, but this technique does give adequate results with only one group of animals.

By itself, the measurement of a lethal temperature is of limited value either to the ecologist or to the physiologist. By extending the analysis to include a wider range of lethal temperatures and exposure times, the complete thermal resistance of a given sample of animals can de defined. As the exposure period is extended the sigmoidal graphs of dosage mortality are shifted leftwards to give a progressively lower LD_{50}. This is more clearly illustrated in Fig. 6.4, which shows the LD_{50} eventually becomes constant at long exposure times. This means that there is a critical temperature above which 50% of the sample is killed in a discrete period and below which 50% mortality does not occur (the 'zone of tolerance') even for extended exposure periods. Fry (1971) has termed this lowest lethal temperature the *'incipient lethal temperature'*, a value which is most easily obtained by the determination of dosage mortality after a long exposure time, say 48–96 h. Exactly equivalent procedures for cold exposure provides the incipient low lethal temperature. Beyond the incipient lethal temperature there lies a 'zone of resistance' in which more than 50% will succumb to the effects of exposure. Events in the zone of resistance are best studied by time–mortality experiments.

An important observation is the very high temperature dependence of the destructive effects of temperature. In practice this means that experiments can only be performed over small temperature ranges. Whilst it is common in measurements of rate effects of temperature to experiment over a 20–30°C range of temperatures, the maximal range over which lethal temperature may be observed is usually less than 5°C. A variation of a

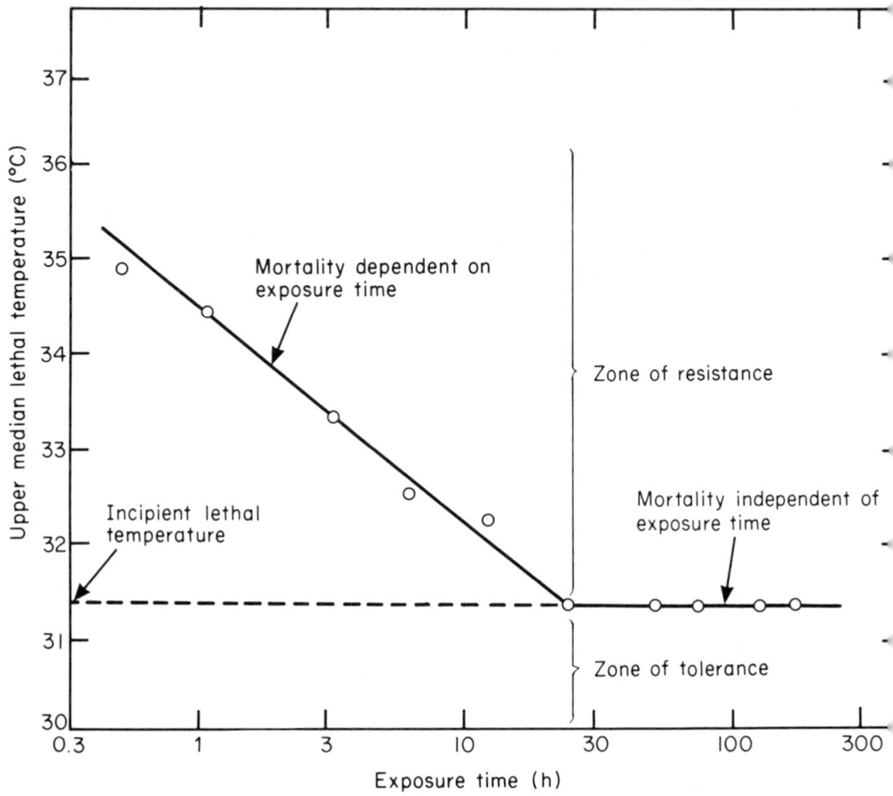

Figure 6.4 The effect of exposure time upon the median lethal temperature (LD_{50}) as established from dosage mortality curves. The LD_{50} decreases as exposure time increases, until it reaches a plateau. This minimal LD_{50} represents the 'incipient lethal temperature' which separates the zones of tolerance and resistance. (Data from Doudoroff, 1945.)

fraction of a degree in the exposure temperature may have a dramatic effect upon the time to median mortality, and it is necessary to control temperature during these experiments to within 0.1°C of the desired value. This point is well illustrated in Fig. 6.2, where a shift of 1.0°C caused a doubling of the LD_{50}.

6.3 Resistance adaptation

The resistance of an animal to the effects of lethal temperature is rarely constant. Brief conditioning of an animal to temperatures which are just sub-lethal often induces a degree of hardening, so that a previously lethal temperature may be tolerated. Similarly, winter animals are usually more cold tolerant and more heat sensitive than summer animals. Collectively, these alterations of thermal resistance are termed *'resistance adaptations'*, but it is necessary to distinguish between adaptations which result from laboratory conditioning (*resistance acclimation*) and those which may be observed in animals in their natural environment (*resistance acclimatization*), since these are not necessarily the same. This adaptive ability is a widespread phenomenon in the animal kingdom and is probably of great significance in increasing the ability of animals to survive the likely extremes of temperature that may occur during the different seasons of the year.

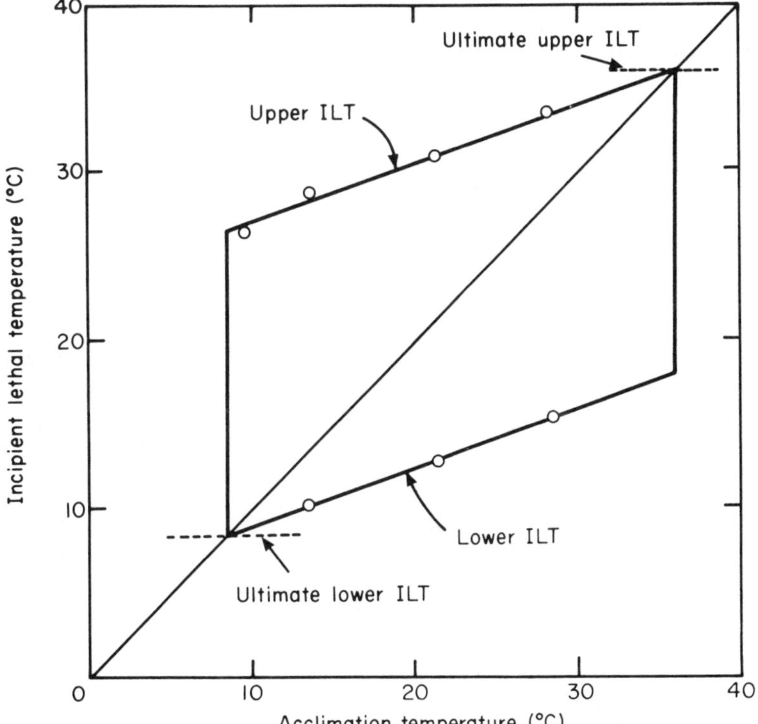

Figure 6.5 The effect of thermal acclimation upon the upper and lower incipient lethal temperatures (ILT) of the marine fish *Spheroides maculatus*. By joining the graphs for the upper ILTs and the lower ILTs where they cross the diagonal line (where ILT = acclimation temperature) a 'tolerance polygon' is created. The area within the polygon represents the combination of test and acclimation temperature that is survived by at least 50% of the sample. The lowest and highest ILTs achieved by acclimation are termed the ultimate ILTs. (After Hoff and Westman, 1966.)

A complete picture of the adaptive ability of a particular species can be obtained by measuring the incipient high and low lethal temperatures for animals that have been acclimated to a range of constant temperatures. Figure 6.5 shows how the incipient high and low lethal temperatures are increased when acclimation temperature is increased. These values can be plotted as a function of acclimation temperature to give a complex graph which, because of its shape, is termed a 'tolerance polygon'. The area between the two curves for the incipient high and incipient low temperatures represents the combination of acclimation and lethal temperatures that 50% of the sample can withstand for an indefinite period (i.e. the zone of tolerance). Outside this tolerance polygon one can superimpose the lines for tolerance for specified exposure times to give a nearly complete description of the lethal effects of temperature on a sample of adult animals. Brett (1958) has extended the idea of a tolerance polygon for adult lethality to include the corresponding graphs which designate the thermal limits for other processes such as growth and reproduction (Fig. 6.6). This gives

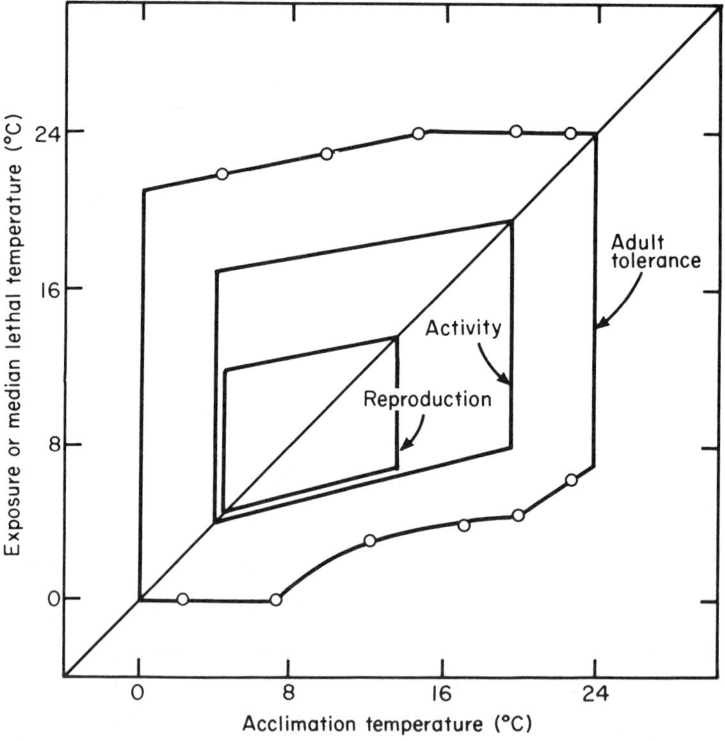

Figure 6.6 The tolerance polygon for the sockeye salmon (*Onchorhynchus nerka*) together with the more limited temperature zones within which activity and reproduction can take place. (After Brett, 1958.)

perhaps the most complete representation of the thermal tolerance of a species throughout its entire life-cycle.

Incipient lethal temperatures may increase by as much as 3°C for a 10°C increase in acclimation temperature. Sometimes the increase is linear, as in Fig. 6.5, but more usually it is sigmoidal or plateaued (Fig. 6.7). However, this process cannot proceed indefinitely, and there is a point at which the incipient lethal temperature is the same as the acclimation temperature. This temperature is termed the '*ultimate incipient lethal temperature*' (UILT).

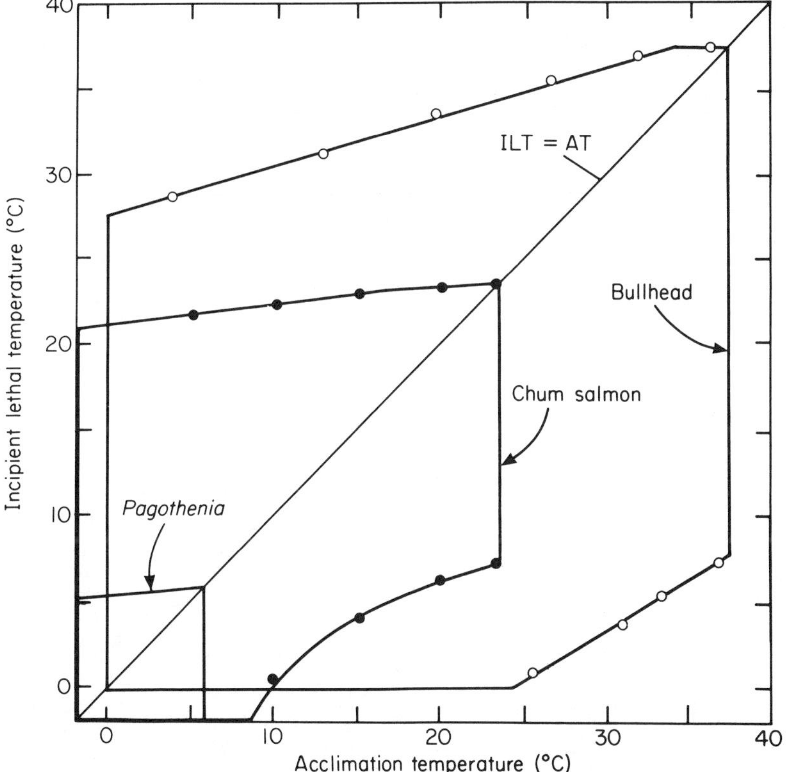

Figure 6.7 A comparison of the tolerance polygon for a eurytherm (the bullhead, *Amieurus nebulosus*), a cold stenotherm (the chum salmon, *Onchorhynchus keta*) and an extreme stenotherm (the Antarctic rock perch, *Pagothenia bernachii*). The area within each polygon represents the temperature 'space' available to each species. The values were 1200°C^2, for the bullhead, 650°C^2 for the salmon and 28°C^2 for the Antarctic fish. Note that the bullhead displays appreciable resistance acclimation and a zone of tolerance that extends down to the freezing point of fresh water. The chum salmon shows more limited acclimation of heat resistance, but more significant cold acclimation. The curve for *Pagothenia* has been estimated from Somero and DeVries (1967). Other curves are from Brett (1944, 1952).

It represents the fixed limit for thermal tolerance for that sample of animals, and probably corresponds rather closely to the CT_{max} of Cowles and Bogert (1944). The low UILT is often limited by the freezing point of water. However, the true low tolerance may be even lower, since many fish species, such as trout, are active at the freezing point of water.

The extent to which acclimation can extend the thermal tolerance range of a species is related to the lifestyle of the species and to the seasonal range of temperatures that it usually encounters. Species which have evolved in continental climates tend to have very marked abilities in this respect (i.e. eurythermy) compared with those from less variable climates (stenothermy); compare, for example, the large tolerance polygons of the bullhead with that of the chum salmon and the Antarctic *Pagothenia bernachii* in Fig. 6.7. Whilst resistance acclimation extends the zone of tolerance to encompass the likely seasonal variations temperature, this adaptive range

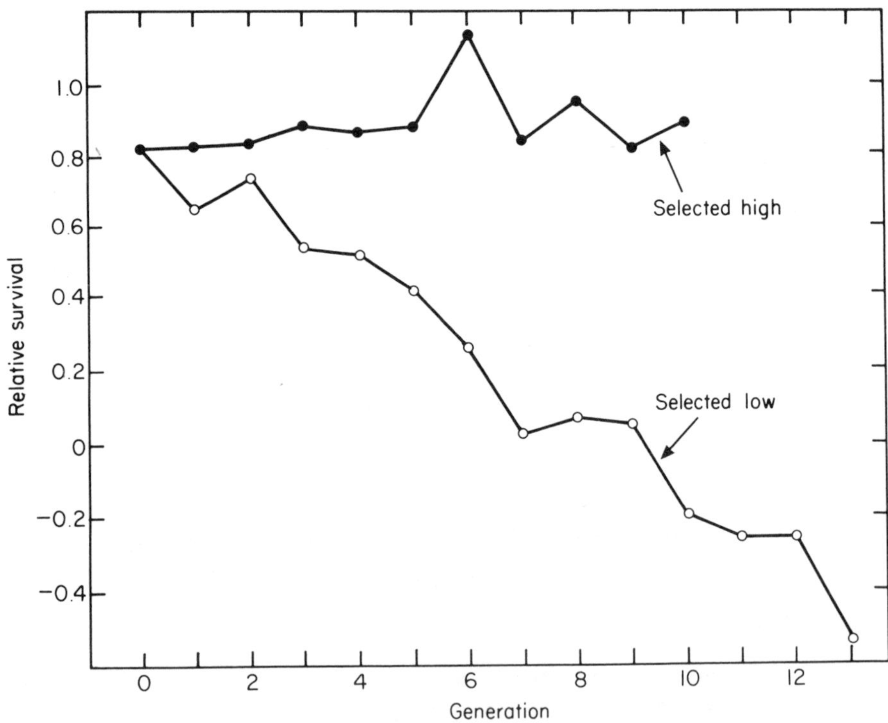

Figure 6.8 Changes in survival of *Drosophila* populations in response to artificial selection for increased and decreased resistance to heat shock. The ordinate is an index of the proportion surviving heat shock. Selection for decreased heat resistance was very rapid. (After Morrison and Milkman, 1978.)

does not usually match that displayed by the evolution of animals in different habitats. For this reason evolution often produces differences in thermal tolerance of a species throughout its geographical range. Because terrestrial animals may avoid temperature extremes by microclimate selection, they generally possess a poorer ability to modify their thermal resistance.

We have been careful not to define the UILTs as the genetically fixed limits of resistance for the species as a whole, since it has long been recognized that tolerance polygons may be altered to some extent by factors other than temperature. Chief amongst these is season and photoperiod, in which summer animals may have an enhanced heat resistance compared with winter animals, even though they have had identical laboratory conditioning before the experiment (Bulger and Tremaine, 1985). Thus, resistance acclimatization may exceed resistance acclimation to extend the ecological tolerance beyond that represented by conventional tolerance polygons. The effect of season may be profound and of critical importance in extending the zone of tolerance to suit the seasonal conditions. For example, the LD_{50} for a 12-h exposure in the summer-acclimatized minnow, *Chrosomus*, was 3–4°C above that of winter fish (Tyler, 1966).

Resistance to thermal stress is certainly of great selective advantage to animals and is subject to modification over evolutionary time. Indeed, provided that selection pressure is sufficiently great, then thermal resistance can be modified over relatively few generations. Morrison and Milkman (1978) have shown that heat resistance could be lost from a culture of *Drosophila subobscura* over five to ten generations, though selection for increased heat resistance was not achieved (Fig. 6.8).

The rate at which thermal resistance may be altered by acclimation depends, in part, upon the direction of temperature change during acclimation. Increasing acclimation temperature tends to elicit more-rapid adjustments in thermal resistance than does reducing acclimation temperature (Fig. 6.9(a)), and heat resistance is altered more rapidly by a given temperature increment over the upper range of temperatures than over the lower range (Fig. 6.9(b)). This is because changes in thermal resistance are dependent upon metabolic and other cellular processes whose rates are themselves temperature dependent. As a consequence of this, the gain in heat tolerance in a natural situation would be rapid enough to follow the daily maximum temperature, whilst the loss of heat tolerance during the autumn and winter may lag behind a rapid fall in environmental temperatures. Thus, a sudden exposure to cold poses a greater problem than heat exposure. Figure 6.10 shows the change in heat resistance of some lake fish during a summer. Because of the time delay in the process of acclimatization, the changes in resistance were smoothed relative to the changes in water temperature. Nevertheless, the gentle dips and troughs in

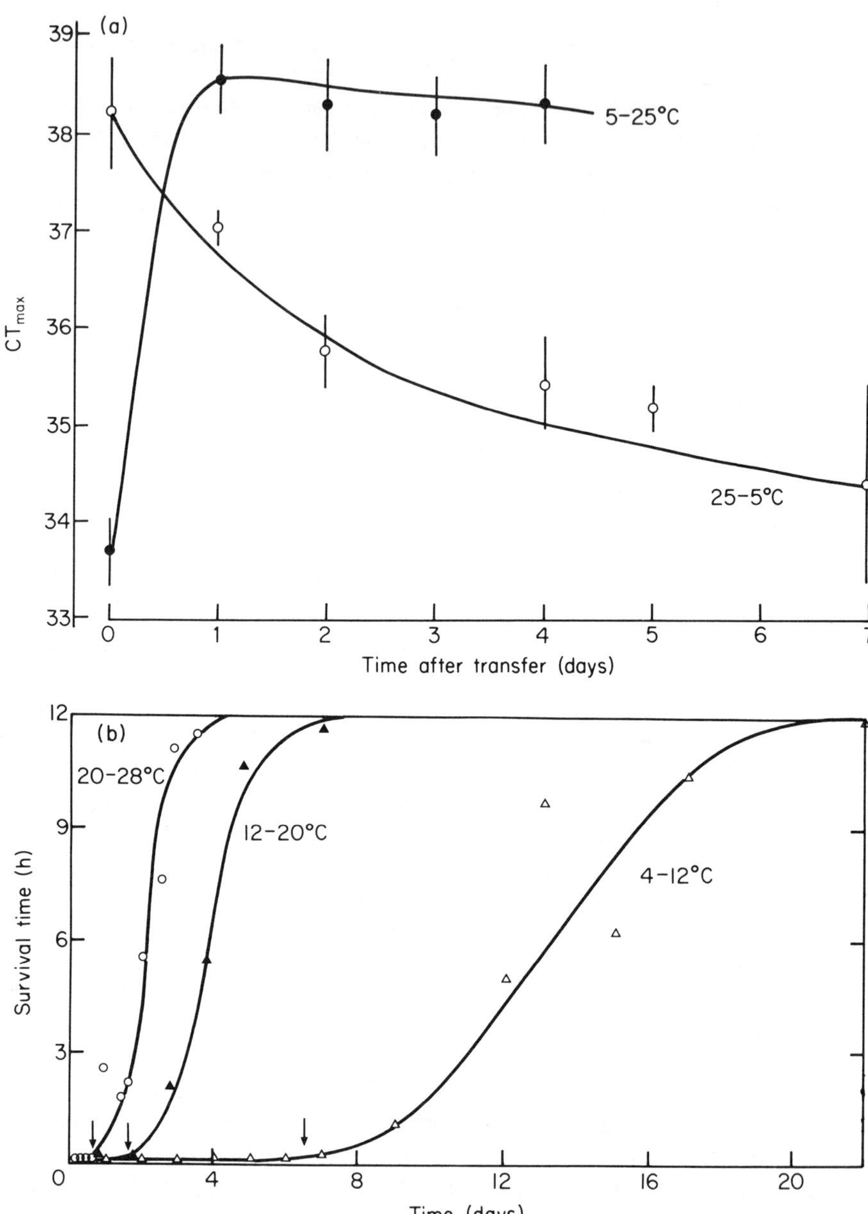

Figure 6.9(a) The time-course of changes in the CT_{max} of the crayfish, *Orconectes rusticus*, after transfer of 5°C-acclimated crayfish to 25°C and vice versa. The times for half-completion of the changes in resistance were 0.45 and 2.53 days, respectively. (After Claussen, 1980.) (b) The time-course of changes in heat resistance of the goldfish after increases in temperature of 8°C from acclimation temperatures of 4, 12 and 20°C. Note the extended lag times (arrows) before changes in resistance were observed, especially at the lowest acclimation temperature. The temperatures used to test for resistance were 30.3, 34 and 37°C, respectively. (After Brett, 1946.)

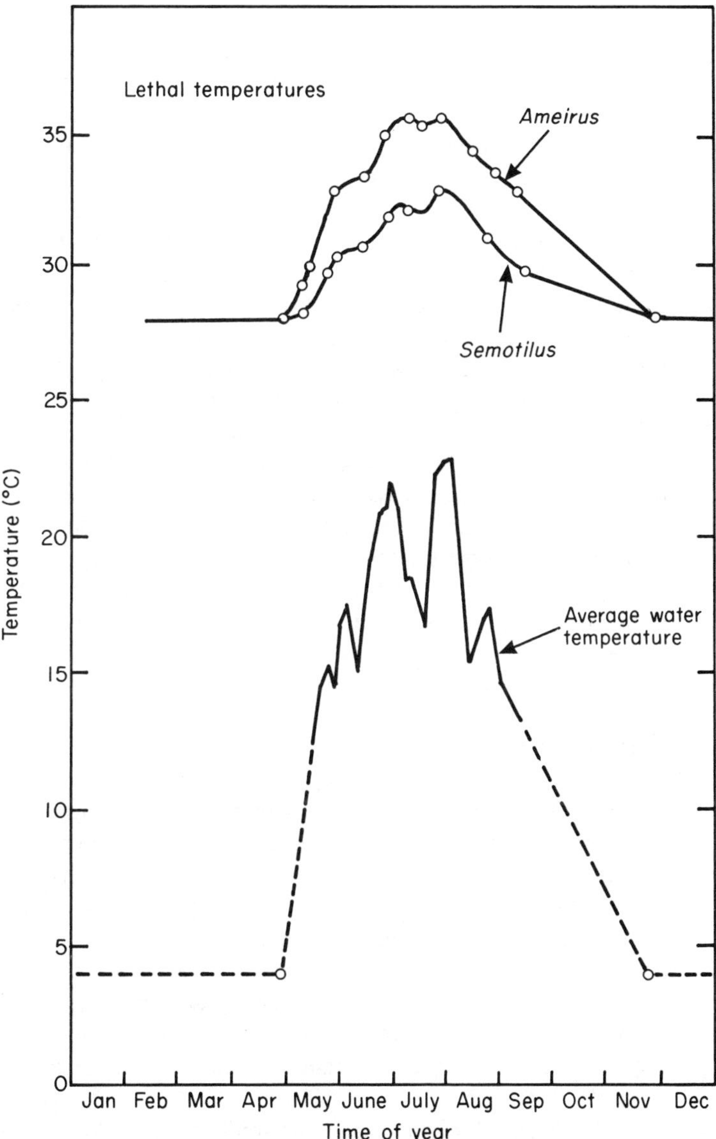

Figure 6.10 Seasonal variation of median upper lethal temperature for two fish species in relation to the average water temperature of Lake Opengo, Ontario, Canada. The time-course of changes in thermal resistance closely follow the changes in water temperature. (After Brett, 1944.)

the LD_{50} curve indicate a high degree of sensitivity to environmental conditions.

Other points to note are, first, that the shape of the time-course curves are often sigmoidal with pronounced lag phases before any change in thermal resistance occurs. The significance of the lag phase is not understood, but it may represent an inability to modify thermal resistance until certain other time-dependent processes, such as repair to thermal injury, have taken place. Secondly, the changes in heat resistance may follow a somewhat different time-course from the corresponding changes in cold resistance, which suggests that these two processes are not identical.

Resistance acclimation is undoubtedly a complex process, and some progress has been made in identifying the physiological mechanisms by which it is brought about. Unfortunately, there is no such clear understanding of the control processes which regulate the adaptive processes. There are two obvious possibilities: a regulation by control processes at the cellular level of organization or centrally by the central nervous or hormonal systems. The possible involvement of photoperiod as an environmental cue for seasonal resistance acclimatization suggests a regulatory role for the central nervous system which, incidently, may permit anticipatory changes to thermal resistance. On the other hand, Bowler *et al.* (1983) have shown resistance acclimation in a cultured cell line from the fathead minnow which indicates that cellular regulation also plays an important role.

6.4 Hardening

This is a term used to describe a brief, transitory increase in thermal resistance to extreme temperatures, which results from a short exposure to sub-lethal temperatures. Hardening is distinguished from resistance acclimation not only in the manner of its induction, but also because it is more rapidly gained and lost than resistance acclimation. In practice the division between these two processes may be more arbitrary than real, and may represent extremes in a continuum of adaptive responses. This is particularly true for heat acclimation, because its rate of attainment is temperature dependent and resistance acclimation at high sub-lethal temperatures may be particularly rapid.

Clear evidence for heat-hardening under field conditions comes from the desert pupfish, *Cyprinodon macularis* (Lowe and Heath, 1969). This fish lives in water that can rise in temperature from 25°C at dawn to 40°C in mid-afternoon. Fish frequently enter water at 40–41°C even though water at 30°C is available nearby. These higher temperatures are uncomfortably close to their summer CT_{max} (43°C), and well above the winter CT_{max} (37°C)

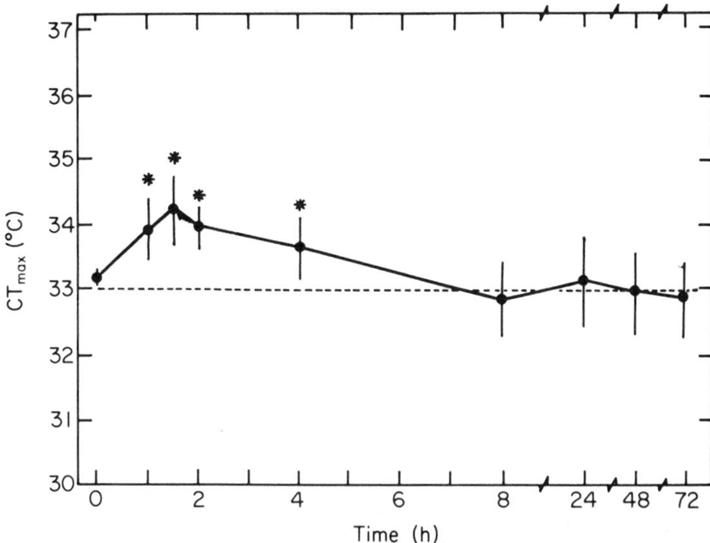

Figure 6.11 Heat-hardening in the fathead minnow, *Pimepheles promelas*. The CT_{max} was measured by heating the fish at 1°C min^{-1} until the onset of spasms. Immediately following, exposure animals were returned to their acclimation temperature (15°C) for the time indicated until the second CT_{max} determination. The broken line represents the CT_{max} of unexposed animals. Note that 1–4 h after the initial exposure the animals displayed significantly greater resistance to the second exposure. Vertical bars represent two standard errors of the mean, and the asterisks indicate a significant difference between the initial or unexposed CT_{max} and the second exposure. (After Maness and Hutchison, 1980.)

but, more importantly, are higher than the fish can be acclimated to in the laboratory. It is suggested, therefore, that the ability of these fish to enter and spend time at 40°C may be a heat-hardening effect of considerable adaptive value. In consequence one might speculate that heat- or cold-hardening may be found in animals which experience large diurnal fluctuations in temperature. Maness and Hutchison (1980) have reported variations throughout the day in hardening in a number of amphibians and fish (see, for example, Fig. 6.11). The periods for peak hardening were found to correspond with the periods of highest environmental temperature which supports the adaptive role for heat-hardening.

The physiological basis for hardening is obscure. It has been demonstrated not only at the organism level, but also in isolated tissues and in cultured cells of invertebrates, vertebrates and mammals (Fig. 6.12). An interesting recent observation in this respect has been that a variety of cells, ranging from yeast to mammalian cell lines, synthesize a family of proteins in response to sub-lethal heat shock. The function and significance of these

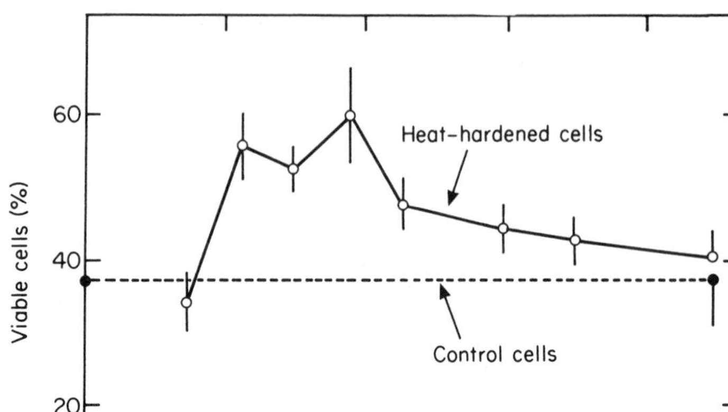

Figure 6.12 Heat-hardening in a cultured cell line from the fathead minnow, *Pimepheles promelas*, Cells were given a 10-min 'shock' at 35°C and then returned to 18°C for the time indicated before receiving a test of resistance at 43°C for 5 min. Control cells received only the resistance test. Cell survival increased from 34% to 50% during the 40 min after the initial shock. Values represent the means ± s.d. (After Schmidt *et al.*, 1984.)

'heat-shock proteins' is not understood, but their presence and persistence does seem to correlate with the development of cellular thermotolerance following a brief heat shock (Li and Werb, 1982). Other manipulations of culture conditions, such as the addition of ethanol or sodium arsenate or the reduction of oxygen tension, also induces a cross-tolerance to high temperature concurrently with the synthesis of heat-shock proteins. Thus, these proteins may represent a general response to enhance cellular survival. Whole-organism hardening may be related to these various cellular hardening responses, since in *Drosophila* at least, artificial selection for heat tolerance is associated with the elaboration of heat-shock protein synthesis (Alahiotis and Stephanou, 1982; Alahiotis, 1983). On the other hand, at least part of the effects of exposure to stressful temperatures is due to behavioural shock, and a brief pre-exposure may habituate the animal to the shock response.

6.5 Causes of heat death

Understanding the mechanisms of resistance acclimation is largely a problem of identifying the causes of thermal injury and death, since it is by altering these that resistance is altered. Read (1967) has proposed two criteria by which the physiological events which lead to thermal death may by identified. First, the events must occur over the same combination of temperature and time as thermal death in the whole animal. Secondly, they must change by a similar amount and with a similar time-course to the changes in thermal resistance of the whole animal during acclimation to an altered temperature.

The major problem in deciding for or against the various theories to account for the mechanism of heat death is that little evidence exists which relates resistance of the whole organism to that of its cells. It is often found that enzymatic activity is significantly more thermostable than the organelle or cell from which the enzymes are extracted, that cells are more thermostable than organ functions and that these are more thermostable than the whole animal (Ushakov, 1964). This hierarchical sequence indicates that heat death is not a simple result of protein thermostability, even though they both possess an exceptionally high Q_{10}. It is more probably a result of a loss of integration at a higher level of organization. It is clear from a number of studies that heat death is not accompanied by a general breakdown of the overall metabolic processes of animals or their cells, since oxygen uptake may subsequently proceed at normal levels in isolated tissues of heat-damaged animals (Bowler, 1963; Grainger, 1975).

The general problem of cause and effect makes the identification of the mechanism of heat death difficult, and the separation of those processes which initiate heat death from secondary and tertiary consequences of those processes is often unclear. For example, several studies on fishes suggest that osmoregulatory failure is an important early event during heat death, due to pathological effects at the gill epithelium or to the difficulties in satisfying an increased oxygen demand in an environment with a lower oxygen concentration. The damage to the respiratory surface may then lead to hypoxia, which then influences normal brain function. According to this scheme brain damage is the ultimate cause of heat death, but is not the primary lesion.

A different view is that the central nervous system is itself particularly vulnerable to temperature extremes, and that a direct perturbation of neuronal function causes a general loss of nervous integration and control of both autonomic and neuromuscular activity.

Perhaps the most powerful evidence in favour of the particular sensitivity of the central nervous system is that localized heating of the brain, using thermodes, induces the same sequence of behavioural deficits as those observed during heat stress (Friedlander *et al.*, 1976). This sensitivity to heat is probably due to the lability of synaptic transmission, since both central synapses and neuromuscular junctions are blocked by smaller temperature changes than those which cause axonal blockage. In addition, the spontaneous activity of some central neurones declines dramatically

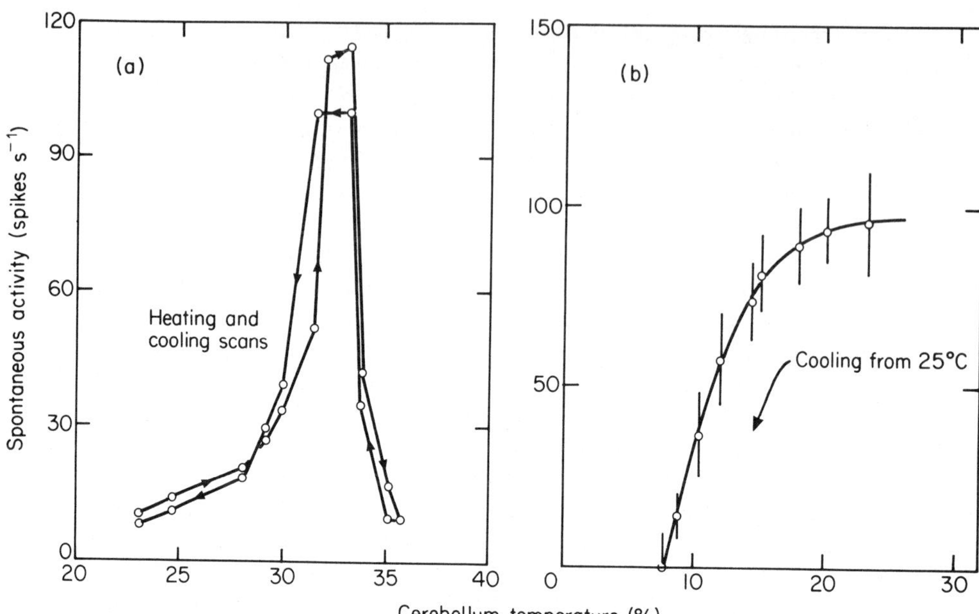

Figure 6.13 The effect of (a) local heating and (b) cooling of the cerebellum of goldfish upon the spontaneous electrical activity of the Purkinje cells. In (a) the activity of a single neurone is shown as it was heated and cooled (5°C-acclimated fish). In (b) the data points and bars represent the means $\pm$ s.e.m. for 15 cells. (After Friedlander *et al.*, 1976.)

above certain critical temperatures which correspond closely with those temperatures which cause behavioural and motor deficits. Figure 6.13(a) illustrates this point clearly, and also shows that this effect is reversible provided the exposure to high temperature is brief. The temperature of maximum spontaneous activity in the Purkinje cells of goldfish cerebellum was increased by increases in acclimation temperature from 32°C for 5°C-acclimated fish, to 39°C for 25°C-acclimated fish, which again correlates with changes in the temperatures for motor dysfunction.

A common observation, particularly in invertebrates, is that haemolymph Na^+ concentration decreases and K^+ concentration increases during heat death. Figure 6.14 shows a scheme for heat death in the freshwater crayfish, where the primary lesions and secondary effects are considered. The dramatic changes in haemolymph cation concentrations are probably due to the loss of the normal barrier properties of the plasma membrane of a bulk tissue, such as muscle, which allows the ionic gradients across this membrane to collapse. The subsequent rise in haemolymph K^+ is known to disrupt the normal bioelectrical properties of excitable cells, and results in a progressive loss of neuromuscular co-ordination. Death is probably caused

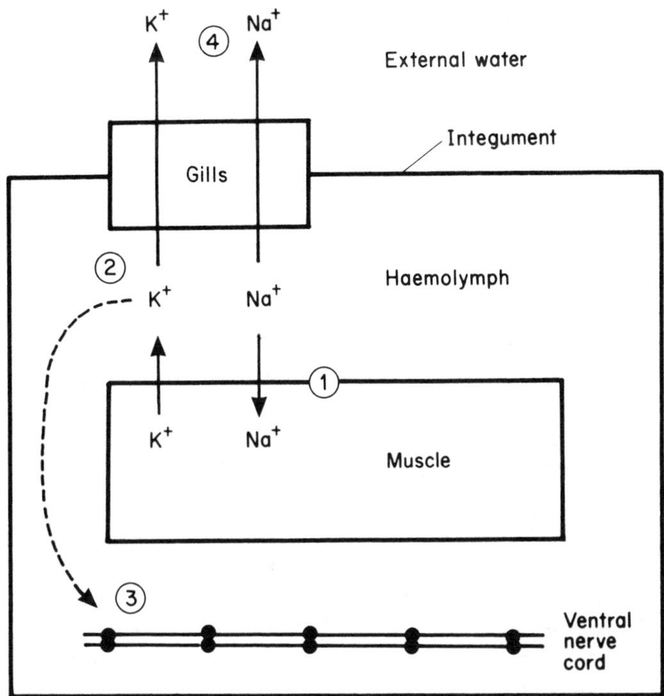

Figure 6.14 The proposed sequence of events during heat death in the freshwater crayfish, *Austropotamobius pallipes*. 1. Exposure to high temperature causes a progressive collapse of the normal permeability barriers between the haemolymph and the cellular compartment. 2. Haemolymph Na^+ concentration decreases whilst K^+ increases as the concentration gradients run downhill. 3. High K^+ in the haemolymph causes abnormal neuronal activity and a progressive loss of neuromuscular co-ordination. 4. Progressive loss of cations to the surrounding water. (After Bowler *et al.*, 1973b.)

by the loss of gill irrigation (Bowler *et al.*, 1973a and b), which is viewed as a tertiary consequence of the initial perturbation of membrane properties.

6.6 Cold injury and death

Until recently, the lethal effects of cold were less well studied than those of heat. In part this was because the point of death is less easy to establish, as many organisms simply become quiescent at low temperatures. Cold presents two distinct problems for organisms, depending upon their sensitivity. Cold injury and death may occur at temperatures well above 0°C due to the effects of cooling in the absence of ice formation, a phenomenon known as *chill coma*. On the other hand, death in more cold-resistant organisms may occur only at temperatures at which their body fluids become frozen. Resistance in this latter group is associated with an ability either to withstand the effects of freezing of body fluids (the so-called '*freeze-tolerant*' animals) or to supercool and thereby avoid ice formation (the so-called '*freeze-sensitive*' animals, reviewed by Baust and Rojas (1985) and by Zachariassen (1985)).

6.6.1 *Chill injury and coma*

A number of different causes of chill coma have been proposed. These may occur sequentially during cold exposure, and Doudoroff (1942) has been able to separate the early death in some sensitive individuals (primary chill coma) from death in more-resistant individuals after more prolonged exposure (secondary chill coma). Perhaps the most obvious cause is that low temperature suppresses the normal cellular processes to a level below that required for maintenance purposes. Although there is little direct evidence in support of this as a general mechanism of chill injury, it is difficult to exclude critical cells from being particularly sensitive.

In fish and other aquatic organisms osmoregulatory and respiratory failure has been proposed as the ultimate event in chill injury and death. Some evidence exists for structural damage to the gill epithelium of fish in the cold, which could certainly lead to impairment of both osmoregulatory and respiratory functions, and ultimately death. In some cases an increase in the oxygen content of water improves survival (Pitkow, 1960), though in others this has no effect. In another case immersion of fish into an isosmotic solution improved survival of fish in the cold (Doudoroff, 1945), suggesting that cold suppressed osmoregulatory function. This has been confirmed by Maetz and Evans (1972), who showed that cold affected the active extrusion of Na^+ more than the opposing leak. As a result the fish were progressively less able to maintain normal blood ion concentrations.

The disruptive effects of cold may become apparent with only moderate cooling even in eurythermal organisms. As with heating, an obvious effect in fish is upon their behavioural competence with a progressive loss of conditioned responses, excitability and motor ability. This may be associated with a general depression of nervous activity, which leads to a general immobilization, or perhaps with effects at rather specific sites within the central nervous system, such as the respiratory centre. Prosser and his colleagues have again produced convincing evidence that these behavioural symptoms are associated with the central nervous system, since the localized cooling of the cerebellum with a thermode led to chill coma (Friedlander *et al.*, 1976). Goldfish acclimated to 25°C showed distinct behavioural impairment below 15°C, yet nervous conduction of peripheral nerves continued down to −0.3°C (Roots and Prosser, 1962). Friedlander *et al.* (1976) examined the effects of progressive cooling upon a

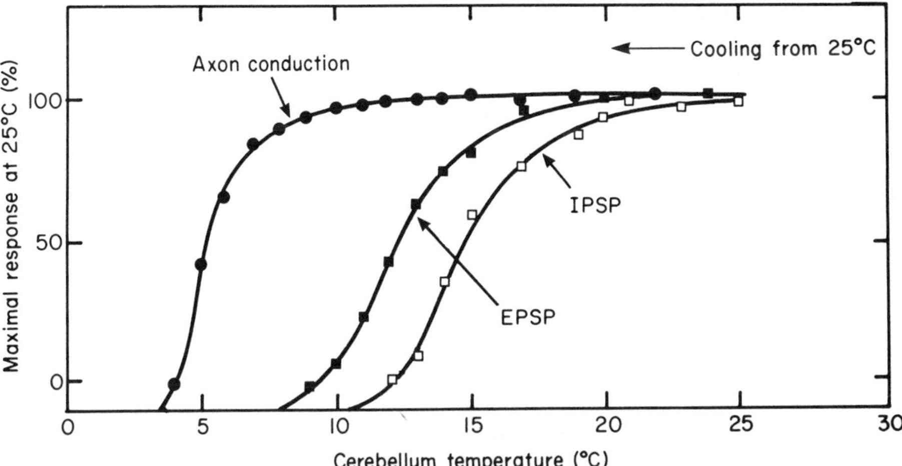

Figure 6.15 Effects of localized cooling of the goldfish cerebellum upon the electrical responses of Purkinje cells elicited through various neuronal pathways. ●, Antidromic response due to stimulation of Purkinje axon (i.e. axonal conduction); ■, excitatory post-synaptic potentials (EPSP) produced through a monosynaptic pathway; □, inhibitory post-synaptic potentials (IPSP) produced through a disynaptic pathway. (After Friedlander *et al.*, 1976.)

well-characterized neuronal network, the cerebellum of goldfish. Figure 6.13(b) shows how the spontaneous activity of Purkinje cells in goldfish cerebellum was depressed below 15°C. This preparation provides a convenient means of comparing cold-block of axonal conduction with transmission across synaptic junctions. Figure 6.15 shows that axonal conduction was affected only below 6–7°C, whilst stimulation of the Purkinje cell by stimulation of adjacent neurones (thus involving a synaptic route) was suppressed below 15°C. It is important to remember that these effects occur over the same range of temperatures as the motor and behavioural defects noted earlier for the whole animal.

These experiments show that synaptic function is particularly cold sensitive, and that inhibitory synapses are somewhat more cold sensitive than excitatory synapses. A particularly clear-cut example of the peculiar sensitivity of synaptic transmission is provided by the neuromuscular junction of the frog sartorius muscle. Jensen (1972) showed that muscle contraction in response to nerve stimulation was blocked by cold, even though axonal conduction and muscle contraction in response to direct electrical stimulation were not affected. The obvious inference is that transmission of the signal at the neuromuscular junction was specifically interrupted in the cold. The temperature for cold-block of neuromuscular transmission was modified by acclimation; for example, cold-block was approximately 1°C in 10°C-acclimated frogs but 4°C in 25°C-acclimated frogs.

6.6.2 *Freeze injury*

Freezing occurs either by the growth of a crystal from a particle or structure which organizes water molecules to form an ice nucleus (innoculative nucleation) or by the chance and spontaneous aggregation of water molecules to form a nucleus (spontaneous nucleation). Thus, nucleation is the key event in ice-crystal formation, and freezing can be avoided by inhibiting spontaneous and inoculative nucleation.

The freezing point of a fluid is defined as the temperature at and below which a seed ice crystal will grow. However, many fluids may be cooled well below this temperature because spontaneous nucleation does not occur, a phenomenon known as supercooling. In many cases it is not simply the freezing point of body fluids which is critical to animal survival in cold environments, but the extent to which the fluid may be cooled before spontaneous nucleation occurs.

Freezing generally occurs more readily in the extracellular fluids than in the intracellular fluids. Microscopic examination of tissues from frozen intertidal animals reveals an exceptional degree of distortion and shrinkage of cells, with ice crystals present only in the spaces between cells (Kanwisher, 1959). The plasma membrane seems to act as a physical barrier for the growth of extracellular ice crystals into cells, but as temperature is lowered it becomes progressively less effective. Within a few seconds of thawing the tissues assume a normal appearance.

The formation of crystalline water in the extracellular fluid leads to the formation of pockets of unfrozen and hypertonic fluid. This osmotically draws water out of the cells and gradually dehydrates and shrinks the cell. By contrast intracellular freezing increases the solute concentration in the unfrozen cellular fluid. This leads to an osmotic influx of water from the extracellular compartment, cell swelling and eventually cell rupture. Osmotic stresses have long been thought to be involved in extracellular freezing injury, though it is not clear whether the critical injury is a consequence of high solute concentration, as proposed by Lovelock (1953a), or by volumetric contraction of cells *per se*, as proposed by Meryman (1974). Lovelock (1953b) showed that glycerol can protect against freeze injury by 'buffering' against osmoconcentration when extracellular fluids are frozen. The induction of osmotic water movements during extracellular freezing is fundamental to the growing field of cryopreservation (Mazur, 1984).

Freeze-tolerant species appear to have a greater percentage of intracellular water that is osmotically inactive (i.e. bound to intracellular organelles or macromolecules), so cell volume does not fall so dramatically even when 80% of extracellular fluid is frozen. It is therefore tempting to suggest that freeze tolerance is related to an ability to withstand high salinity and desiccation. This idea is supported by the increased freeze resistance of intertidal animals when they are acclimated to high salinity (Murphy and Pierce, 1975). Furthermore, Hinton (1960) was able to transfer chironomid larvae to −270°C without harm, provided that their water content was reduced to 8%. Overwintering resting phases of insects usually have a low water content, and this may contribute to their increased resistance to cold.

Freeze injury may also arise from processes other than cellular dehydration. First, the growth of ice crystals into the cell may cause mechanical damage to cellular structures. Secondly, the ratio of free to bound water within cells decreases during freezing of the extracellular compartment and this, together with the associated increase in solute concentration may denature or inactivate enzymes. Finally, metabolism is very depressed at freezing temperatures, due in part to the reduced diffusional rates for metabolites, ions and respiratory gases in ice compared with water. Nevertheless, energy metabolism does continue in freeze-tolerant animals in the frozen state. For example, in the frozen gall-fly larve *Eurosta*, adenylate energy charge is maintained by glycolysis, at least for a few weeks (Storey and Storey, 1985).

6.7 Mechanisms of cold tolerance

6.7.1 *Freeze tolerance*

Freezing of intracellular water is invariably lethal, whereas a few invertebrates can survive the presence of extracellular ice. In general, these animals live in environments where freezing conditions are frequently encountered, such as in terrestrial and intertidal habitats in high latitudes (Murphy, 1983). Sessile invertebrates in the intertidal zone are particularly at risk and regularly go through freeze–thaw cycles with no ill effects. Calorimetric studies have shown that up to 70–80% of the body water of *Balanus balanoides* (Fig. 6.16) and the periwinkle *Littorina littorea* may be in the form of ice (Murphy, 1979). Recently, the first case of freeze tolerance in the vertebrates has been reported. Schmidt (1982) has found that several species of tree frog in North America survive extracellular freezing during the winter.

Survival of freezing decreases both with the extent of freezing and with the length of time frozen. *Balanus balanoides* can withstand −10°C for several days but −20°C is usually lethal even after a brief exposure. *Littorina littorea* can survive −8°C for approximately 8 days, but survival at −12°C and −13°C is reduced to 6 h and 2.3 h, respectively (Murphy and Johnson, 1980). Freeze tolerance usually increases during the autumn. In *Balanus* the median low lethal temperature falls during the autumn and winter from −7°C to −15°C for an 18 h exposure. The development of

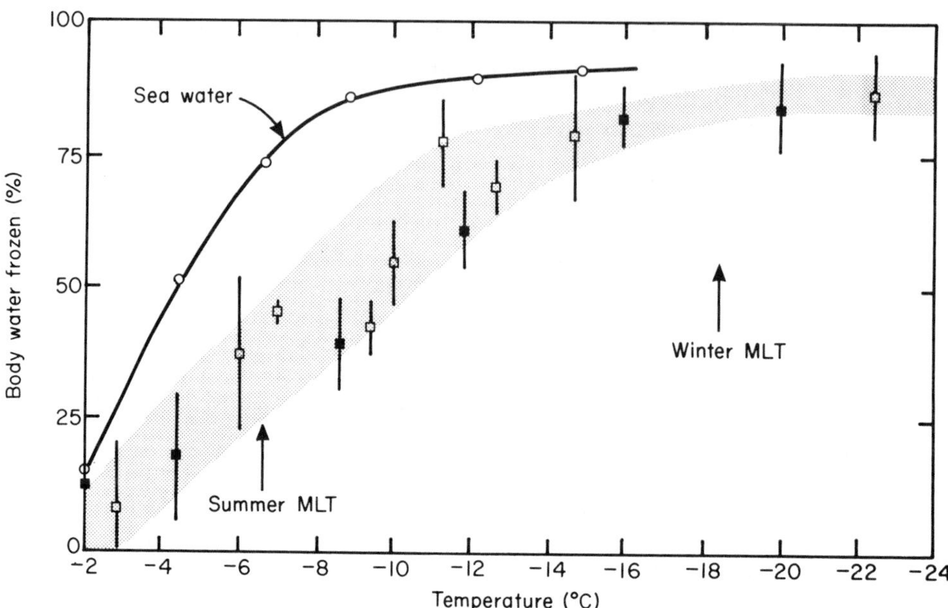

Figure 6.16 The formation of body ice in the barnacle, *Balanus balanoides*, in the summer (□) and winter (■). Note that the proportion of ice formed at any freezing temperature was substantially less than that formed in sea water. This difference seems to reflect the presence of 'bound' water. The winter median low temperature is shown; the proportion of frozen body water was 80%. (After Crisp *et al.*, 1977.)

tolerance is often associated with quiescence, though in *L. littorea* it parallels the attainment of reproductive condition.

The seasonal fluctuations in freeze resistance can be demonstrated in isolated tissues as well as in whole animals, but its physiological basis is not well understood. Several species of freeze-resistant insects have been shown to accumulate glycerol and other polyhydric alcohols during the autumn and winter. In some insects there is a close correlation between the degree of freeze tolerance and the concentration of alcohol in the haemolymph (Miller, 1978). However, other insects can survive freezing without glycerol, and other species cannot survive freezing even in the presence of large amounts of glycerol (Miller, 1969). Freeze tolerance in wood frogs is associated with a dramatic (1000-fold) increase in blood glucose concentration which is rapidly induced with the onset of freezing. There appears to be no anticipatory synthesis of cryoprotectants during autumn and winter (Storey and Storey, 1986).

The rate of freezing and thawing is particularly important to the survival of animals, especially in the intertidal zone where rapid freeze–thaw cycles may be experienced (Murphy, 1983). Rapid cooling (i.e. above 0.4°C min^{-1}) is more likely to result in the formation of intracellular ice, and rapid thawing is then usually essential for survival. Slow cooling enables the full potential for freeze tolerance to be realized, since the progressive and uninterrupted dehydration of the cellular compartment leads to a maximal depression of the freezing and supercooling points of the intracellular fluids. Since large specimens cool more slowly, they are able to survive exposure to freezing conditions more successfully than smaller individuals (Murphy and Johnson, 1980).

An interesting recent development is the discovery that many, if not most, freeze-tolerant insects have a very limited ability to supercool because of the inclusion of nucleating agents in the haemolymph (Duman, 1982). These agents ensure that ice formation occurs in the extracellular fluid at relatively high sub-zero temperatures. During slow cooling this facilitates the gradual and controlled formation of extracellular ice, and thereby may prevent the formation of damaging intracellular ice (Baust and Edwards, 1979) that may occur with rapid freezing of an extensively supercooled insect. With few exceptions their presence is seasonal and limited only to the cold part of the year. Duman and Patterson (1978) have shown that the nucleation substance in some insects is non-dialysable, heat sensitive and inactivated by a bacterial protease, all of which suggests a proteinaceous molecule. Indeed, Duman *et al.* (1984) have recently purified an ice-nucleating protein from a hornet.

It has become apparent that the distinction between freeze-sensitive and freeze-tolerant species is more apparent than real. Baust and Lee (1981) have noted that northerly populations of the gall fly, *Eurosta solidagini*, tend to induce ice formation in the cold by the elevation of supercooling points, whilst southern populations depress supercooling points as a means of avoiding ice formation. Presumably each strategy has advantages in its respective environment.

6.7.2 *Mechanisms of freeze avoidance*

Freeze avoidance relies upon the inhibition of spontaneous nucleation (i.e. supercooling) or upon the prevention of inoculation. In some instances the growth of environmental ice into animals is prevented by special properties of the integument. In at least one species of limpet a copious mucus is secreted which prevents the spread of ice down to $-10°C$ (Hargens and Shabica, 1973). The prevention of inoculative nucleation seems to be why overwintering forms of insects stop feeding and evacuate the gut well before sub-zero temperatures are experienced, since gut contents are important sites of nucleation (Fig. 6.17). Some fish may also exist in a supercooled state. For example, juvenile specimens of the cunner, *Tautogolabrus adspersus*, passes the winter in the inshore areas of Nova Scotia in a dormant state (DeVries, 1982). If such torpid, supercooled fish are touched with an ice crystal they rapidly freeze and die. Supercooling is not an uncommon strategy in fish which inhabit ice-free, deep-water environments, and the

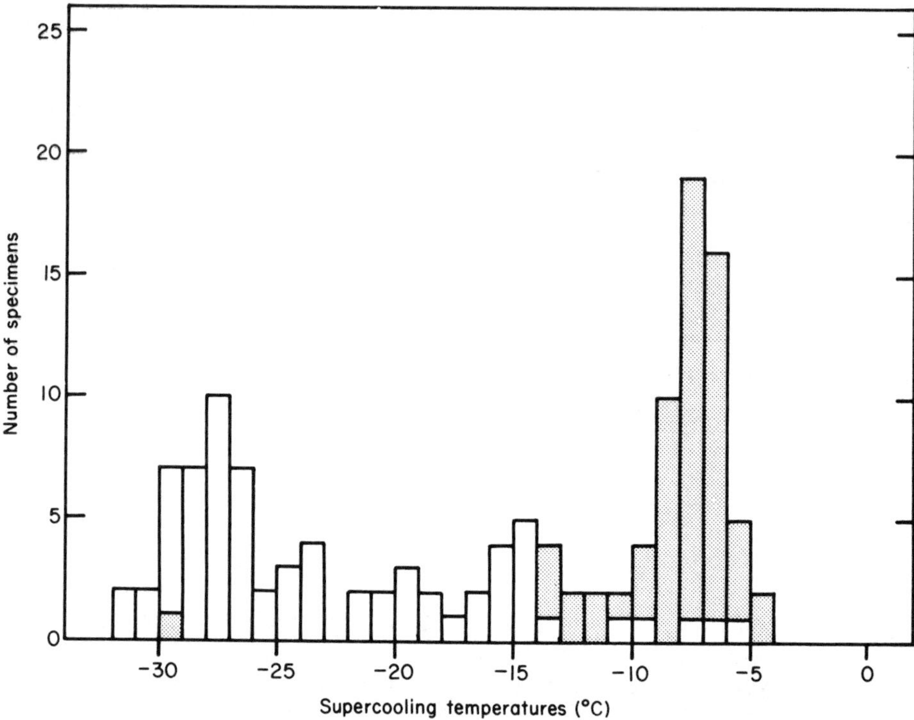

Figure 6.17 The frequency distribution of supercooling temperatures in a population of the collembolan, *Tetracanthella wahlgreni*, with (shaded bars) and without (open bars) gut contents. (After Sømme and Conradi-Larsen, 1977.)

supercooled state is sufficiently stable to permit survival without additional adaptations. However, the fact that supercooled fish may freeze in contact with ice crystals indicates that their integument is unable to resist inoculative freezing.

An obvious means of freeze avoidance is to increase the solute concentration of the body fluids and to depress their freezing point by a colligative mechanism. Indeed, the transfer of the winter flounder to the cold increases the plasma concentration of sodium by 18%. In *Fundulus* the concentration of glucose and other solutes may increase by over 400%. However, these adjustments only decrease the freezing point by 0.2–0.3°C at best, which is small compared, for example, with the difference in freezing points of fish plasma (– 0.7 to – 1.0°C) and that of sea water (– 1.86°C) (DeVries, 1974). Although this increase in plasma osmotic pressure is not sufficient to prevent inoculative freezing, it may inhibit spontaneous nucleation and permit a degree of supercooling. Whether this response is adaptive or merely the result of a fortuitous loss of osmoregulatory control is not known.

A growing list of animals, principally insects and many other terrestrial

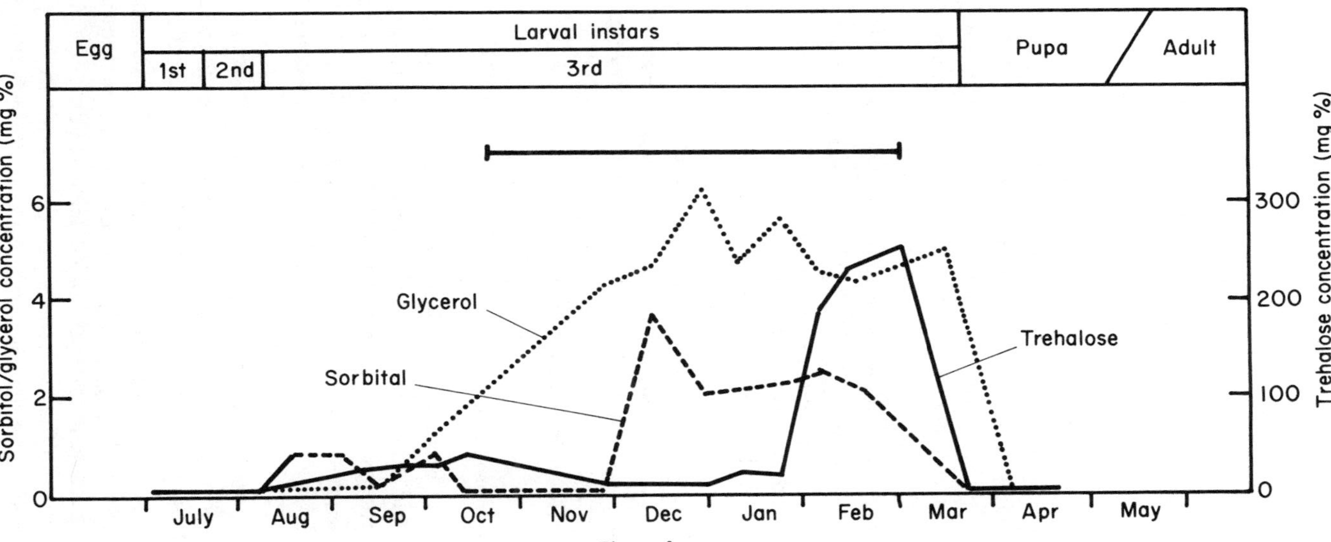

Figure 6.18 The accumulation of haemolymph cryoprotectants during the autumn and winter months in the gall fly, *Eurosta solidaginis.* The third larval instar becomes freeze-tolerant following the accumulation of a three-component cryoprotectant system. The horizontal bar indicates the period of 90% survival following freezing. (After Morrissey and Baust, 1976.)

arthropods, produce substances in their blood which prevent freezing of their blood or haemolymph, the so-called '*antifreezes*'. These animals produce large quantities of polyhydric alcohols such as glycerol, sorbitol, threitol and mannitol (Baust, 1981). Overwintering larvae of the sawfly *Bracon cephi*, for example, are reported to possess 25% glycerol by weight to give a haemolymph concentration of 5 mol l^{-1}, or 30% by weight (Salt, 1959), although this is unusually high. These cryoprotective substances may be present in species-specific, multicomponent mixtures which are usually synthesized from glycogen stores in late autumn and remain throughout the cold period (Fig. 6.18). In some insects they are synthesized only during diapause, so may be present in the egg, larvae, pupae or adult stage, depending on which is the overwintering form. At the termination of diapause or when environmental temperatures rise, glycogen is resynthesized and the polyol concentration of the haemolymph decreases.

Whilst these substances certainly lower the freezing point of body fluids in a colligative manner, they may also have other beneficial effects. First, the increased concentration of polyols may greatly enhance the supercooling properties by inhibiting spontaneous nucleation of water (Lovelock, 1953b). Indeed, recent studies have shown that for every degree Celsius depression of the freezing point caused by glycerol the supercooling point is depressed by two degrees Celsius. Secondly, by making extensive hydrogen bonds with supercooled water they may reduce the speed of ice-crystal growth or modify the structure of ice crystals. Thirdly, polyols may act as solvents for electrolytes and so lessen the damaging effects of cell dehydration. Finally, they may penetrate the intracellular compartment, where they have similar protective effects.

6.7.3 *Thermal hysteresis*

It has recently become clear that other factors may promote survival in extremely cold and ice-laden environments by interfering with or even preventing the growth of ice crystals. This means that the freezing point of the fluid is depressed to lower temperatures than the colligative properties would predict, yet the melting point is unaffected since this depends purely on the normal colligative properties of the fluid. The separation of the freezing and melting points of fluids is termed 'thermal hysteresis' and factors that promote this unusual property have been found in some insects (Duman, 1982) and in polar fish (DeVries, 1974, 1980). Several glycoproteins have been purified from the plasma of some polar fish species, which appear to be responsible for the ability of these fish to resist inoculative freezing in the surface, ice-laden waters of the Antarctic. On a molal basis the antifreeze glycoproteins depress the freezing point of water 200–500 times more effectively than sodium chloride (Fig. 6.19), but also exhibit the phenomenon of saturation. Thus a 2% solution of glycoprotein

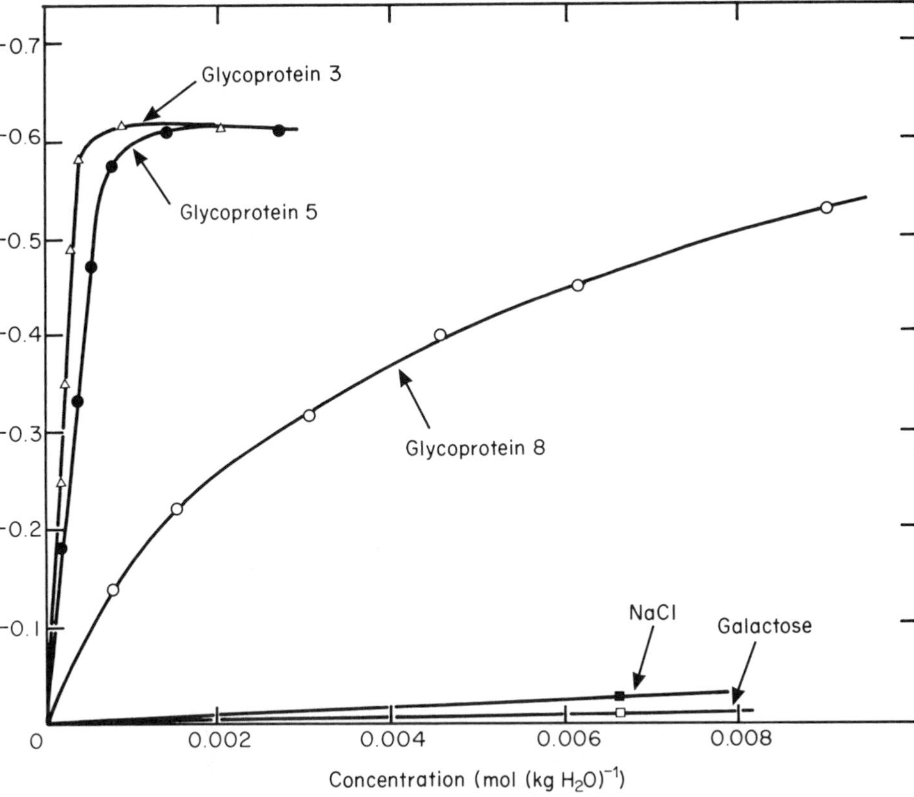

Figure 6.19 The freezing points of aqueous solutions as a function of the molal concentration of a sugar (galactose), a salt (NaCl) and some fish glycoproteins. The glycoproteins are significantly more effective on a molal basis than either NaCl or galactose, and this, together with their non-linear graphs, suggests that they operate by non-colligative mechanisms. (After DeVries, 1971.)

in water will freeze at $-0.9°C$ but will not melt until the temperature is raised to $-0.02°C$. This clearly provides considerable protection against freezing but does not have the disadvantage of a high molar concentration and the osmoregulatory problems that this would create. The inhibitory effects of glycoproteins upon ice crystal growth is a stable property. In one study the growth of a seed ice crystal remained inhibited even after 10 days (DeVries and Lin, 1977).

A good example of the use of glycoprotein antifreezes is in the Antarctic fish *Pagothenia borchgrevinki* whose plasma freezes at $-2.07°C$. After removal of salts by dialysis the plasma freezes at $-1.1°C$, so almost half of the freezing point depression in this species is due to a non-dialysable molecule at a plasma concentration of approximately 3% by weight. A number of glycoproteins have been purified from *Pagothenia* plasma with

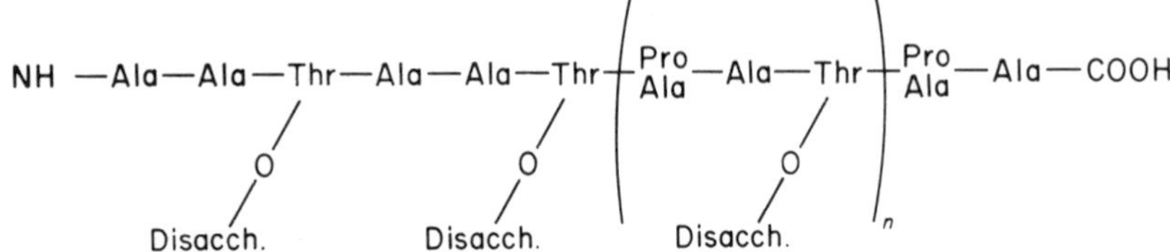

Figure 6.20 The repeating structural unit of an antifreeze protein isolated from the plasma of the Antarctic fish *Pagothenia borchgrevinki*. Different glycoproteins differ in size by including different numbers of repeating units: glycoprotein 7, $n = 4$; glycoprotein 8, $n = 2$. Disaccharide (Disacch.) consists of galactose and *N*-acetylgalactosamine. (After DeVries, 1974.)

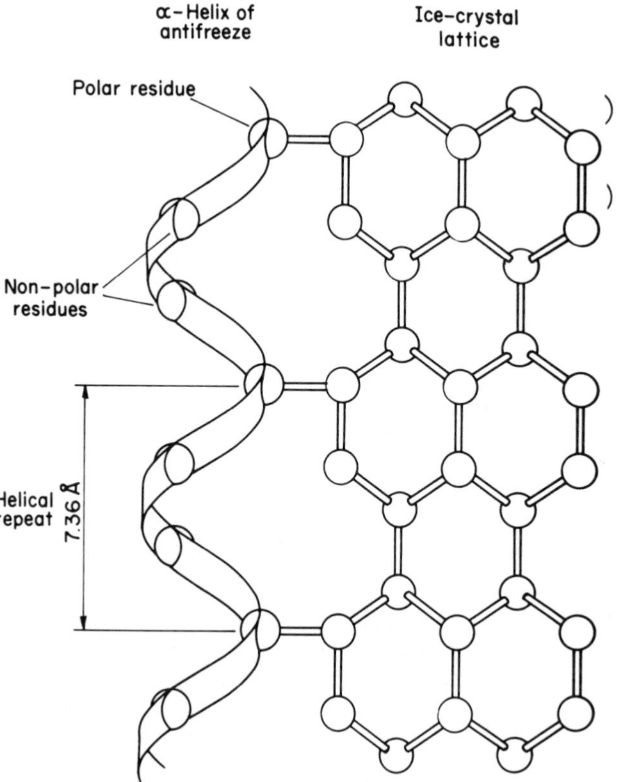

Figure 6.21 A model showing how the regular peptide repeat of the glycoprotein antifreeze molecule could form hydrogen bonds with the oxygen atoms of the ice-crystal lattice. If the repeat distance of the helix is similar to that of the ice crystal (i.e. 7.36Å) hydrogen bonding would be enhanced. Hydrogen bonds are indicated by solid bars. (After DeVries and Lin, 1977.)

molecular weights ranging from 2600 to 34000. Their molecular structure is typically composed of repeating units of glycotripeptides (Fig. 6.20). The secondary structure of these molecules is not known with any precision, but there is some evidence for a rod-shaped molecule arranged in a helical configuration. Raymond and DeVries (1977) and DeVries and Lin (1977) have suggested that they form an extended α-helix in which the hydroxyl and carbonyl groups are presented at a regular repeat that corresponds closely with the crystalline lattice of ice (Fig. 6.21). Their 'adsorption-inhibition' hypothesis suggests that these molecules recognize the crystal structure of ice and physically bind to the growing edge of the ice lattice in order to prevent access of free water molecules. Inhibition of ice-crystal growth apparently results from the inability of the advancing ice front either to overgrow the adsorbed antifreeze molecule or to pass between them. This notion of a stereospecific recognition of ice structure is supported by the loss of antifreeze activity of glycoproteins and their binding affinity

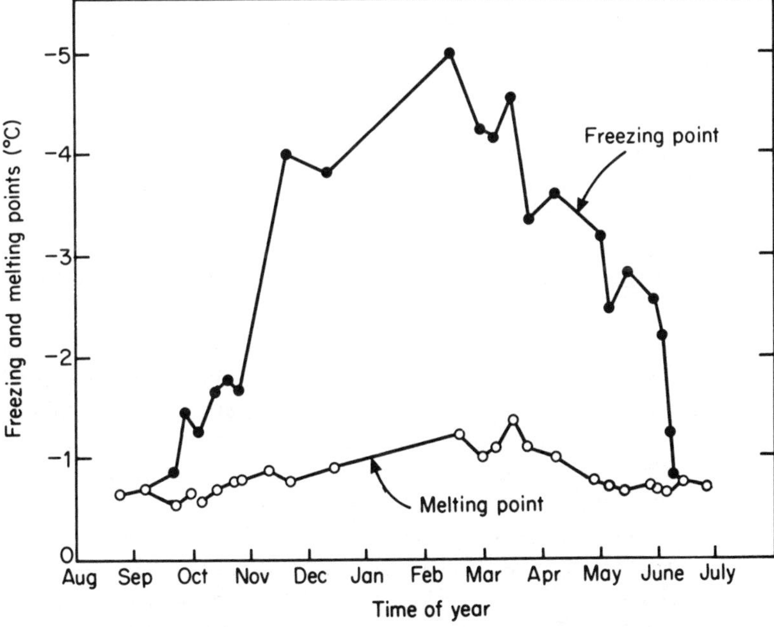

Figure 6.22 Seasonal variation in the freezing (●) and melting (○) points of the haemolymph of larvae of the beetle *Meracantha contracta*. The difference between the lines represents the extent of thermal hysteresis and the protection against freezing. (After Duman, 1977.)

for ice when the hydroxyl groups of their sugar residues are converted into carboxyl groups.

Arctic and northern fish species may experience a large seasonal range of temperatures (−1.7°C to 22°C) and, unlike *Pagothenia* and other Antarctic species, experience freezing temperatures during the winter months only. As a result there is a large seasonal fluctuation in the concentration of antifreeze compounds in the blood. For example, exposure of the Arctic flounder, *Pseudopleuronectes americanus*, to 10–12°C results in a loss of all serum antifreeze within 3 weeks. Acclimation studies suggest that these striking changes in antifreeze concentration during the spring and autumn are influenced by seasonal changes in both photoperiod and temperature. Recently, Lin (1979, 1983) has demonstrated that antifreeze production in the Arctic flounder occurs in the liver. Indeed, during the winter months few proteins other than antifreeze are synthesized. Antifreeze synthesis is controlled at the level of DNA transcription and translation, since mRNA for antifreeze appears during the autumn months about 4 weeks before antifreeze is found in the blood. Similarly, the mRNA disappears about 4 weeks before antifreeze is lost from the blood.

Figure 6.22 shows the seasonal variations in thermal hysteresis in the haemolymph of the beetle *Meracantha contracta*. The thermal hysteresis proteins appear in late September or early October and peak in midwinter. Hysteresis does not disappear until late May, when the likelihood of frost is negligible. Experimental studies suggest that both photoperiod and low temperatures are involved in the seasonal induction of these proteins (Duman, 1982).

6.8 A cellular basis for thermal injury and death

It is clear that thermal injury and death of animals follows from the failure of one or more of its homeostatic systems, which may vary from species to species. Just what constitutes the most thermolabile cellular component underlying the failure of the system in any one instance is not known with any certainty, although there are several good candidates. Any theory of cellular thermal injury and death must account for the fact that these phenomena have a high temperature coefficient, they operate over a very restricted temperature range and that this range can be modified by thermal acclimation. Thermal lability of cells is probably a property of two groups of macromolecules: the proteins and the membrane lipids. Most hypotheses concerning the aetiology of thermal injury emphasize one or other of these effects.

It is well known that proteins denature at high temperature, usually with a Q_{10} of 10–600 (Johnson *et al.*, 1974). The Q_{10} for thermal injury is also of this order, and this correspondence constitutes important evidence for the critical role of protein denaturation in heat injury. It is often observed that proteins of more thermophilic organisms denature at higher temperatures than those of cold-adapted organisms (Ushakov, 1964). On the other hand, protein thermostability seems to be unaffected by thermal acclimation. This, together with the fact that these enzymes are usually inactivated at temperatures somewhat above those that cause heat injury, means that gross denaturation of cellular proteins is probably not involved in heat death. This conclusion must be qualified to the extent that those enzymes which have been studied may not be the most thermolabile or the most critical. Indeed, there are two groups of proteins whose thermolability roughly matches that of the cells and animals from which they were obtained: membrane-bound enzymes and muscle actomyosin ATPase.

The *in vitro* inactivation of proteins is greatly affected by conditions of pH, ionic strength and the concentration of substrates, products and allosteric modulators. Thus, a major difficulty in establishing the most relevant inactivation temperatures is in choosing the *in vitro* conditions which most closely resemble the *in vivo* intracellular environment during heat injury.

The relationship between enzyme thermostability and whole-animal thermotolerance noted earlier probably arises from the evolution of proteins with a conformational freedom appropriate for function at their particular environmental temperature regimes. Thus, warm-adapted species possess more 'rigid' proteins whilst cold-adapted forms possess more 'flexible' proteins. Although these differences may not neccessarily have much relevance to heat injury or to resistance acclimation, it is nevertheless a reflection of an important molecular adaptation.

The idea that tolerance to heat is limited by the stability of cellular lipids has its origins in the work of Heilbrünn (1924). He observed that organisms adapt to changes in temperature by altering the saturation of their cellular lipids, and proposed that this acts to preserve the physical state of these lipids in a particular condition. Heat injury was thought to result from a change in state of the cellular lipids, the so-called 'lipoid-liberation' theory. The crucial role of phospholipids in the structure and function of cellular membranes was not then fully appreciated. More recently this idea has been applied more specifically to cellular membranes. The idea that cells maintain their membrane lipids in a critical physical condition or 'fluidity' in order to support normal function has already been introduced (Chapter 5). This means that membranes of cold-acclimated animals are generally more fluid than those of warm-acclimated animals, so that the former will attain a critical, and perhaps damaging, value at a lower temperature than the latter (Cossins, 1983). This idea is supported by the similar membrane fluidities of brain synaptosomes (sealed vesicles formed from synapses during homogenization) of cold and warm-acclimated goldfish at their respective heat-coma temperatures (Cossins *et al.*, 1977). In other words the difference in fluidity between cold- and warm-acclimated goldfish was exactly that required to account for their different heat-coma temperatures.

Several other compelling pieces of evidence supporting the idea that heat injury follows from excessive membrane fluidity comes from studies of mammalian cell cultures. First, the heat resistance of a mammalian cell culture increased with increased culture temperature (Anderson *et al.*, 1981). This was accompanied by significant changes in the ratio of cholesterol to phospholipid and in membrane fluidity. Cress and Gerner (1980) found an impressive inverse relationship between the sensitivity of different mammalian cell cultures to heat exposure and their cholesterol content. The importance of cholesterol in both of these cases is related to its marked stabilizing effect upon membranes, by reducing molecular motion and increasing the overall order of the hydrocarbon chains.

Secondly, manipulation of membrane fluidity by agents other than temperature produce predictable changes in the thermotolerance of cells. Thus, alcohols and clinical anaesthetics, both of which increase fluidity, caused a decrease in the thermotolerance of some cell lines (Yatvin, 1977). On the other hand, hydrostatic pressure, which tends to decrease membrane fluidity, has been shown to increase thermotolerance (Minton *et al.*, 1980). Finally, the dietary manipulation of membrane lipid composition of bacteria and some cells of higher animals leads to changes in membrane fluidity, in maximal growth temperature and in sensitivity to heat exposure.

Precisely which aspect of membrane function fails as a result of excessive molecular motion has not been unequivocally shown, though it is most probably related to the stability of particular membrane-bound proteins. For example, in freshwater crayfish a breakdown in the normal permselectivity of the sarcolemma occurs early in heat death. The normal permeability characteristics of plasma membranes are properties either of the passive, electrodiffusional leak or of specific protein-mediated mechanisms, the so-called gating proteins, which may become damaged during heat death. Some membrane-bound enzymes of crayfish muscle are, in fact, inactivated over the same range of temperatures which cause heat death. Furthermore, the thermostability of these enzymes may be altered in a predictable manner by thermal acclimation (Cossins and Bowler, 1976; Gladwell, 1975).

In another example, heat injury and death in the blowfly, *Calliphora erythrocephala*, are closely related to the loss of function and the appearance of ultrastructural damage in flight muscle mitochondria (Davison and Bowler, 1971). The mitochondria lose respiratory control, which means that oxidation of substrates is not properly coupled with the phosphorylation of ADP. Recovery from heat-induced damage is related to repair of these mitochondria and the re-establishment of adequate respiratory control (Bowler and Kashmeery, 1979). Again, the susceptibility of these structures to high temperature is altered in a predictable way by thermal acclimation.

The lowering of temperature produces several damaging effects. Metabolic pathways are generally slowed, but the different Q_{10} for different pathways may also lead to an imbalance in overall cellular metabolism. Membrane fluidity is also reduced, but more significantly may induce a change in the physical state of the bilayer from a liquid-crystalline state to the more rigidified gel state. Lyons (1973) has suggested that chill injury in plants and in homoiothermic animals results from such a phase transition of their membrane lipids at the critical temperatures. This leads to increased membrane permeability and to large increases in the Arrhenius activation energy of membrane-bound enzymes, both of which were thought to contribute to a loss of viability in the cold. This idea has been supported by the observation of Arrhenius discontinuities of membrane-bound enzymes and membrane order of mitochondria at temperatures which cause chill injury in chill-sensitive plants and homeotherms. These discontinuities were thought to represent phase-transition temperatures. They were absent in the mitochondrial membranes of chill-resistant plants and most poikilothermic animals, probably because of their increased unsaturation of membrane lipids (see page 42).

Because of its obvious importance to the study of chill injury in crops and in the preservation of cooled mammalian organs during surgery, this original hypothesis has been subjected to intense scrutiny and found to be wanting in several respects (Wilson and McMurdo, 1981). First, the concept of a single and clearly defined bulk phase transition in natural membranes now appears to be a gross oversimplification. It is a complex event and probably involves only a small proportion of the membrane lipids. Secondly, the original spectroscopic techniques employed by Lyons and his colleagues to define discontinuity temperatures has been questioned. Thirdly, the existence of sharp discontinuities, as opposed to more-gradual curves, is doubtful in some cases, though not in others. Finally, other workers have found enough exceptions to the original observations to limit greatly their usefulness. Nevertheless, the fact that lipid-perturbing agents protect mammalian cells against chill injury does provide strong evidence for a role of membranes in this process (Kruuv *et al.*, 1983).

6.9 Thermal tolerance in multivariate conditions

In laboratory studies of the resistance of animals to extreme temperatures it is usual to maintain other environmental factors as constant as possible. In nature, however, this is rarely the case, and a number of environmental factors – biotic and physical – vary, often in a correlated way. The enormous complexity of these interactions makes it difficult to predict the potential consequences of a given set of circumstances. Similarly, the relative importance of any one lethal factor may change through interactions with other factors.

In the terrestrial environment the factor most likely to affect the way in which temperature causes mortality is humidity. The threat of desiccation occurs not only in arid habitats, but also in the intertidal zone. This specific interaction has been most thoroughly studied using insects in which exposure to high lethal temperatures is usually tolerated better in dry air than in moist air. This is because the insect can cool by evaporation from its surface. It is estimated that a large insect can afford to lose up to 20–40% of its body weight before desiccation becomes a crucial factor. By contrast, small insects such as houseflies have too little body water for evaporative cooling to be an effective protection against high temperature, and they tend to have higher survival times in moist than dry air.

In the aquatic environment oxygen tension and salinity are the most obvious factors which may interact with temperature to cause mortality. The resistance to combinations of lethal factors tends to reflect the evolutionary and acclimation experience of the organism. For example, stenohaline organisms tend to have a maximal thermal tolerance in a salinity that is usual and which does not constitute an additional stress. *Clymenella torquata*, an estuarine polychaete worm, can tolerate low temperature better at low salinity than high temperature at low salinity (Kenny, 1969), these being conditions which this organism normally experiences. Similarly, *Gammerus duebeni* can withstand high temperatures better at high salinities, again a combination that is frequently encountered in nature (Kinne, 1964).

Acclimation to one factor may influence the resistance to another lethal factor. In nature, acclimatization to these different factors occurs simultaneously, and so optimizes the resistance of the animal to the likely combinations of lethal factors. The classical study by McLeese (1956) on lobsters illustrates these interactions particularly well. He acclimated lobsters to 27 different combinations of temperature, oxygen tension and salinity, then determined the upper lethal temperatures and the lower lethal salinities and oxygen tension for each combination. The upper lethal temperature was increased by raising the acclimation temperature, but was lowered by acclimation to decreased salinity and oxygen tension. Similarly, the effect of salinity acclimation was not constant, but depended upon acclimation temperature.

These interactions between different lethal factors in lobster are illustrated graphically in Fig. 6.23, in which the ultimate lethal levels of the three factors have been determined and integrated into a three-dimensional surface which represents the boundary for 50% mortality for lobsters exposed to the three lethal factors simultaneously. The surface has three nearly flat areas where one factor alone is lethal. In each case acclimation to different levels of the other two factors has no effect upon resistance to that factor. The curved section of the surface represents an area of mixed lethal effects, the most obvious being the combination of high temperature and low oxygen tension to produce mortality at normally sub-lethal conditions when each is acting separately.

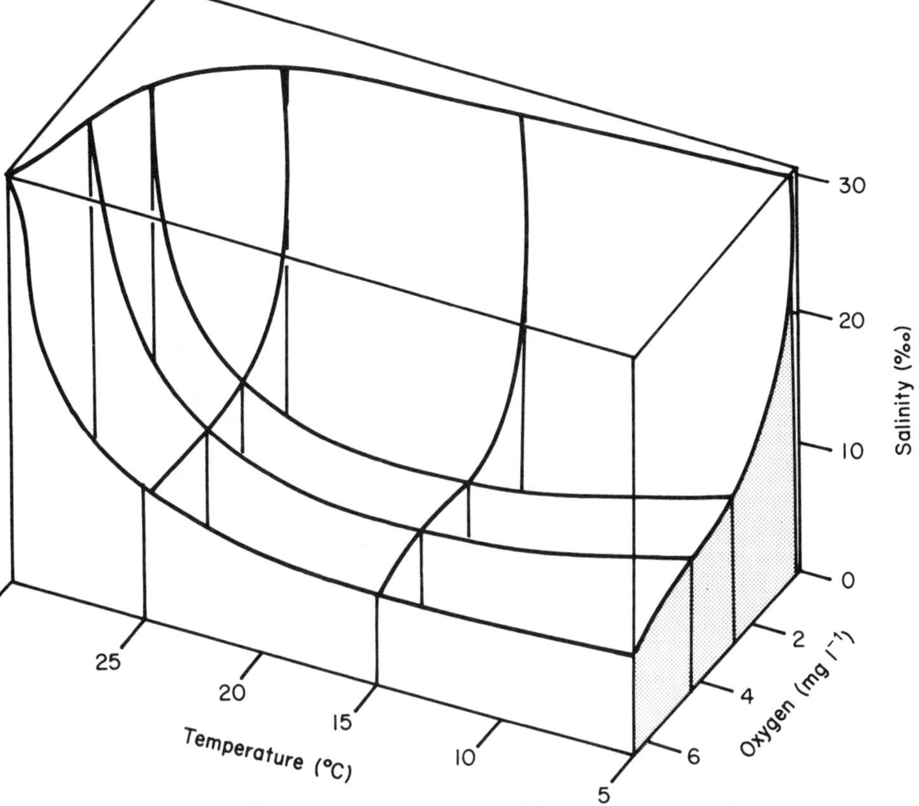

Figure 6.23 A three-dimensional representation of the boundaries for median lethality for temperature, salinity and oxygen concentration for the American lobster *Homarus americanus*. The figure was constructed by drawing templates for the minimal lethal salinities at different acclimation conditions of oxygen and temperature. The template for 5°C-acclimated lobsters is shown on the right front face with shading. Six templates were constructed, and these were placed on a base drawn from a graph of temperature against oxygen to provide a three-dimensional surface. Above this surface at least 50% of the lobsters survived. (After McLeese, 1956.)

McLeese's diagram was an early attempt at a multifactorial study. Recent advances have provided a more powerful empirical description of the complex interactions between several factors which act both to acclimatize and as lethal factors, namely response-surface methodology (Alderdice, 1972). The results of complex factorial experiments such as those of McLeese may be modelled with polynomial equations in order to generate three-dimensional images relating, for example, mortality to two lethal factors (see Fig. 6.24). Although these descriptions are empirical and do not help with the interpretation of the physiological events wihch underlie

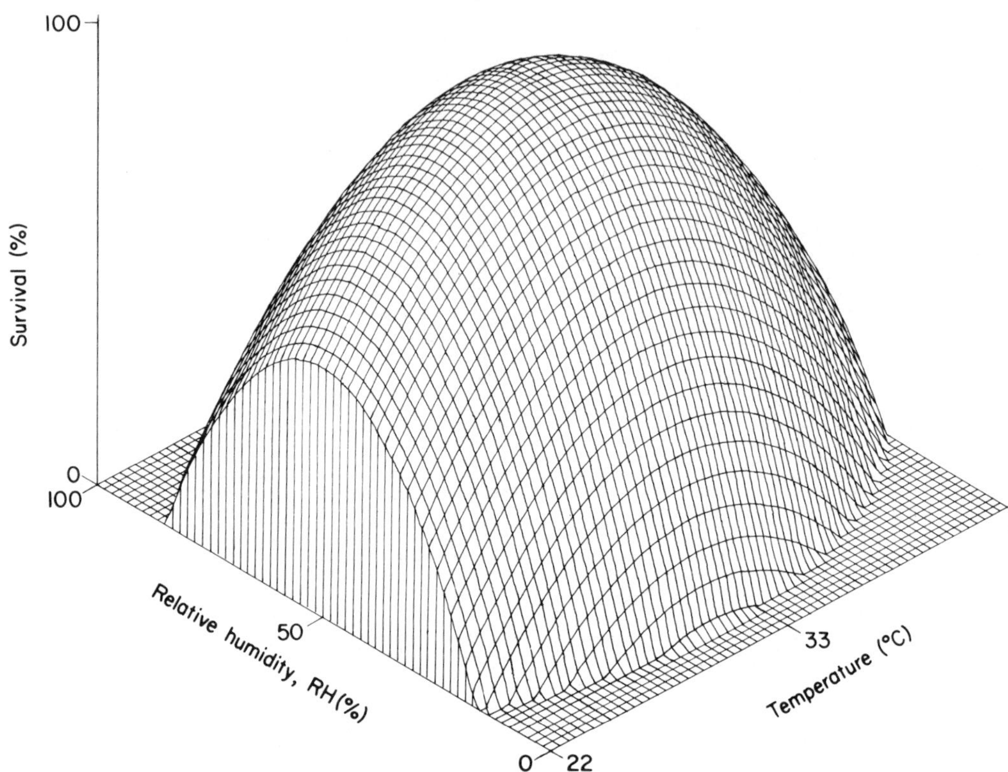

Figure 6.24 A response surface for the interactions of humidity and temperature upon mortality of *Schistocerca* nymphs. The data were obtained from Hamilton (1956) and modelled on a computer by K. Hodgkinson.

mortality, they do enable a more ready appreciation of the interactions which occur and how they are affected by experimental treatments, life cycle, and so on. The complex computer-generated surface such as that shown in Fig. 6.24 is usually represented by a two-dimensional graph of factor intensity (i.e. temperature on one axis and humidity on the other), with contours representing the different levels of response (i.e. survival). This format has a utility in allowing an easy and more precise assessment of the likely consequences of a given combination of factors (Fig. 6.25).

We have seen that animals appear to optimize their physiological resistance to those combinations of lethal factors that are most likely to occur in their respective habitats. No such optimization may be expected, at least in the short term, in response to many of man's pollutants. Of course, pollutants such as pesticides and heavy-metal ions may themselves cause significant mortality but, more importantly, these pollutants may cause a reduced ability to tolerate other environmental stresses such as extreme

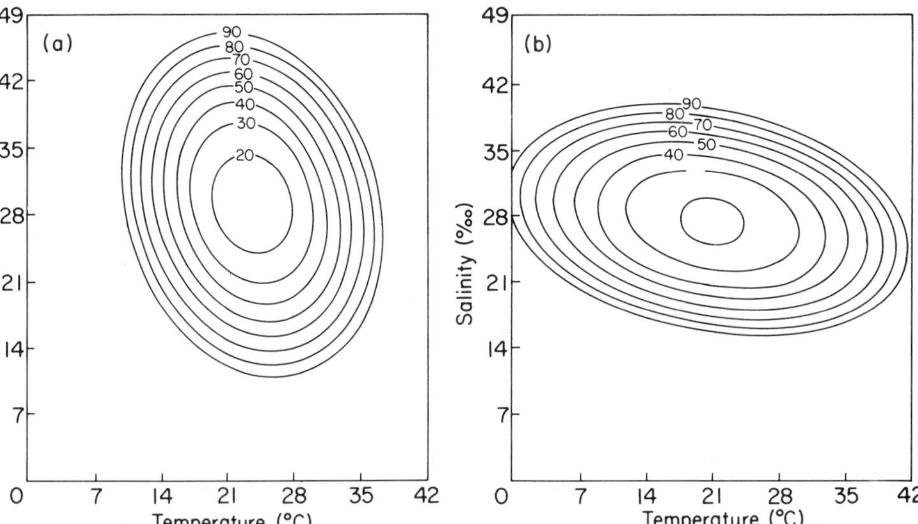

Figure 6.25 The response surfaces (in % mortality) calculated for larval mortality of the fiddler crab, *Uca pugilator*, in (a) the absence and (b) the presence of cadmium at 1 ppm. The curves were fitted to the observed mortality using 13 combinations of salinity and temperature. (After Vernberg *et al.*, 1974.)

temperature or salinity, especially during sensitive stages of the life-cycle. For example, Vernberg *et al.* (1974) have shown a pronounced effect of very low cadmium and mercury concentrations upon the tolerance of zoeae larvae of the fiddler crab, *Uca pugilator*, to combinations of temperature and salinity. Larvae normally survived combinations of 21–28°C and 23–34% salinity, but in the presence of cadmium at 1 ppm the survival limits were reduced to 18–23°C and 24–29% salinity. The isopleths in Fig. 6.25 show that temperature was the dominant cause of mortality in control larvae but salinity was more important in cadmium-treated larvae. Exposure of fish fry to arsenic led to a progressive reduction in CT_{max}, even though there was no outward sign of the physiological effect of this poison. Finally it is not safe to assume that toxicity of a pollutant rises with increased temperature, since other factors such as increased rate of passage through the gut, rate of detoxification and excretion of the pollutant may interfere.

6.10 Lethal limits, temperature and animal distribution

Temperature is commonly believed to be an important factor in limiting the distribution of animals. In view of the sensitivity of animals to extreme temperatures and the manifold strategies employed by them to survive exposure, it can hardly be denied that it is an important factor. Yet the evolution of species, communities and ecosystems is the result of a great variety of influences, both biotic and abiotic, which operate in all sorts of subtle ways, so we might expect the role of temperature to be less than clearcut.

Reports of a major role of temperature in limiting distribution usually rely on a circumstantial relationship between a climatological description of the environment and species abundance. However, this is only the start of the problem, because as we have seen macroclimate is not necessarily closely related to microclimate and, secondly, temperature may have its effect not from a direct lethal effect of temperature upon the adult or sub-adult form of a species, but indirectly through effects upon the food supply or upon a predator, parasite or competitor. Thus, although many animals live close to their lethal thermal limits, this is seldom the primary determinant of their distribution. The geographical distribution of many species correlates more closely with the temperatures that occur during the breeding season. This is probably because gametogenesis and embryogenesis are generally the least-tolerant stages of the life-cycle and therefore constitute the 'weakest link' (see Chapter 7). For example, it is probably the thermal tolerance of corals during breeding and development that results in their being confined to waters above 20°C. In other animals it may be a combination of the heat tolerance of adults and the thermal requirements for reproduction that limit distribution. Hutchins (1947), referring to Northern-hemisphere marine organisms, suggested that the latitudinal expansion of their range is influenced by either the cold tolerance of adults in winter and heat tolerance in summer or by the thermal requirements for reproduction and repopulation. The northern limit of the barnacle *Balanus amphitrite* appears to be set by their inability to survive mean monthly temperatures below 7.2°C combined with the need for a mean monthly temperature in the summer of at least 18°C for reproduction.

A good example of the complexities that exist is the agreement between the southerly limit for the barnacle *Balanus balanoides* and the winter sea-surface isotherm for 8°C (Lewis, 1964). Although the adult is killed above 25°C there is no great impairment to growth and gonadal growth up to 20°C during the summer months. Final maturation is usually delayed until the autumn, but nevertheless can take place at sea temperatures of 15°C. The apparent connection between distribution and the 8°C isotherm is due to the fact that these sea temperatures are usually preceded in the autumn by the appropriate air temperatures to release the maturation barrier. The isotherm has no specific meaning except in the context of the normal seasonal progression of environmental temperatures.

In some species, however, the routine occurrence of severe thermal conditions at any one stage of their life-cycle has resulted in the production of resistant stages, such as freeze-tolerant insects. Perhaps the most bizarre is the larvae of the West African chironomid *Polypedilum* (Hinton, 1960).

This organism inhabits temporary pools which frequently dry out. The larvae become desiccated and they can withstand unusually high temperatures. Many overwintering arthropods produce diapausing stages. In desert and arid-zone species there may be a summer dormancy called aestivation. These various resting stages allow survival during particularly stressful periods.

For temperature to have its effects as a direct lethal factor it need only act infrequently. The risk from such unexpected extremes is naturally greatest at the edge of a geographical distribution. For example, following a hard winter in Florida the northern limit of the shore crab *Uca rapax* retreated about 100 miles. Such occurrences are usually termed 'thermal catastrophes'. Records show both heat and cold catastrophes have occurred in a variety of unlikely environments, including the sudden cold spells in Florida and the Gulf of Mexico which occur at intervals of about 10 years. One of the best-documented cases is the severe cold weather experienced by southern Britain during the winter 1962–63. Sea temperatures fell to −1.5°C for several weeks, and rock pools and beaches were frozen. Exposed intertidal species were subjected to very low air temperatures and became frozen. A detailed account of the effect of this weather on the British fauna is given by Crisp (1964), who describes how huge mortalities were experienced by some important components of the fauna. These authors concluded that the palearctic species, which here are at the southerly end of their range, suffered the lowest mortality. The most severely affected were those species with predominantly southern distributions, which were at the northerly end of their ranges on the south and south-west coasts of Britain. Mass mortality was observed in many species, and the abundance of many species was dramatically reduced. Very rarely, however, was a species totally eliminated from a habitat; only in the case of the barnacle *Balanus perforatus* was the species distribution clearly more restricted after that winter.

Natural catastrophes such as these emphasize the care needed to ensure that the wider use of fresh and sea water for cooling purposes in power and desalination plants do not cause localized environmental damage (Coutant, 1972; Barnett and Hardy, 1984). Summer is a particularly hazardous period because the added environmental heat load together with thermal pollution may be lethal to the local fauna. The problem is exacerbated by the synergistic effects of temperature with chemicals used in cooling waters and mechanical damage caused at intake screens and by water turbulence. These situations are most likely to occur in tropical and subtropical regions where estuarine and marine animals live close to their lethal limits. Any effects are likely to be localized close to the release point of the warm-water effluent, and here a permanent change in species composition is possible and has been recorded (Barnett and Hardy, 1984). Cold-stenothermal forms may be replaced by more eurythermal species which benefit from the higher temperatures by having increased developmental rates and a longer breeding season.

A potentially more hazardous situation would predictably arise if the warm water effluent was intermittent. The sudden temperature shocks caused by start-up and cessation of power station operations may cause a number of mini-catastrophes and, by limiting reproduction, may lead to an impoverished environment. The biological effects of artificially raising water temperature are likely to be more profound in freshwater and estuarine situations than on the coast, because the sea provides an enormous thermal bin where cold and warm waters quickly mix. Nevertheless, it seems that a concentration of power stations around seas of limited size may cause a significant change in the average annual temperatures and in the seasonal maxima and minima.

6.11 Conclusions

Undoubtedly, the lethal effects of temperature pose a major problem for all living organisms, but particularly for ectotherms. This problem is compounded not only by the predictable daily and seasonal variations in temperature, but also by the unpredictable and sudden onset of unusually extreme conditions. Survival thus requires a suite of adaptive responses ranging from daily heat hardening and seasonal resistance acclimatization to the intervention of resistant stages of a life-cycle. Each combination of responses can only be fully appreciated within the context of the lifestyle of the animal, its evolutionary history and the life history, energetic and reproductive strategies employed by the species.

At a simple level these responses all enhance survival in the face of extreme conditions. They may also allow individuals to minimize the sublethal and debilitating effects of thermal extremes, so the zone of physiological competence extends closer to the absolute tolerance limits. A remaining problem, however, is the extent to which adult tolerance limits are actually tested in nature and the degree to which selection in these instances operates to fix that characteristic in a species. There are many clear-cut examples of animals living sufficiently close to their tolerance limits that differential mortality occurs. However, Fig. 6.10 shows a case where the upper lethal temperature of some lake fish comfortably exceeds the daily maximum water temperature. Yet the general order of increasing thermal tolerance correlates with the ecological characteristics of each species, and all species employ resistance acclimatization. So if resistance limits in this instance are not tested in nature, then it seems that some other associated property restricts distribution. One possibility is that resistance adaptations are, at least in some cases, manifestations of adaptive responses for life processes over the normal range of temperatures (i.e. capacity adaptations).

The relationship between these two aspects of temperature adaptation is not clear, but since both ultimately depend upon the same cellular and molecular machinery there is good reason to expect that physiological adjustments during seasonal acclimatization and over evolutionary time will produce changes in both the capacity for normal functions and in resistance limits.

This linkage also provides an explanation for another problem. Some species, such as the mummichog *Fundulus heteroclitus*, are unusual in surviving exposure over the full biological range of temperatures. This impressive eurythermicity raises the question of why tolerance limits for other species become sufficiently restricted that thermal injury and death occur in nature. Why, for example, are salmonids largely restricted to temperatures below 25°C when heat death is a real problem in certain localities? One answer is that there are costs associated with eurythermy or, to put it another way, there are benefits to stenothermy which outweigh, in selective terms, the need for wider tolerance. These benefits of adopting a thermal-specialist approach may be rather subtle and need not directly relate to the thermal environment. Thus, physiological performance (growth rates, conversion efficiency, locomotory abilities, etc.) over a limited range of temperatures may be generally improved over that of a more eurythermal species. Of course, each strategy has its particular benefits in its respective environment.

7 Effect of temperature on reproduction, development and growth

7.1 Introduction

Temperature influences the complex processes of reproduction, development and growth in several ways. It may act as a rate function (Q_{10} effect), it may be the signal that initiates the process or, finally, it may be a threshold factor such that a particular temperature(s) is necessary for continuation of the process.

In nature temperature is rarely a static factor, acting at a particular intensity for a certain time, it also interacts with other environmental variables. This complicates studies on the effect of temperature on reproduction, development and growth, but a further complexity is that these processes are not constant with time. Laudien (1973) considered these to be 'changing' reaction systems, and this emphasizes that temperature effects on these processes would not be constant and would also vary with different stages in the process concerned.

In view of such complexities it is difficult to distill a set of principles that encompass the diverse strategies for reproduction, development and growth that animal species follow with respect to temperature. One consistent feature that does emerge is that these three complex processes are successful only in a particularly narrow range of temperatures, a range much narrower than that which permits survival in adults. This is well illustrated by the resistance polygon of Brett for the sockeye salmon, shown in Fig. 6.6.

7.2 Reproduction

Sexual reproduction is a complex set of processes which culminate in the production of a zygote. It involves gonad growth, germ-cell maturation, gamete release and fertilization. In some species complex behavioural events such as mating, nest building and maternal care also occur.

7.2.1 *The thermal trigger?*

Reproduction is only successful if the offspring survive to reproduce. Thus, it must be timed to coincide with the season most propitious for the survival and development of the offspring. The physiological, morphological and behavioural changes necessary in the adult for reproduction are often brought about by hormones produced by specialized neurosecretory cells in the central nervous system (CNS). These neurosecretory cells will secrete their hormones, and so initiate the train of events leading to mating, in response to stimuli by other CNS centres that process information about seasonal environmental changes.

For reproduction to be so entrained the environmental factor must vary appreciably throughout the year, and the variation must be predictable from year to year. In temperate terrestrial climates the seasonal changes in day-length would be an excellent 'trigger', whereas temperature would not. However, temperature may make an additional suitable 'trigger' for reproduction in organisms living in surface freshwater and marine habitats.

Little direct evidence exists to implicate temperature change, or even the attainment of a specific temperature, as the 'trigger' for the initiation of reproduction. The problem is complicated by the difficulty in separating rate–temperature effects from those of a trigger. For example, Kinne (1970) reports that in many marine invertebrates reproduction seems to occur only when a particular temperature has been reached. He emphasizes that some workers argue that this indicates a 'trigger' function for temperature, whilst others consider it merely a rate–temperature effect, with gametogenesis occurring at low temperatures, but only slowly.

Nevertheless, temperature and photoperiod often interact to determine the onset of reproduction. In the stickleback, *Gasterosteus aculeatus*, Baggerman (1957) found that gonadal development was stimulated by temperature as a rate function, but only when the animals experience a long-day photoperiod. The freshwater crustacean *Hyalella azeca* behaved similarly (De March, 1977). These animals are reproductively quiescent on a short-day photoperiod, irrespective of temperature, but when on a long-day photoperiod regime reproduction is stimulated by rising temperature.

In many species of fish, temperature and photoperiod together modulate all but the final stages of reproduction. In some cyprinids, however, temperature has been shown to have the predominant role in stimulating gametogenesis (Bye, 1984). However, Scott (1979) stresses the difficulties inherent in interpreting such data; vitellogenesis in the minnow *Phoxinus phoxinus* is stimulated by temperature and photoperiod, and a change in water temperature and photoperiod synchronize reproductive cycles. This occurs because the change in temperature promotes a behavioural change in the fish before which the fish were not experiencing conditions where they could respond to the photoperiod clues. In this case temperature is acting indirectly on reproduction by promoting changes in behaviour.

7.2.2 *Gonadal growth and development*

Gamete production makes heavy nutritional demands on reproductive individuals. It is for this reason that gonad growth and somatic growth are generally not compatible. Reproductive activity usually begins after the major feeding season and when growth has slowed or even stopped.

It is because of this that, in many species, gametogenesis occurs not when seasonal temperatures are at their highest, but when they are falling. This pattern has been reported for a wide variety of organisms, including perch and the barnacle *Balanus balanoides*. In both cases ovarian growth starts in September or October. In perch yolk is laid down in November and the developing ova continue to grow until the spring spawning. In the barnacle ovarian development is restricted to the season when temperatures have fallen below 10°C (Barnes, 1963).

In many species, within a narrow permissible range, increasing temperature usually speeds gonad development. Above that range gonadal growth

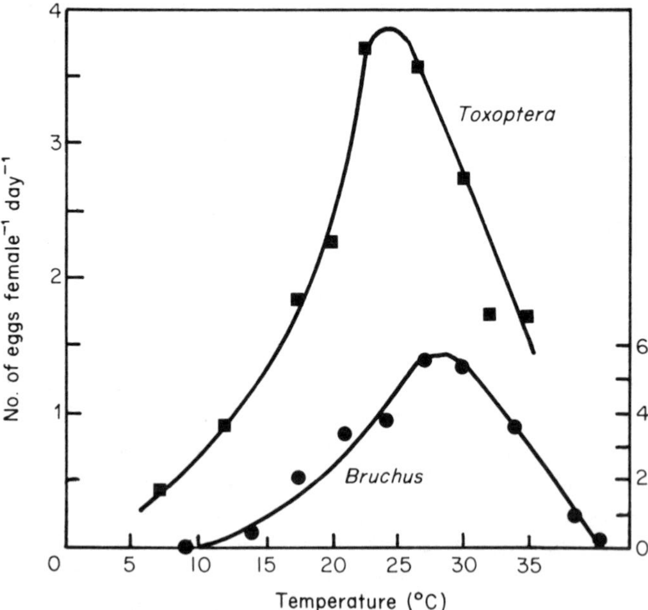

Figure 7.1 The effect of temperature on the numbers of eggs laid by two species of insect, *Toxoptera graminum* and *Bruchus objectus*. Note that both high and low temperatures inhibit egg laying and that 'optimum' temperature is that at which the largest daily oviposition occurs. (After Bursell, 1974.)

is progressively impaired. Thus, both high- and low-temperature 'limits' are reported, and reproduction is shown to have an 'optimum' temperature. This is shown for two insect species in Fig. 7.1. The rise in egg production with temperature is clearly a rate response to rising temperatures, and the fall in egg production toward the upper threshold occurs because developing eggs are resorbed. A similar optimal temperature response has been observed in a classical study of the desert-pupfish, *Cyprinodon nevadensis*, where the number of eggs per spawning and the percentage of eggs hatching have a maximum value at 30°C, whilst adults survive up to 42°C (Shrode and Gerking, 1977). Temperature also dramatically affects the quality of the eggs produced, since an inverse relationship was described for yolk content and temperature. Above 32°C a high proportion of soft-shelled yolkless eggs were produced. Thus, the temperature range for oogenesis in *C. nevadensis* is much more restrictive than for hatching of healthy eggs (Fig. 7.2).

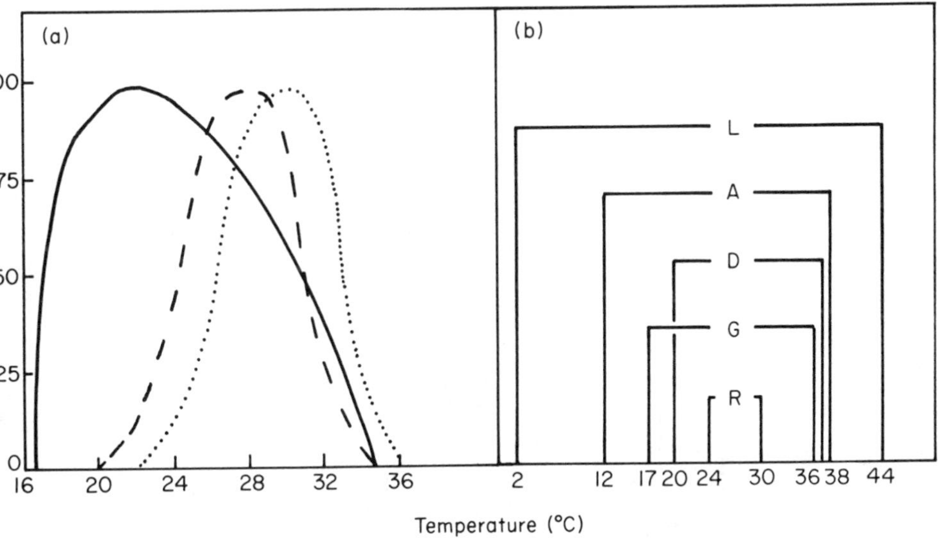

Figure 7.2 Thermal limits for reproductive, development, activity and critical thermal limits in *Cyprinodon n. nevadensis*. (a) Limits for growth (percentage change in dry weight week^{-1}) on a ration of 4.5% of body weight day^{-1} (——); and egg viability (— —) and egg production (.....). Note limits for growth are wider than those for reproduction, and also that growth is 'optimal' at 20–24°C when reproduction is severely inhibited. (b) Diagrammatical representation of thermal limits for reproduction and other functions. Note that reproduction has the narrowest limits of about 6°C, whereas adult tolerance has the widest, some 42°C. L = lethal limits; A = activity limits; D = development limits; G = growth limits; R = reproduction limits. (After Shrode and Gerking, 1977; Gerking and Lee. 1983.)

Few organisms, however, experience constant diurnal temperature regimes, so information about the effect of regularly fluctuating temperatures on gonad growth will be important. Shrode and Gerking (1977) found a fluctuating temperature regime in the laboratory did not accelerate egg production in the desert-pupfish compared with rates at the calculated median temperature (see Fig. 7.3). On the other hand, Hoffman (1974, 1986) found that temperatures fluctuating on a day–night cycle stimulated egg production in the cricket by up to three-fold. Determinations of the activity of the neuroendocrine system suggest that this effect of fluctuating temperatures acts via the endocrine system, rather than as a direct effect on

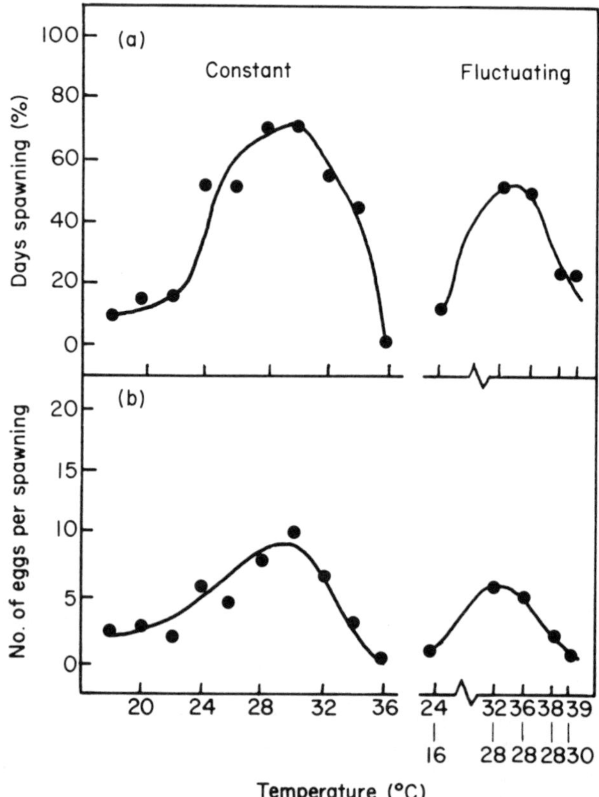

Figure 7.3 The effect of constant and fluctuating temperatures on reproduction in *Cyprinodon n. nevadensis*. (a) The proportion of days on which spawning occurred and (b) the numbers of eggs per spawning in fish acclimated and reared at the test temperatures 18–39°C, or under a daily fluctuating temperature regime. Spawning, and the number of eggs laid, were optimal from 24°C to 32°C, and at fluctuating temperatures of 32–28°C and 36–28°C. Thermal conditions outside those sharply limited reproduction (After Schrode and Gerking, 1977).

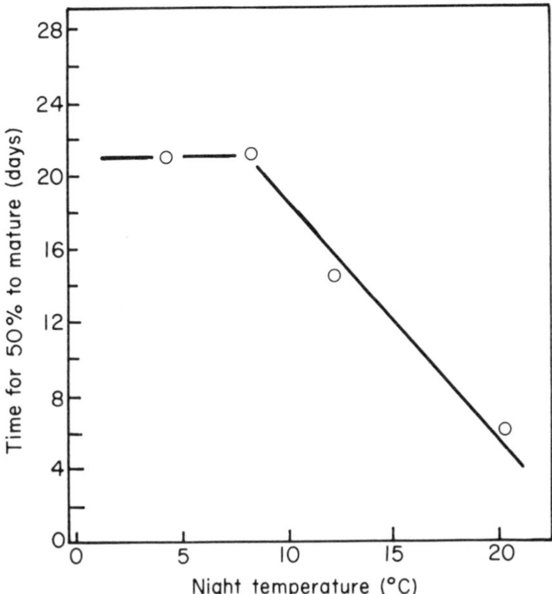

Figure 7.4 Time taken for egg maturation in different daily regimes of alternating temperatures. Day temperature was 20°C (8 h) in all cases but night temperatures (16 h) were either 4, 8, or 12°C. Development time increased linearly with the decrease in night temperatures to a limit at about 8°C. This is lower than the 13.5°C limit found in constant conditions, and so development can occur at temperatures below that threshold if they alternate with higher temperatures (After Meats and Khoo, 1976).

the ovary. The importance of such studies is emphasized by the work on the fruit fly, *Dracus tryoni*. The low-temperature threshold for egg development was calculated to be 13.5°C. However, development took place at lower temperatures if low temperatures alternate diurnally with higher temperatures (Fig. 7.4). At a constant temperature of 20°C egg development took 190 h, but when 20°C (for 8 h) alternated with 8°C (for 16 h) development took 570 h. During this time the adults spent 190 h at 20°C, so no development apparently took place in the time the females were at 8°C (for 300 h). However, when 20°C alternated with a temperature higher than 8°C development continued at the lower 'night' temperatures even when below the 13.5°C threshold. It seems that the response of aquatic organisms to fluctuating regimes is closer to that predicted from the median temperature. This is presumably a reflection of their more thermostable environment where fluctuating temperatures are not experienced.

Temperature may also exert a longer-lasting effect. In the beetle *Tribolium*, for example, exposure of females to low temperatures for several days increased oviposition when females were returned to higher temperatures. The interpretation of the effect of fluctuating temperatures on egg production is made difficult by such responses, and a simple relationship with the mean temperature may not hold in natural conditions.

In species in which both sexual and asexual phases occur in the life-cycle the pattern of reproductive activity may be influenced by temperature. This is the case in two coelenterates *Rathkea octopuntata* and *Coryne tubulosa*, (Kinne, 1970). In the former species, medusae bud asexually when water temperatures are below 7°C, but at higher temperatures gametes are produced. In *Coryne* asexual production of new hydranths and stolons occurs at temperatures above 14°C; however, after transfer to 2–3°C medusae producing gametes are formed. In both cases this switch between sexual and asexual reproduction is determined by temperature. A similar temperature-dependent switch between asexual and sexual development has been described for *Daphnia* sp. (Fries, 1964) and aphids (Lees, 1959). In general, many such species reproduce sexually in warmer conditions and by parthenogenesis in cooler parts.

Thus, for many organisms the effect of temperature on reproduction is such that gonad growth and maturation takes place in the season most appropriate for the subsequent survival of the offspring. The seasonal resurgence of reproductive activity requires an environmental trigger of which temperature or changing seasonal temperature may be one, but in most instances it is the more predictable photoperiod clue. However, once initiated gonadal growth and gametogenesis continue within a particularly narrow range of temperatures, within which rising temperatures are usually stimulatory.

7.3 Development

The rate of development is usually described as the time taken to change from one stage to another at a particular temperature (duration of development). Alternatively, it can be expressed as the reciprocal of the time taken in the form of a rate process (Fig. 7.5). Development usually, but not necessarily, involves growth, and in all cases concerns a system undergoing change.

The effect of temperature on the rate of development is a simple one in principle: within the permissible temperature range an increase in temperature speeds development. However, at some higher temperature a deleterious effect on development will occur, so that a plot of the rate of development as a function of temperature has a sigmoid shape (Fig. 7.5). Still higher temperatures will progressively slow the rate to produce a graph with an 'apparent' optimum temperature for development. The designation of such maxima as optima is spurious and misleading, for as Bursell (1974) points out they have neither functional nor ecological significance to the organism. For example (Fig. 7.6), development in *Calandra onzae* is fastest at 30°C. However, at that temperature mortality is also high. Thus, the

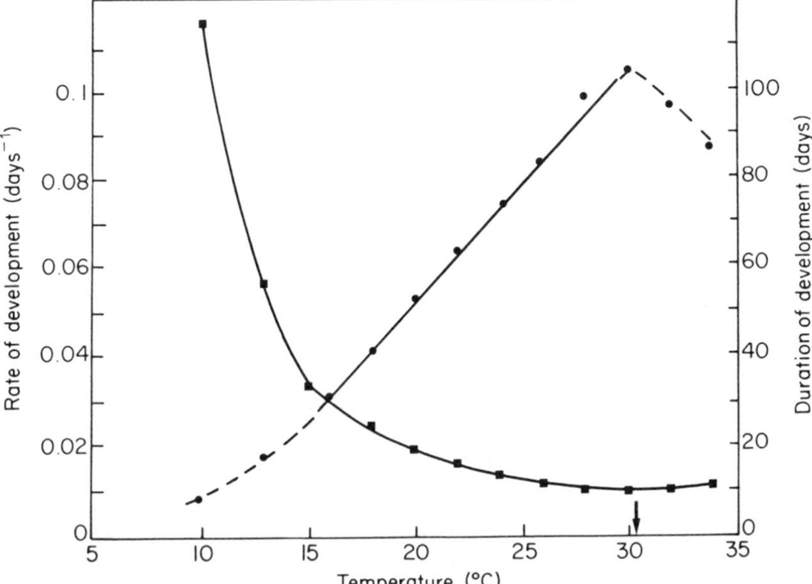

Figure 7.5 The effect of temperature on the rate of development (●) and on the duration of development (■) of the egg stage of *Fasciola hepatica*. (After Grainger, 1986.) The curve showing the duration of development at different constant temperatures is typical for many species, except that their location on the temperature axis will widely differ. When these data are re-plotted to show the rate of development a typical sigmoidal curve is obtained. In some cases the central (solid) part of the curve may show an approximately linear relationship with temperature.

range of temperature over which development can continue is wide, but the range over which rapid development is possible, without increasing mortality significantly, is much narrower.

Many studies on development have resulted in the presentation of temperature optima for a species. However, little consideration has been given to how such optima are reached. It is clear that a number of other biotic and abiotic factors, apart from temperature, can be limiting (e.g. nutrition, accumulation of wastes, crowding, hierarchy and photoperiod). Not only will the intervention of these factors shift the temperature optimum, but different factors may become limiting at different temperatures. For these reasons it is best to define the optimum conditions for development to be those at which mortality is lowest, rather than those at which the rate is fastest.

Before optimum conditions are specified it should be shown that they fit the ecological requirements of the organism. For some species it may be most advantageous to undergo slow development with low mortality,

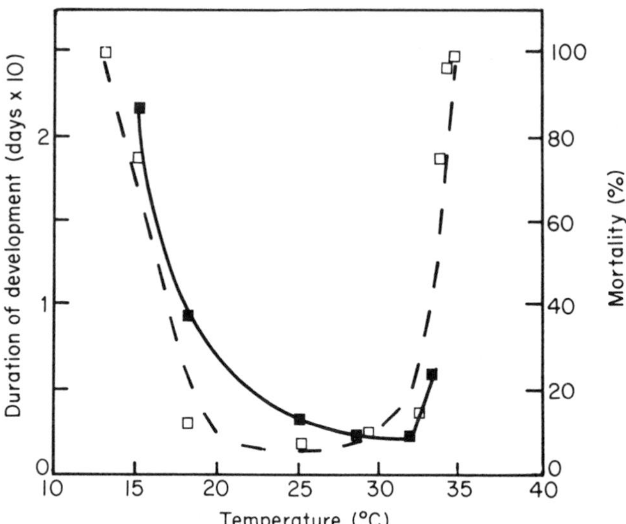

Figure 7.6 The effect of temperature on the duration (—■—) mortality (--□--) of the developmental stages of *Calandra onyzae*. Note that the development can continue over a wide range of temperatures, but that towards the extremes of this range mortality is very high. At the apparent 'optimum' temperature for development (about 30°C) mortality is already increasing. (After Bursell, 1974.)

whereas for others a rapid development with high mortality might be adaptively more advantageous.

7.3.1 *Mathematical formulation of the rate of development with temperature*

The accelerating effect of an increase in temperature on development is well known. The fact that it is possible to represent this relationship graphically has led to many attempts to formulate the relationship mathematically. It was hoped that this would allow a temperature characteristic or coefficient to be obtained, from which the speed of development at any temperature could be calculated. Furthermore, it was considered that once the relationship between temperature and the rate of development was known it may be possible to provide an explanation into the causes underlying that relationship.

This latter intention has not been realized. Development results from a sequential pattern of changes, each of which involves both physical and chemical processes. So complex is this phenomenon that it is surprising that development can be related to temperature in a simple manner at all. Consequently, the biological significance of the derived empirical relationship remains obscure.

One of the simplest treatments used is the thermal summation, or day-degrees rule:

$$yT = a \tag{7.1}$$

This predicts that the product of the developmental period (y) and temperature (T°C) was constant. A modification of this equation takes into account that at temperatures above 0°C the development becomes imperceptibly slow, so the terms *effective temperature* and *null-point* were introduced. The effective temperature becomes the temperature above that for the zero rate of development ($T - T_0$), where T_0 is the temperature for a zero rate of development or the null-point. Thus Equation 7.1 becomes

$$y(T - T_0) = a \tag{7.2}$$

In practice the null-point is difficult to determine, since in many instances it is approached asymptotically, as is shown in Fig. 7.7, and the temperature at which development stops completely is hard to determine. For this reason

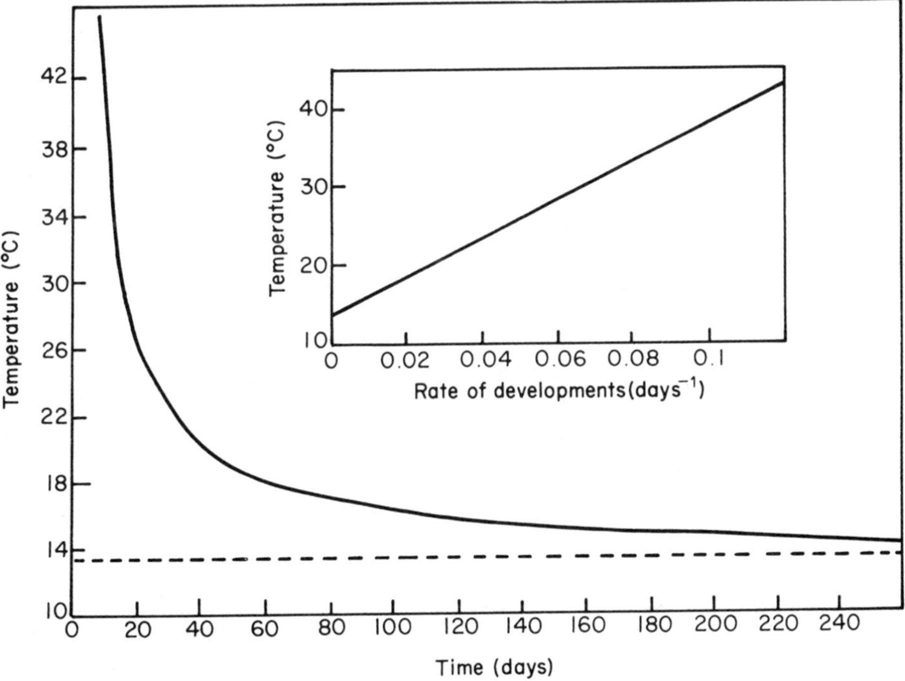

Figure 7.7 Duration of development in the insect *Ceratibis capitata* as a function of temperature. The hyperbolic relationship can be transformed into a straight line by plotting the reciprocal of developmental time against temperature (inset). From the extrapolation of this curve to the y axis, the null-point can be determined. (After Uvarov, 1931.)

some workers have chosen that temperature at which development is just measurable as the null-point. However, Equation 7.2 is the equation for a hyperbola, so the relationship can be transformed into a straight line by plotting the reciprocal, which allows a more acceptable means of estimating the null-point. An example is given in Fig. 7.7. The major problem in applying this thermal summation rule is the accurate determination of the null-point. A further difficulty arises because of the variability of the null-points due to interaction with other environmental factors, such as salinity, humidity, oxygen tension and photoperiod.

The application of this rule relies on the relationship between developmental rate and temperature being linear, which when determined

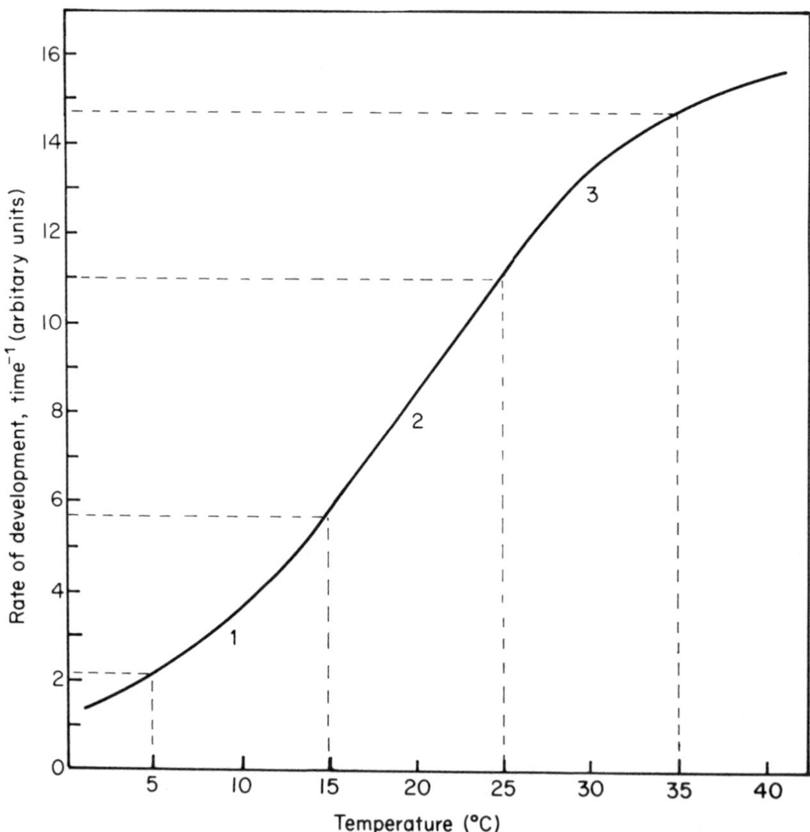

Figure 7.8 Generalized diagram showing the relationship between temperature and the rate of development. Only in phase 2 is the relationship approximately linear, where the 'thermal summation' rule could be applied. The broken lines indicate the decrease in Q_{10} for the process over the ranges 5–15°C (2.6), 15–25°C (1.93) and 25–35°C (1.34).

experimentally is rarely found to be the case. The relationship found is more usually sigmoidal or curvilinear, often with an inflexion at high temperatures, as is shown in Figs 7.5 and 7.8. However, it is often found that over a central range of temperatures the rate behaves as a nearly linear function of temperature. In these instances the rule has some utility. It is a poor predictor at low temperatures, when the estimated rate may be too low, and also at high temperatures – above the inflexion – when the estimate is too high.

Notwithstanding these criticisms, the summation rule has been successfully used by applied biologists. A classical study has been made by Touzeau (1966) on the emergence of the codlin moth, *Laspeyresia pomonella*. The null-point was 10°C and the difference between it and the daily maximum temperature was used as the 'effective temperature'. He was able to predict that the moth would emerge when the sum of effective temperatures reached 400.

The relative inadequacies of the summation rule to describe the effect of temperature on development has led to the introduction of more complex relationships being proposed, of which there are three principal ones. Bělehrádek (1957) suggested the following relationship between the rate of development (v) and temperature (T):

$$\log v = \log a - b \log T \tag{7.3}$$

where a and b are constants. This equation, in a modified form, fits the rate of development in amphibian eggs:

$$\log v = \log a + b \log (T - x) \tag{7.4}$$

where a, b and x are constants. The significance of the constants is not clear; McLaren (1965) found that his constant a was inversely proportional to egg diameter and that x was closely correlated with spawning temperature.

The second relationship is the so-called logistic curve of Davidson (1944),

$$v = \frac{z}{1 + e^{a - bT}} \tag{7.5}$$

where v is the rate of development, T is temperature and a, b and z are constants, but are different from those in Equation 7.3. If v is plotted against temperature a sigmoid curve results, with z as the upper asymptote. The equation transforms into a straight line

$$\log_e\left(\frac{z - v}{v}\right) = a - bT \tag{7.6}$$

As can be seen in Fig. 7.9, the data obtained for the development of *Drosophila* eggs fit Equation 7.6 better than Equation 7.4, but nevertheless a deviation still is seen at the upper temperatures. The values derived for the

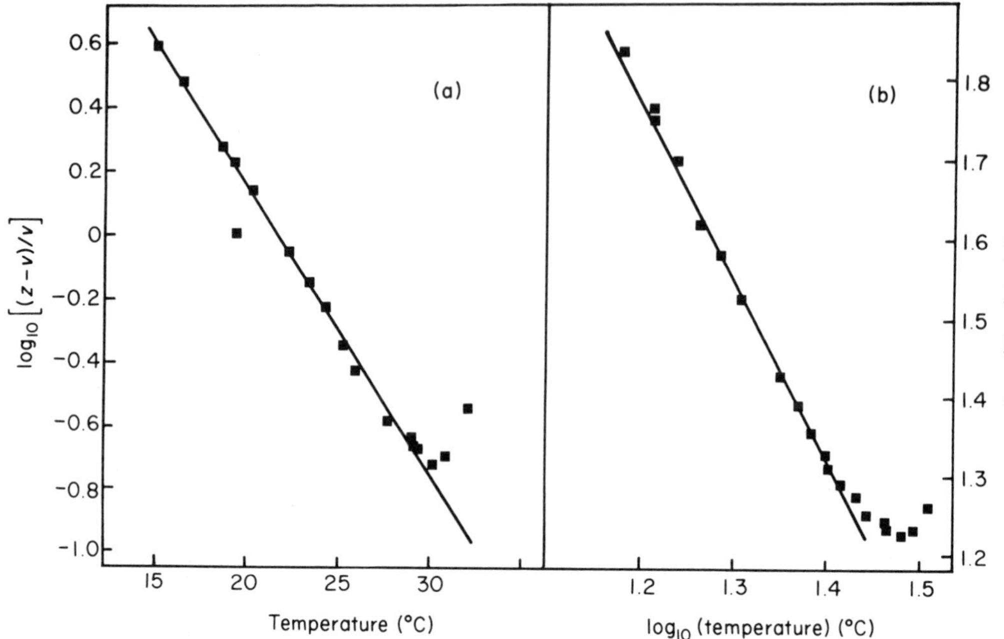

Figure 7.9 Two empirical methods of obtaining a linear relationship between temperature and development of *Drosophila* eggs. (a) Data treated according to the 'logistic' curve $\log_e[(z-v)/v] = a - bT$ (see text), and (b) the same data treated according to the logarithmic relationship of Bělehrádek; $\log v = \log(a - b) \log T$. (After Davidson, 1944.)

constants appear to have little biological meaning. For example, the value of z for *Drosophila* eggs was 42°C, whereas the 'optimum' temperature for their development determined experimentally is close to 30°C; 42°C would be lethal.

Finally, the catenary curve of Janisch (1927) has also been used, it has the form

$$y = (m/2)\,(a^T + a^{-T}) \qquad (7.7)$$

where y is the duration of development, m is the duration of development at the 'optimum' temperature, T is the difference between the environmental temperature and the optimum temperature, and a is a constant.

The catenary curve has also provided good fits for data on the effect of temperature on development, but the requirement of determining the 'optimum' temperature makes it very cumbersome to operate.

It is evident that the mathematical formulation of the relationship between temperature and development will be dependent upon the constancy of other environmental parameters such as salinity. These formulations should also consider such biological factors as thermal history, a point emphasized by Kuramoto (1976). Indeed, it is noteworthy that the precision adopted for the theoretical considerations is not always matched with exactness in experimental procedures. Whilst such treatments have often successfully described the developmental process, they have done little to reveal the underlying principles of the temperature effect.

7.3.2 *Fluctuating temperatures*

Much of the experimental work on the effect of temperature on development and growth has been restricted to studies at constant temperatures. In view of the fact that constant temperatures are rarely experienced in nature, such work is of questionable ecological significance.

Relatively few studies exist that have dealt with the effect of variable temperatures on development. In principle a change in temperature may have three effects. First, the rate of development may alter during the period of, and as a direct effect of, the temperature change (an immediate effect). Secondly, exposure to a particular temperature regime may change the development rate at a subsequent temperature (an acclimation effect) and, thirdly, the fact that a temperature change occurs may cause a stress effect (Grainger, 1986). A good example is provided by regeneration in *Ambylostoma* embryos, which was more rapid at 18°C if the embryos had previously been kept at 5°C rather than at 14°C (Piatt, 1971). The 'immediate' effect can be demonstrated by comparing the rate of development under conditions that cycle regularly between two temperatures with that at some computed 'mean' temperature. It is precisely this latter calculation that has brought problems of interpretation of experimental results, because this again assumes a linear relationship between developmental rate and temperature, which is not usually the case. Suppose that development, to a certain point, takes 8.5 days at 10°C, 4.5 days at 15°C and 2 days at 20°C, then if an egg is placed at 10°C for 1 day, 20°C for the next day, followed by the next day at 10°C, and so on, after 3 days it will be seen that some three-quarters of the development is over and less than one further day is necessary to complete development. This will be an apparent acceleration over the 4.5 days taken at the mean temperature of 15°C. This apparent acceleration occurs because the increase in the rate of development which results from the time spent at the higher temperature (15°C↔20°C) is not cancelled out by the retardation during equivalent period of time spent at the lower temperature (15°C↔10°C). This example can be used to highlight another potential source of error. If the egg had been first placed for 1 day at 20°C, followed by 1 day at 10°C, 62% of the total development would then have been completed and less than a further day at 20°C would be required to complete development. Thus, the measured rate of development depends on the actual sequence of temperatures chosen (10°C→20°C→10°C or 20°C→10°C→20°C). Keen and Parker (1979) and Grainger (1986) have discussed the importance of considering both the lack of linearity in the temperature–rate curve, and the influence of the point in the cycle at which the egg begins development, in studies using fluctuating temperature regimes.

Different reports exist that show that developmental rate is retarded, accelerated or unaffected by a fluctuating temperature regime. It is likely that these conflicting reports over the existence of an 'immediate' effect are due to incorrect computation of the mean temperature. Reference to Fig. 7.6 will again make this clear. If the temperature regime chosen for study was 20°C↔30°C, then compared with the rate at 25°C an apparent retardation would be seen. This occurs because the time spent at 30°C will cause only a small increase in development (over that at 25°C) in contrast with the marked reduction in rate when at 20°C as compared with that at 25°C. The reverse will be true if the temperature varies about a low mean towards the null-point. In the central part of the range it is likely that the relationship between temperature and development is more nearly linear, so development will appear to be unaffected by the fluctuating temperature compared with that at the mean. Such difficulties in the interpretation of these experiments was pointed out by Kaufmann (1932) and the apparent acceleration effect of fluctuating temperatures is called the *Kaufmann effect.*

In consequence a better estimate of the development at reference temperature(s) is to sum the fraction of development that would have been completed in the time spent from the known rates at a series of constant temperatures. This approach also allows studies to be carried out in which temperature fluctuates in an irregular manner. Grainger (1959) has used this approach to study the development of frog's eggs. He found that gastrulation was the most stenothermal stage, and was also most sensitive to temperature change, no Kaufmann effect was evident.

Relatively few studies exist in which laboratory work on the rate of development is used to model the rate of development under field conditions. Grainger and co-workers (see Grainger, 1986) have attempted to do this for the liver fluke *Fasciola hepatica* in Ireland. They produced mathematical models for each developmental stage, the egg as well as intrasnail stages, as a function of temperature, which allowed confident estimates to be made for the duration of development at any temperature in field conditions. However, in the field temperatures vary, so an estimate of the 'effective' temperature must be made. As the study spanned an annual cycle it was impracticable to work at smaller than daily intervals. For most of the year the mean daily temperature was used, given by the relationship $(T_{max} + T_{min})/2$. However, development is negligible below 10°C, so when T_{min} falls below 10°C the estimated 'effective' temperature becomes $(T_{max} + 10°C)/2$ for that proportion of the day (a) when temperature exceeds the null-point. An estimate for the proportion of the daily period (a) for which the temperature exceeds the null-point is thus obtained:

$$a = \frac{T_{max}\,^{\circ}\mathrm{C} - 10^{\circ}\mathrm{C}}{T_{max}\,^{\circ}\mathrm{C} - T_{min}\,^{\circ}\mathrm{C}}$$

When both the maximum and minimum temperatures fall below 10°C, then (a) is defined as zero, and clearly development does not occur.

From this it was possible to compute the daily fraction of development at each daily effective temperature. Thus, if a stage requires D_i days for complete development at T_i, then the proportion of development completed in 1 day at T_i is $1/D_i$. In this way the daily increments of development can be summed, and when the sum adds up to unity development will be complete.

The aim of the study was to provide, in the field, a means of predicting the prevalence of when the parasite was at its infective stage. In consequence other factors, such as moisture, which may influence the rate of development, and also survival of both the parasite and the molluscan hosts, had to be taken into account. This added considerably to the complexity of the algorithm. Figure 7.10 shows the observed, compared with a predicted, fluke uptake into sheep in Ireland in 1973. The expected pick-up was computed from the laboratory-derived models for the rates of development and survival under the estimated field conditions.

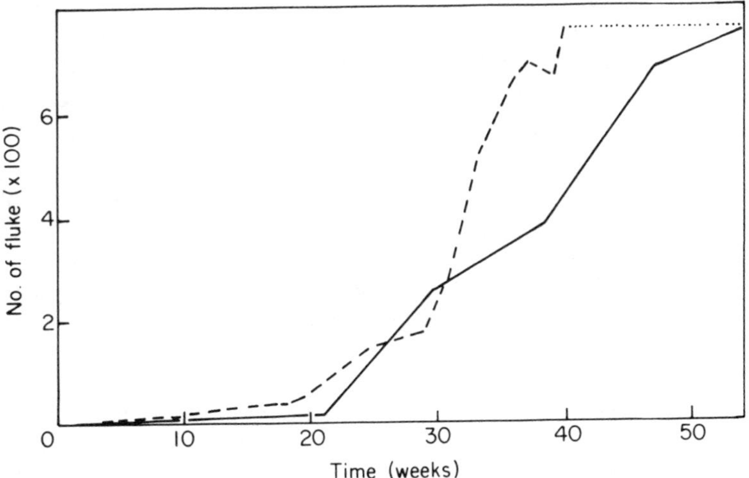

Figure 7.10 A comparison of predicted (——) and observed (---) take up of liver fluke into tracer sheep in Ireland in 1973. The predicted accumulation was computed from laboratory-derived models of the effect of temperature on fluke development and survival under the estimated field conditions of temperature and ground water levels. (After Hope-Cawdrey *et al.*, 1978.)

The model clearly provides an adequate prediction of the developmental rates of the parasite in the field. This provides an aid to the application of efficient measures to control transmission, as animals can be moved from infested pastures at times of maximum risk. It seems the most difficult aspect of the modelling found was the definition of the appropriate microclimatic conditions of temperature and moisture experienced by the parasite and molluscan host (Hope-Cawdrey *et al.*, 1978). Grainger (1986) has also recently provided a valuable guide to the methods used to measure and record field-temperature data, as well as the methods for reproducing those field temperatures in the laboratory. These studies stress the role that mathematical modelling has in dealing with complex ecological problems, and has also focused attention on the gaps in our understanding of the biological problems faced.

7.4 Thermal limits of development and temperature-sensitive stages

A large number of studies show that the temperature range over which development is possible for a particular species is relatively limited, but it is wider than that for gamete production (see Fig. 7.2, for example).

To determine the mechanism by which temperature operates as a threshold factor limiting development, both direct and indirect effects must be considered. First, the temperature may be too low to allow development to start or continue, or so high that damage occurs. These are direct effects. Alternatively, temperature may act indirectly by influencing the available food resources necessary to support development to completion.

There are clear indications that some developmental stages are more temperature-sensitive than others. Gastrulation in *Rana* eggs has been shown to be more sensitive than other stages between fertilization and the internal gill stage of the tadpole (Grainger, 1959). At temperatures below 7°C and above 25°C many eggs either failed to complete gastrulation or did so abnormally. For most other stages development took place most rapidly at 25°C. Kinne and Kinne (1962) report similar observations for the embryonic stages of the fish *Cyprinodon macularis*. In this case the stages between fertilization and the end of gastrulation were the most sensitive. In *Mytilus edulis*, however, cleavage of the egg was very temperature-sensitive, since it did not occur below 5°C and was abnormal above 20°C.

The conclusions that can be drawn from these various studies are that thermal conditions under which gametogenesis, spawning and early embryonic development will successfully occur are relatively narrow. During the time that these processes are taking place, therefore, organisms tend to be more stenothermal, and consequently strict control operates to ensure that reproduction takes place in the correct season when suitable thermal, and other, conditions exist for development. It is obvious that the placing of such restrictions will be one factor operating to limit the distribution of a species.

7.4.1 *Developmental null-points and their ecological significance*

The operation of null-points in development (see page 257) can be thought of as a traffic-light effect, preventing development until the appropriate environmental conditions prevail. They explain, for example, the synchronous emergence of blackfly, even though the eggs are laid over a 5-month period. Pupation has a higher null-point than larval development.

In some cases the different instars have been shown to have different null-points that reflect the changing seasonal temperatures, Zwölfer (1933) in a now classical study on the moth *Lymantria monacha*, found that the five instars and the pupal stage show a series of ascending null-points, 3.2°C, 5.7°C, 7.2°C, 7.6°C, 7.8°C and 8.4°C, respectively, that are consistent with field temperatures.

The most definitive work in this field has been carried out by Corbet (1957) on British dragonflies. *Anax imperator* emerges in the spring with a closely synchronized emergence (Fig. 7.11). This occurs because in *Anax* the final instar is a resting one and metamorphosis is inhibited until the spring. Adults emerge in the period May–August, egg-laying takes place in June–September and hatching and larval growth is rapid until October, when the lower temperatures again become inhibitory. Growth resumes in April in the following year, so by midsummer larvae of a wide size-range exist in the population, depending upon the time for growth in the previous summer and individual growth rates. Entry into the final instar takes place between June and August, and further development in these summer temperatures is very slow, it is only when temperatures fall that diapause development takes place. Thus, when spring comes in the following year all the overwintering larvae are capable of responding, which results in a synchronized emergence. A few larvae, however, that enter the final instar early in the summer avoid the diapause stage and, by metamorphosing, they emerge during the same summer. Thus, whilst most of the *Anax* emerging are 2-year-old individuals, emergence in the rapidly developing individuals occurs after 1 year.

7.4.2 *Resting stages and diapause*

An important adaptation for species that experience adverse seasonal conditions is the inclusion in the life-cycle of a resistant stage. There are many well-documented accounts of species that survive low environmental temperatures by the adoption of a resistant stage, but fewer that do so at

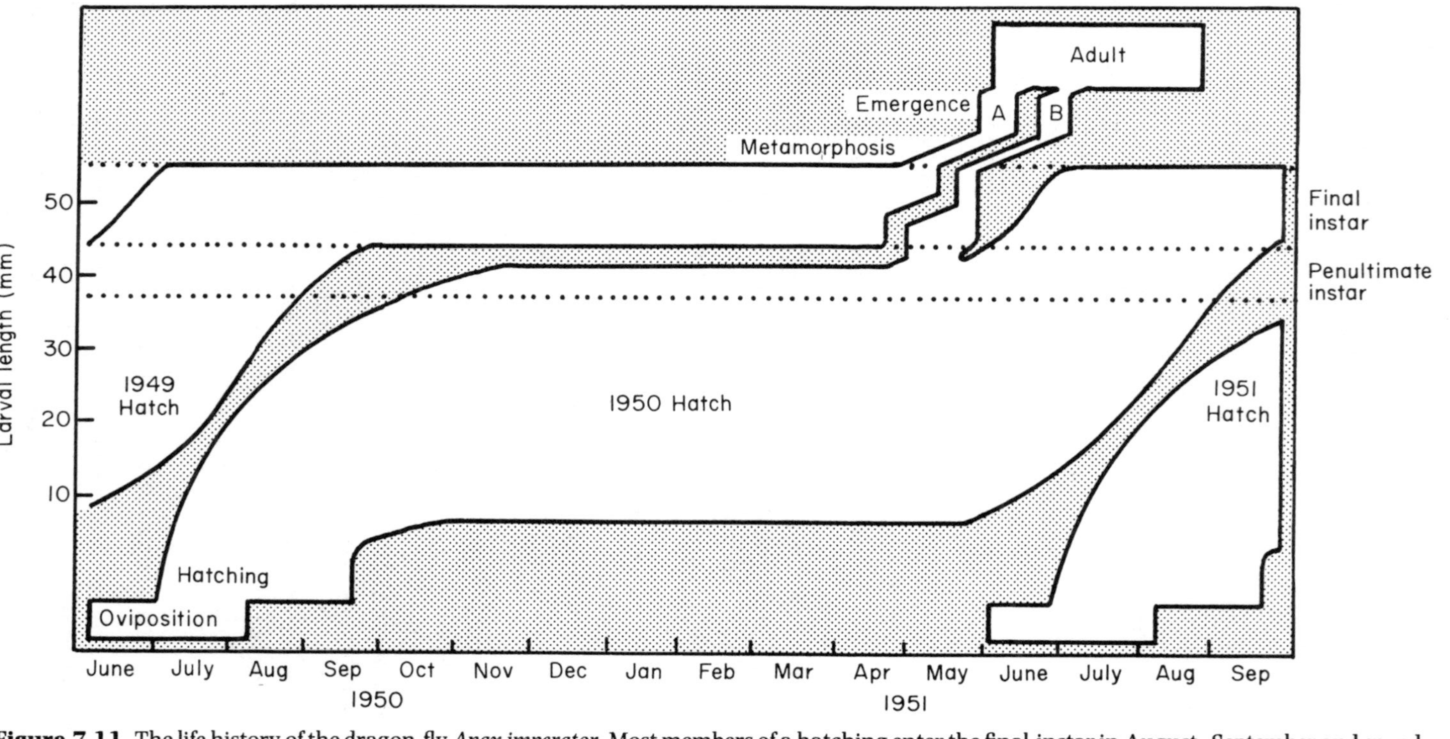

Figure 7.11 The life history of the dragon fly *Anax imperator*. Most members of a hatching enter the final-instar in August–September and spend the winter in diapause, emerging as adults during the following May (Group A). Some precocius larvae, which enter the first larval stage before June, complete development without diapausing and emerge as adults in the same year (Group B). (After Corbet 1957.)

high seasonal temperatures. The tropical insect *Polypedilium vanderplancii*, which inhabits temporary pools in West Africa, does so. During the dry season these pools dry up and the organisms resist the combined effects of desiccation and high temperatures in the larval stage. Larvae in the resistant form are able to withstand exposure to temperatures of 60°C for several hours, and do so, it seems, by a reduction in tissue water and a corresponding reduction in metabolism (Hinton, 1960).

It is important to distinguish between simple quiescence and true diapause. Many examples are cases of quiescence, since the resting stage can return to normal levels of activity as soon as conditions become favourable. True diapause, by contrast, is a pause in, or inhibition of, development. It is characterized by the development of sophisticated physiological mechanisms to control its initiation and termination. The utility of having such carefully controlled resting stages in development has been outlined in an earlier section (see page 265) and ensures that the adult emerges in the season that is most favourable.

Diapause is commonest and best documented in the insects, and is a state induced in most instances by external factors such as temperature and photoperiod. In a few cases entry into diapause has been shown to be obligatory and, being independent of external factors, is endogenously controlled (Braune, 1971). In other cases diapause is facultative, and it occurs when the sensitive stage in the life-cycle is exposed to the set of factors that are inductive. Once the inductive process has occurred it is usual for development to continue so that the diapausing stage of development is reached. Here development stops and diapause has been entered. In some cases the length of the diapausing stage is set by endogenous factors, since the animal only becomes sensitive to the sets of environmental clues that will eventually bring it out of diapause after a specific period. In other cases the diapausing stage must experience a particular temperature regime for a certain time before diapause can be ended (see Behrendt, 1963).

Photoperiod is the most usual trigger for entry into diapause, since it provides the most reliable clue to the changing seasons in temperate regions. Temperature can modify the photoperiod response, as, for example, in the cabbage white butterfly, *Pieris*, where at temperatures above 30°C photoperiodic induction of diapause fails (Danilevskii, 1965).

In a detailed study on *Sarcophaga angyrotoma*, Saunders (1971) provided an explanation for the interactive effects of temperature and photoperiod (Fig. 7.12). In larval cultures maintained at different temperatures in the range 16–26°C and at short day-length (L:D 10:14) diapause induction depended on the length of the larval period and the number of light–dark cycles experienced. The duration of larval development was very temperature dependent, being only 9 days at 20°C but rising to 22–23 days at 16°C. However, the number of light–dark cycles needed to induce diapause in 5%

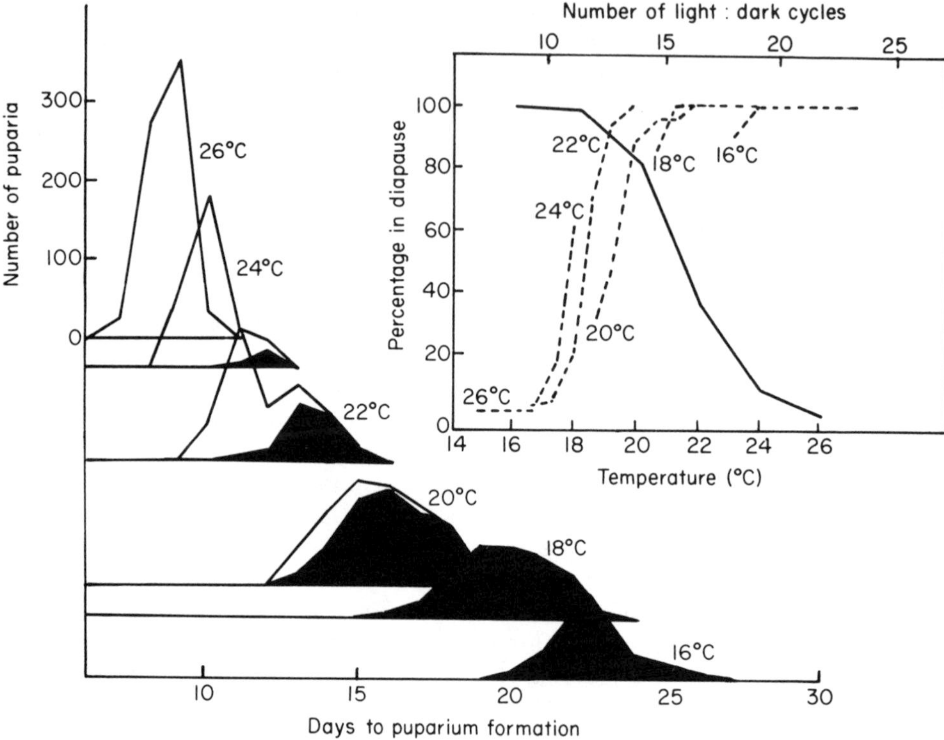

Figure 7.12 Effect of temperature on induction of pupal diapause in *Sarcophaga argyrostoma*, at a light:dark ratio of 10:14 h. In the main graph solid lines show the numbers of larvae forming puparia and the solid areas show the number of those pupae entering diapause. Inset shows the effect of temperature (---) on the proportion of diapause pupae. Also shown is the effect of temperature on the summation of light:dark cycles (L:D, 10:14 h) required to raise the proportion of diapause pupae to 50% (——). At high temperatures (24–26°C) the larvae form puparia before the required day number has been achieved and so development is not interrupted by diapause. At lower temperatures (16–18°C) the necessary day number has been achieved before puparium formation cuts short the sensitive period, so all enter diapause. (After Saunders, 1971.)

of the cases was much less temperature dependent ($Q_{10} = 1.4$). Consequently larvae held at high temperatures formed puparia before they had experienced sufficient light–dark cycles, so the incidence of diapause was low.

Such interactive effects of temperature and photoperiod on development permit a flexibility of response that has ecological significance. Some multivoltine species may be allowed to complete another generation should seasonal temperatures be warm. Furthermore, it may account for the ability of some Arctic and sub-Arctic species to adjust the length of their life-cycle from 1 to 2 or more years, if in one year summer temperatures are not high enough to permit complete development.

For diapause to be effective it must last long enough to ensure that development, when it resumes, does so at the appropriate seasonal time. In many instances this is achieved by the induction of an endogenously controlled, temperature-insensitive stage of a specific duration. In *Aphis* eggs this lasts for about 50 days, whereas in the plant louse *Leptoptera dolobrata* this period lasts 180 days and is followed by a temperature-dependent period. Diapause is often broken by the exposure of the diapausing stadium to a specific temperature(s) for a particular time. In *Acheta commodus* Hogan (1960) reports that the lower the temperature is, the less time is required at that temperature to break diapause; 20–60 days at 13°C and 5 h at −7°C.

Temperature appears to be the dominant factor in the termination of diapause, and many different patterns of response have been recorded. The specific requirements of termination suggest that something other than that of a short, sharp temperature shock is required. In *Bombyx*, eggs that had been kept warm for 30 days required only 69–70 days chilling at 5°C to obtain a 90% hatch when placed back at the higher temperature. If the chilling temperature was lowered to 2.5°C some 80 days were required, and if the chilling temperature was higher, at 12.5°C, more than 100 days were necessary. If low temperatures continued after diapause had ended, then development was inhibited or, at best, continued very slowly. In general, however, it is found that if post-diapause development is prevented, then the viability of that stage is reduced. The range of diapausing temperatures that must be experienced to terminate diapause is very variable between species and is clearly closely related to the climatic factors likely to be experienced in nature. In Arctic and high-Alpine species these temperatures are usually sub-zero, whereas for temperate forms it is usually above zero.

The severe and persistent cold of the Antarctic influences the reproductive and developmental strategies available to both terrestrial and marine invertebrates. In the more variable terrestrial environment where the invertebrate fauna is sparse the mite *Alaskozetes* survives the extreme cold of winter in mixed populations of juvenile and adult forms. Growth and reproduction can only occur in the brief summer and, consequently, life-cycles can be prolonged for 2–3 years. Bonner (1979) draws comparison with the polar sea environment, where temperatures are low but relatively stable, and freezing is less of a threat. The Antarctic marine invertebrates are more seriously affected by the extreme seasonal variation in food availability, where algal blooms provide abundant food for brief periods, to be followed by scarcity. This is probably the reason for the absence of planktonic larval stages of benthic invertebrates in polar seas. Many molluscs, for example, produce large yolky eggs with long developmental times, from which emerge tiny snails. This developmental strategy, together with brood protection and vivipary, are common features where low temperatures and slow growth and development would make it impracticable to produce planktonic larval forms that would have only a short feeding period available. All polar marine invertebrate species show adaptive trends towards lower fecundity, producing well-developed and well-protected juveniles instead of an abundance of planktonic larvae (Clarke, 1979).

7.4.3 *Abnormal development and thermal stress*

It is commonly reported that towards the thermal limits for development the percentage of abnormal embryos increases. This teratogenic effect of temperature has already been referred to in the work on frogs' eggs (page 264). It is not surprising that temperature might have such an effect, since development is a complex pattern of events. If development is to be normal the sequence by which the steps of any one chain of events occurs must be precise, but also the relationships between adjacent chains must be synchronized. That a chain may branch so that alternative routes to development are possible adds further complexity. If branch points, or even different steps in two interdependent pathways, have different temperature dependencies, then it is possible to see how temperature might cause a switch, channelling development along an abnormal path, or might disturb the correct timing of adjacent interdependent pathways, throwing them out of synchrony. Either event might produce an abnormal pattern of development.

Low temperatures have also been shown to produce abnormal development in many species. For example, hatching in trout embryos at low temperatures is precocious, and in the slug *Limax* the membrane separating the pulmonary and visceral cavities does not develop, this produces a 'herniation' (Segal, 1963).

Such effects are not restricted to low temperatures. Tsukuda (1960) found that guppies reared at temperatures higher than 30°C developed with crooked vertebral columns. The higher the temperature was, then the greater was the percentage of abnormally developing animals and the earlier the abnormality was manifest. Other work on several species of fish has shown that the exposure, during development, to dramatic changes in temperature produces a number of structural defects. At the extreme twinned 'monsters' were produced, but more commonly abnormalities of the CNS, cranium, branchial arches, fins and muscles were found. The type of deformity that occurred seemed to be related to the stage during development when the thermal shock happened. The thermal stresses applied in this work were within the temperature range experienced by the animals in nature, and so it is possible that such structural abnormalities occur in nature; that they are rarely seen, however, is likely to be because of the strong negative value of carrying a deformity (Garside, 1970).

An interesting aspect of this effect of temperature on the developmental pattern is that in some species the sex ratio is modified by temperature. In some cases this is a direct effect acting through development, but in others it is an indirect effect. This is well known in the ant *Formica rufa*. Gösswald and Bier (1955) found that low temperatures (below 19°C) inactivate the female spermotheca, so the spermatozoa stored there are not available to fertilize the eggs produced by the female. Consequently the unfertilized eggs develop into males, with no females being produced.

More usually, however, the effect of temperature acts directly, and temperature-sensitive sex determination has been described in some turtles and lizards (Bull, 1980). For example in *Chelydra serpentina* females are produced at high and low incubation temperatures, and males are produced at intermediate temperatures. The fish *Rivulus marmotatus* (Harrington, 1967) shows a similar developmental effect of rearing temperature. Here the ovarian part of the ovotestis is destroyed by exposure to low temperatures and only males develop. Conover (1984) reports that low temperature during a specific period of larval development in the fish *Menidia menidia* produces females, whilst high temperatures produce males. He considers this is adaptive for it provides females with a longer season to grow. Reproductive fitness is seen to be related to body size in females, but not in males. Temperature merely acts as the clue to whether the offspring will experience a long or a short growth season, and so provides the mechanism for adaptive sexual dimorphism.

This type of effect of temperature is even more widespread in the insects. Anderson and Horsfall (1965) have described fully the interaction of temperature and the developing mosquito *Aedes stimulans*. Under natural conditions it develops through five instars in temperatures ranging between 5°C and 20°C. Under these conditions half of the population is limited to being female, for half of the individuals are homozygous for female characters. The other half have the potential of developing into males or females for they are heterozygous for sex. The heterozygous component expresses itself entirely as male when reared below 23°C and wholly females when reared above 28°C. At temperatures intermediate between 23°C and 28°C various intersexes are produced, but the higher the temperature is, the more severely are the male characters suppressed. This was shown to be a direct effect on the gonadal anlage, as when the gonad disc was transplanted from *A. stimulans* into *A. vexans*, a species in which the testes develop normally even at high temperatures, the gonad developed into an ovary if the individual was placed at 28°C after transplantation.

These teratogenic effects of temperature suggest that, in some cases at least, the mechanism by which the abnormal pattern is produced is through an effect on gene expression.

7.5 Temperature and gene expression

Much of the available literature in this field comes from work on *Drosophila*. However, as temperature exerts an influence on processes that are common to all organisms, then it is very likely that the results described for *Drosophila* will apply similarly to other ectotherms. First, we can consider the effect of temperature on gene penetrance or expressivity. Under certain environmental conditions some individuals which are homozygous for a particular trait may fail to express this as a character in the phenotype, such genes are said to show incomplete penetrance. They contrast with those genes that show complete penetrance, when all individuals which are homozygous for a trait will show that distinctive trait to the same extent in the phenotype. In genes that show variable penetrance the expression of the gene depends upon the interaction of genetic and environmental factors, of which temperature is important in ectotherms.

Many cases of variable penetrance, with temperature as the permissive environmental factor, are known. For example, in *D. melanogaster* the recessive gene, *cubitus interuptus*, which causes a break in a wing vein, is such a gene. All homozygous flies reared at 19°C are affected, but when reared at 25°C the gene is expressed in only about half of the individuals. In the heterozygous condition the gene is not expressed at all at 25°C, but when reared at 13°C about 10% of flies are affected (Milkman, 1962).

Temperature shocks during development are known to cause morphological alterations to the phenotype, similar to some mutant phenotypes. They are called phenocopies. For example, the mutant *straw* of *Drosophila*, and its heat-induced phenocopy, were both deficient in the same enzyme, one that is involved in melanin synthesis. It is the absence of melanin that gives the mutant and its phenocopy the straw colouration, hence the name. This evidence suggests that the similarity in those phenotypes results from the same deficiency in gene action. This is supported from work by Tissieres *et al.* (1974) which showed that heat shock causes a cessation of the normal protein synthesis. This is followed by the synthesis of specific new proteins which is correlated with the changes in the 'puffing' pattern of the chromosomes. The development of puffs on chromosomes is known to be associated with gene transcription, and these workers concluded that heat shock extensively alters gene transcription and consequently gene expression.

It is probably for these reasons that the period at which the thermal shock is applied determines the type of phenocopy caused. Recent work shows that the type of phenocopy produced depends upon the time after puparium formation that the heat shock was applied. The times during which specific bristle phenocopies can be produced by heat shocks are relatively short, being in most cases less than 2 h long (see Fig. 7.13). Mitchell and Lipps (1978) conclude that the phenocopies originate from an inability to regain

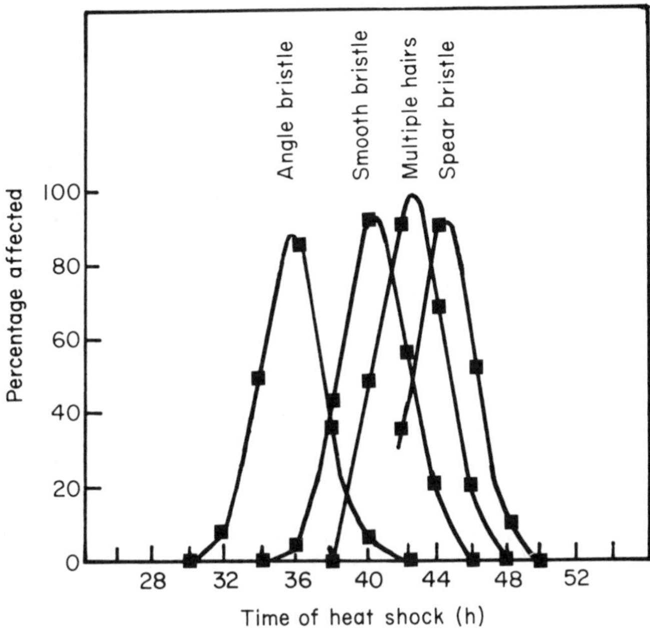

Figure 7.13 Phenocopy frequency as a percentage of emerged flies in *Drosophila melanogaster*, following heat-shock at 40.2°C for 1 h. Note that the temperature-sensitive periods are quite short (about 5 h), but more than 90% of the flies emerging are affected. The application of the shock causes different phenocopies, several more than the four examples given, depending upon the timing of the shock. (After Mitchell and Lipps, 1978.)

gene activity, following the transcriptional block caused by the heat shock, in time to participate appropriately in the continuing pattern of development. This is then equivalent to the lack of that gene activity in a mutant. Consequently, the time of the heat shock must coincide with the time of activity of the normal allele of the mutant gene.

The response of cells to heat shock is to stop existing polypeptide synthesis and greatly increase the transcription of a small family of 'heat-shock' genes, which results in the production of 'heat-shock proteins' (HSP) (see page 220). This phenomenon is of interest because it indicates a line of communication between the environment and the expression of the genome. The question of how the cell senses the heat shock is of particular interest. The fact that this response can be elicited by injurious agents, other than heat, suggests that it is not the genetic apparatus itself that senses the stressor action. Tanguay (1983) has reviewed this field and has proposed a model for the regulation of genetic information transfer during heat shock, shown in Fig. 7.14. The central hypothesis is that the action of the stressors is on a sensitive cytoplasmic factor (x) which releases a protein or proteins

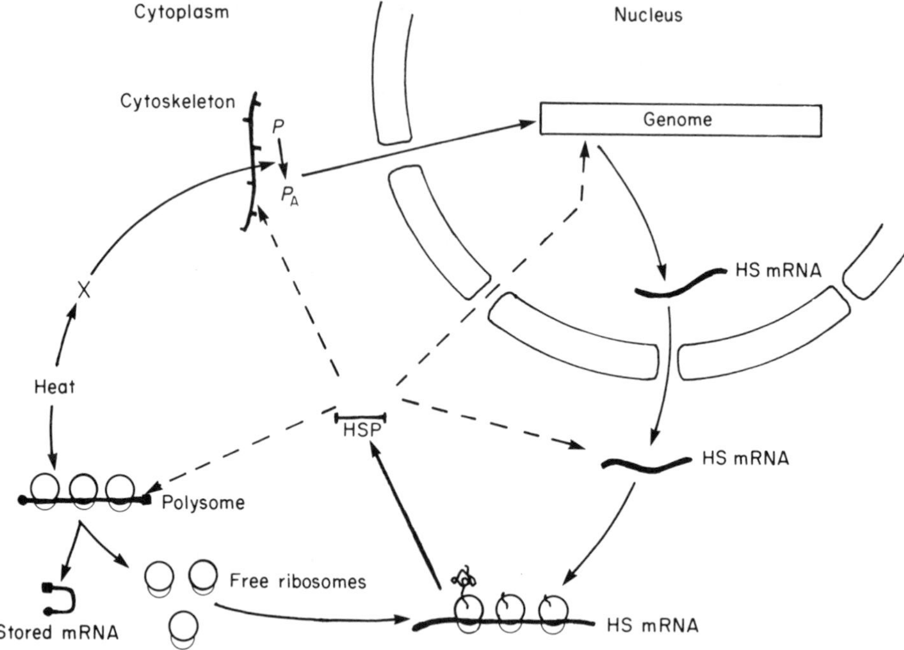

Figure 7.14 A model of genetic information transfer and regulation during heat shock. The model predicts that the shock acts on, or initiates, a cytoplasmic factor *X* which in turn causes the release of protein(s) (P–P_A). This active protein(s) P_A either directly or indirectly causes a transcriptional response at the genome, initiating the production of HSmRNAs. Existing polysomal activity is disrupted by the heat shock, and bound mRNA may be degraded or stored, freeing ribosomes. The HSmRNA is preferentially translated and the heat shock proteins formed are widely distributed to 'sensitive' cytoplasmic and nuclear sites (— →), affording protection from the denaturing effects of the heat shock. (After Tanguay, 1983.)

(P_A) and this is either directly or indirectly involved in the regulation of the transcription of HSP genes. It is clear that this factor exists before the stressor action for the response is very rapid, and requires neither new protein nor RNA synthesis. The mechanism by which normal transcriptional activity is suppressed is not understood, but existing mRNA molecules are either stored in an inactive form or degraded, to allow HSP mRNAs to be translated.

A recent suggestion, reviewed by Monroe and Pelham (1985), is that the stressors cause the release or accumulation of defective proteins and that, in some way, the cell's ubiquitin-dependent degradation system for dealing with such proteins provides the signal. It may be that the system becomes swamped with defective proteins, and this results in a shortage of free ubiquitin. This would require the HS transcription factor to be inactive in the ubiquitinated form, and only when ubiquitin is in shortage would the factor be activated, i.e. the conversion from P to P_A in Fig. 7.14. The involvement of HSP in the transitory increase in heat resistance of cells is persuasive but not yet proven, and this issue will only be resolved when a role for HSP in the development of thermotolerance has been clearly identified, see Chapter 5.

7.6 Temperature and ageing

Just as the duration of the development of an ectothermal animal depends upon environmental temperature, so does the length of adult life; the higher the temperature is, the shorter the life span. This implies that the rate of ageing is temperature dependent, which may not be surprising if, as some workers believe, the ageing process is an integral part of the developmental programme.

The usual method of determining lifespan is to plot a survival curve for a population of animals. Three basic types of curve are found; depending on whether the population shows ageing, the two principal types are shown in Fig. 7.15. When deaths result from random age-independent processes a curve with an exponential decay in numbers of survivors is obtained (Fig. 7.15(b)). In the second case, however, deaths are concentrated in the older age-groups with few occurring in young animals. Natural populations of ectotherms usually show the age-independent pattern of mortality, or a third pattern which has a high infant mortality with lower adult mortality. The age-dependent type of mortality is shown by ectotherms in laboratory cultures. The two extreme life-tables can also be compared from their age-specific mortality curves; in the age-independent curve mortality is constant with respect to age, but this is not so in the laboratory populations where mortality increases progressively with age (Lamp, 1977).

In natural populations ectotherms die before they show signs of senescence. This is not to say they do not age, but that too few life-tables for natural populations of ectotherms have been constructed to enable a conclusion to be reached. In the laboratory the life-shortening effect of a higher maintenance temperature is well proven as, for example, in *Drosophila* (Fig. 7.16) where the expectation of life is shown as an Arrhenius plot. The biphasic shape of the curve suggests death above and below the critical temperature (27°C) results from different processes. Only below 27°C, when the Arrhenius characteristic is low (94 kJ mol^{-1}), would the flies be experiencing ageing. Thus, flies maintained at 3°C have a mean survival of some 220 days, whereas at 25°C this would only be 39 days. Above 27°C the activation energy is increased to values that are characteristic of heat damage.

Pearl (1928) argued that this life-shortening effect of living at a raised

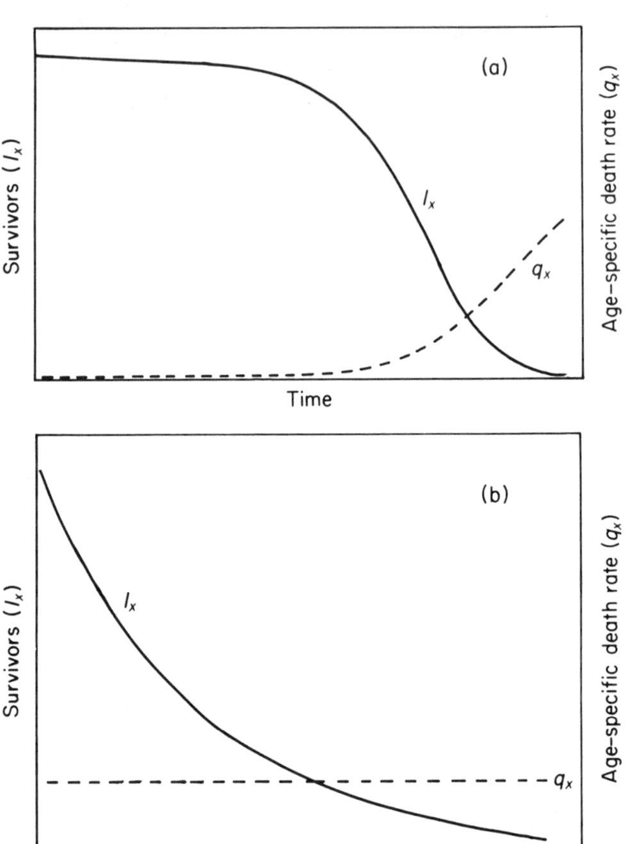

Figure 7.15 Survival curves (——) and age-specific death rates (---) for theoretical populations of animals. (a) Senescing population with low mortality in early age-groups and deaths concentrated in old age groups. (b) Non-senescing population with a random age-independent mortality, resulting in a survival curve with an exponential decay in numbers. This is the form of the curve for natural populations of animals.

temperature was correlated with a higher metabolism, and so was a consequence of a higher rate of living. This theory was held until 1961, when Clarke and Maynard Smith (1961) provided evidence that contradicted it. They kept *D. subobscura* for part of their life at 20°C, and the remainder at 26°C, the rate of living theory predicted that these flies would live longer than those kept throughout at 26°C. In fact, they found that the two groups of flies had the same life span, a clear contradiction of the rate of living theory. In the reverse experiment, flies kept first at 26°C and then transferred to 20°C were found to have the same lifespan as those

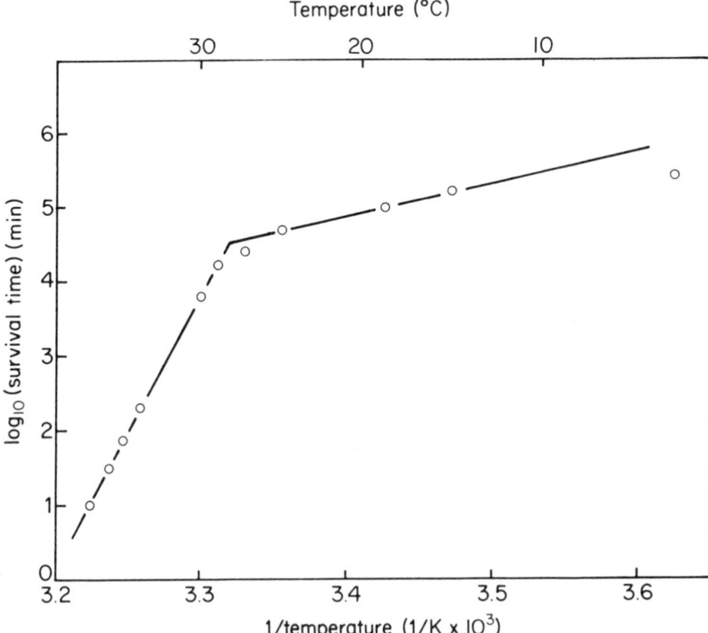

Figure 7.16 Temperature dependence of the length of life of *Drosophila subobscura* in the form of an Arrhenius plot. There is an inflexion at about 27°C. The calculated activation energy for the slopes of the two curves are very different, indicating different causes for death above (heat injury) and below (ageing) the inflexion temperature. (After Hollingsworth, 1969.)

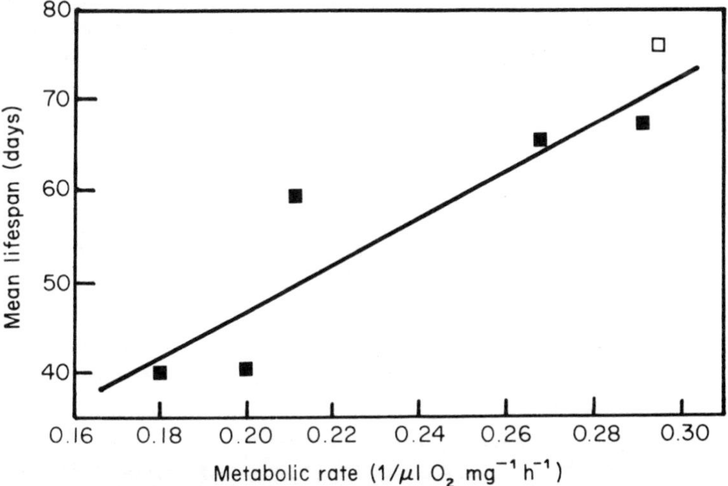

Figure 7.17 Relationship between longevity and metabolic rate in *Drosophila melanogaster*. ■, 'Shaker' mutants; □, wild type. The more-active (higher metabolic rate) flies are found to have a shorter lifespan than flies with lower activity levels. (After Trout and Kaplan, 1970.)

maintained throughout at 20°C, again contrary to the rate of living theory. Maynard Smith proposed a threshold theory to account for his data. He suggested that the rate of ageing was temperature independent, but that a threshold is approached beyond which the animals start to die, the rate of dying he proposed was very temperature dependent. By suggesting that the threshold point was also altered by temperature, he accounted for the observed temperature dependence of the lifespan.

Temperature-shift experiments of this type have been used by several workers with varying results. Some confirm the rate of living theory of Pearl (1928) but others support the threshold theory. Hollingsworth (1969) has proposed that the temperature–longevity relationship is not simple. Life-shortening at high temperatures in *Drosophila subobscura* (above about 27°C) is different from that at lower temperatures (see Fig. 7.16). Consequently, it may be that the higher temperatures used by Clarke and Maynard Smith (1961) may not be quite viable, so may account for the lack of agreement with rate of living theory.

One piece of work which supports the rate of living theory comes from the *shaker* mutant of *D. melanogaster*. These flies are very sensitive to stimuli and are abnormally active. Trout and Kaplan (1970) showed that there was a positive correlation between the lifespan and the reciprocal of the metabolic rate of a number of shaker mutant populations (Fig. 7.17).

Most of our information on the effect of temperature on the length of life of ectotherms comes from studies on relatively few laboratory-maintained animals. These studies have been principally designed to study the ageing process, rather than on the mechanisms by which temperature may act on it, so little can be deduced about the causes of temperature-induced life-shortening. Furthermore, little can be said about the relationship between ageing as a function of temperature in natural populations of ectotherms.

7.7 Growth

Some dispute exists concerning the index used to follow growth. The measurement of a change in length, weight or volume has the advantage of being easy to use and non-destructive, so growth can be followed in an individual. However, in comparative studies care must be taken to exclude the possibility of a different pattern of growth occurring at different temperatures, or with time at the same temperature, such that the relationship between mass and length may change.

There are several ways of depicting growth rates graphically (Fig. 7.18). The simplest is that of a plot of say length against time. This plot typically has an initial phase where growth is exponential, an inflexion and a final phase where the growth rate slows and approaches zero. This changing pattern of growth is clearly shown by plotting the growth increment per

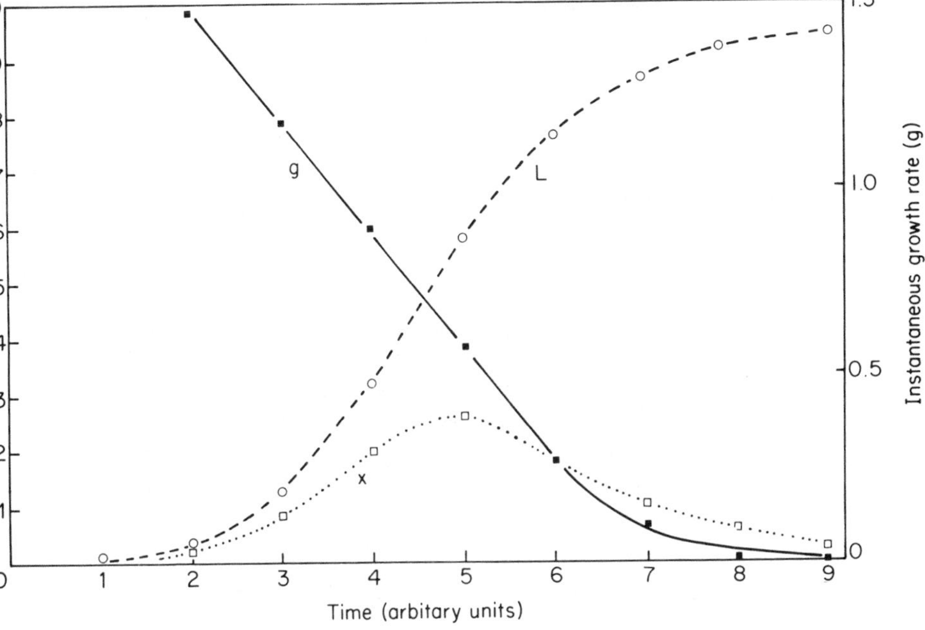

Figure 7.18 Different methods of depicting growth curves. *L*, change in absolute length with time (—○—); *x*, the period increment in length (dL/dt) (····□····); *g*, instantaneous growth rate ($\log_e L_t/L_0$) (—■—).

unit time as a function of time (curve *x* in Fig. 7.18) which passes through a maximum. The most convenient method of showing growth rates is as an instantaneous rate (g)

$$g = \log_e (L_t/L_0) \tag{7.8}$$

where L_t is the size after a stated period of time and L_0 is the size at the beginning of that period. The relationship between finite rates and instantaneous rate is

$$\text{finite rate} = e^g \tag{7.9}$$

The value of using instantaneous rates to follow growth can be illustrated by a simple example. Suppose a 100 kg animal grows at a finite rate of a 10% increment per year. After 1 year it would weigh 110 kg, after 2 years 121 kg and after 3 years 133.1 kg. When this is re-calculated as an instantaneous rate, g becomes 1.1052 so at the end of the first year the animal weighs 110.5 kg, by the second year 122.1 kg and after the third year is 134.9 kg. The instantaneous rate is clearly a more appropriate index for a process, such as growth, which is continuous and usually an exponential rather than a linear process. Clearly the shorter the time intervals over which finite rates are expressed are, the closer they will approximate to the value obtained for the instantaneous rate.

Growth may be indeterminate; that is, a final size is never reached and growth continues so long as other conditions are favourable, as is the case in many fishes. On the other hand, it may be determinate in that a final size is achieved. This usually occurs in the adult in association with the attainment of sexual maturity. This form of growth is found in many insect species.

In both patterns of growth temperature is an important determining factor. Growth is limited to a relatively narrow temperature range, but within that range an increase in temperature leads to an acceleration in the growth rate. This relationship has been shown to hold for a large number of ectotherms that grow continuously. Many studies show a temperature optimum for the growth of a species. It is clear, however, that a number of other biotic and abiotic factors apart from temperature can limit growth. Not only will the intervention of these factors shift the temperature

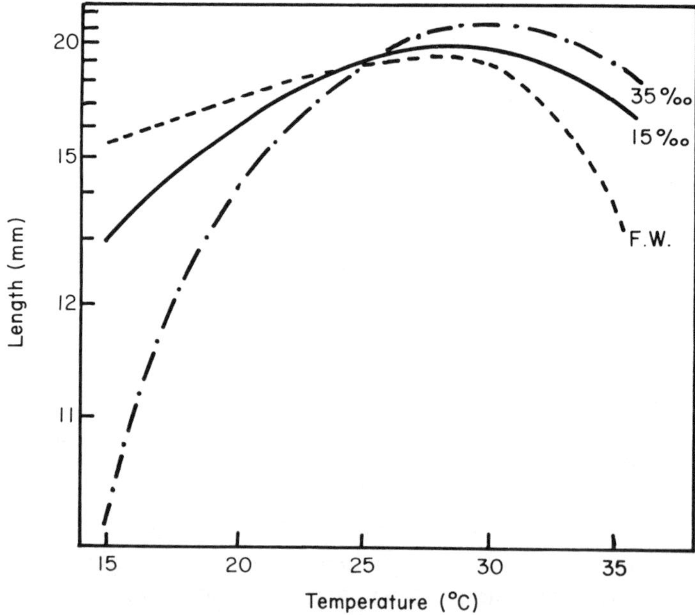

Figure 7.19 The interaction of temperature and salinity in the growth of *Cyprinodon macularius*. Fish living in salinity of 15‰ have a broad optimum of temperature for growth at 20–25°C. The curve is rotated clockwise at lower salinities, giving better growth at lower and poor growth at higher temperatures. The curve rotates counter-clockwise at higher salinities, so growth is retarded at low temperatures but increased at higher temperatures. Salinity 15‰ (——), 35‰ (—·—) and fresh water (---). (After Kinne, 1960.)

optimum (see Fig. 7.19), but different factors may become limiting at different temperatures.

Temperatures outside the permissible range slow down, and may even inhibit, growth. This confines the growth period of many temperate ectotherms to the warmer seasons, and in polar regions it may be restricted to a very narrow thermal range. Some reports suggest that some ectotherms grow throughout the year, as has been found in euphausiid crustaceans. It must be remembered, however, that somatic growth in some species ceases when gonadal growth occurs.

In those arthropods where growth is discontinuous, occurring only at the moult, temperature can affect growth in these organisms either through a change in the moult frequency, or by influencing the size of the increment at the moult. Increasing temperatures have been shown to shorten the intermoult period in many crustaceans (see Fig. 7.20). However, little evidence exists for a temperature dependence of moult increment. Thus, the stimulatory effect of temperature on growth in crustaceans is mainly through an increase in moult frequency.

Few studies have concerned the effect of fluctuating temperatures on growth in crustaceans, but in the euphausiid *Meganyctiphanes norvegica* moult frequency at temperatures fluctuating between 13°C and 18°C was that predicted from the time-weighted mean temperatures, as is shown in

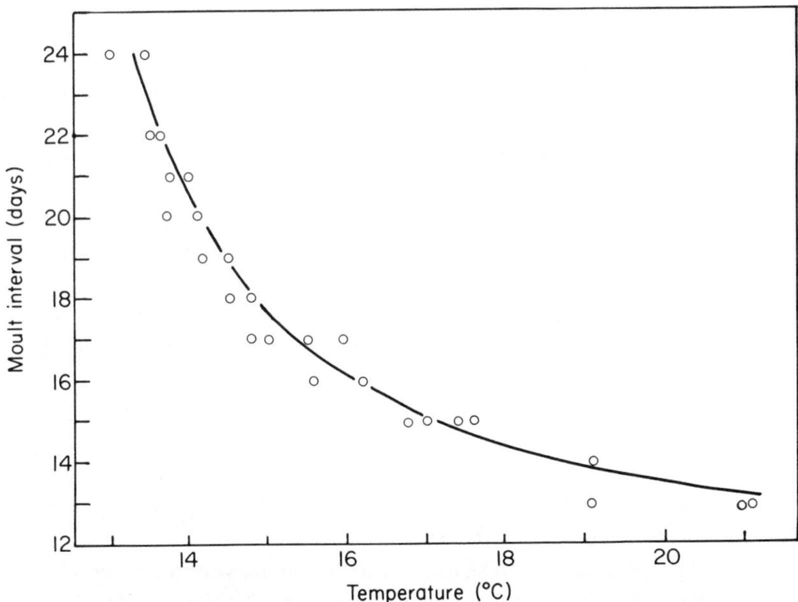

Figure 7.20 Intervals between successive moults as a function of temperature in *Gammarus zaddachi*. (After Kinne, 1961.)

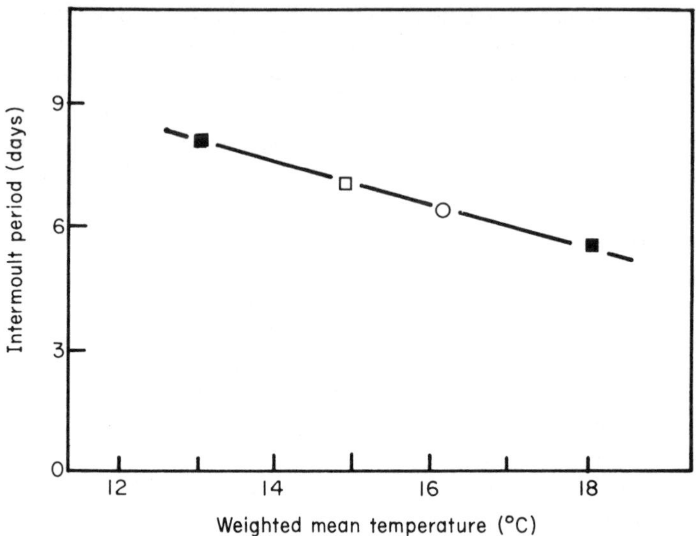

Figure 7.21 Mean intermoult period for euphausiids held at 13°C or 18°C (—■—); or cycled dielly 13°C–18°C (9 h:15 h) (—○—); or 13°C–18°C (15 h:9 h) (-- □ --). Intermoult period shortens with a rise in temperature. (After Fowler *et al.*, 1971.)

Fig. 7.21, so was not affected by temperature fluctuation. Growth in insects and crustaceans is blocked by low temperatures. This is most likely to be exerted through an inhibition of the neurosecretory system that controls the moult process. Not all cases where growth is suppressed by relatively low temperatures can be so explained. In the lizard *Dipsosaurus dorsalis*, Harlow *et al.* (1976) found that at 28°C and below digestive efficiency was highest at the preferred body temperature of 41°C. As well as such effects on digestion, feeding may be inhibited or the food supply reduced by relatively low temperatures.

It must be stressed that, of the many factors known to modify growth rates, nutrition must be dominant. Growth can only occur when food of a sufficient quality is in excess of that required to support maintenance metabolism and routine activity. The relationship between growth rates and ration have been established for several species; Fig. 7.22 shows this relationship for the sockeye salmon. The maintenance ration is that which will just maintain the individual without a weight change. The tangent to the curve describes the ration that gives the greatest growth for the least food (optimum ration), and the asymptote is the ration which just gives the greatest growth rate. Consequently, in order that the effect of temperature on growth be understood a detailed knowledge of its effects upon the processes of ingestion, digestion and assimilation of food, as well as its

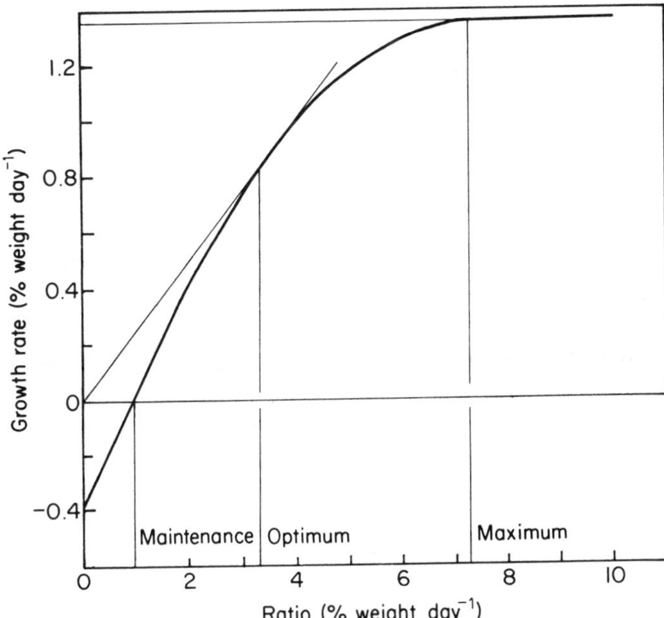

Figure 7.22 Relationship between growth rates and ration in the fingerling sockeye salmon. The derivation of the various parameters of growth are shown. The ration at which the fish show no positive or negative growth is called the maintenance ration. The optimum ration is obtained from the tangent to the growth curve, and is the ration that gives the greatest growth for ration, and the asymptote to the curve is the least ration which supports maximum growth. (After Brett *et al.*, 1969.)

metabolic interconversion, is essential. If these separate processes have different thermal optima or dependences then the resulting effect of temperature on growth may be complex, rather than simple.

Brown (1946), in her study on growth of trout at different temperatures, described two optima; one at 7–9°C and one at 16–19°C. These occurred because temperature had a different effect on metabolism, appetite and activity. The food required shows an S-shaped dependence of temperature. The low-temperature optimum (7–9°C) occurred because basal metabolism and activity were low but appetite was relatively high, so the excess energy intake was channelled into growth. The trough occurred because of the dramatic rise in activity at 10–12°C which took proportionally more of the energy intake. The high-temperature optimum (16–19°C) resulted from a sustained high food intake even though activity had fallen, so allowing more energy available for growth. The decline in growth at higher temperatures still results mainly from a suppression of appetite, as activity is also reduced. It should be remembered, however, that standard metabolism will be several times higher at 20°C than at the lower temperature studied.

The most comprehensive study of this type, however, was carried out by Brett (1971) on the sockeye salmon. A variety of physiological aspects that relate to growth were studied in this fish acclimated to different temperatures between 4°C and 24°C, this upper temperature being close to the lethal limit. (Measurements were made where AT = ET.)

The metabolic studies are shown in Fig. 7.23. Standard metabolism increased continuously with temperature, whereas active metabolism had a marked peak in the 15°C-acclimated group. The decline in active metabolism in the higher-temperature acclimated animals was thought to be limited by the availability of oxygen. The metabolic scope, which is the ability to raise metabolism above the standard level, also showed a marked peak in 15°C-acclimated fish. It is therefore not too surprising that the 15°C group show the highest growth rates. However, if the ration becomes

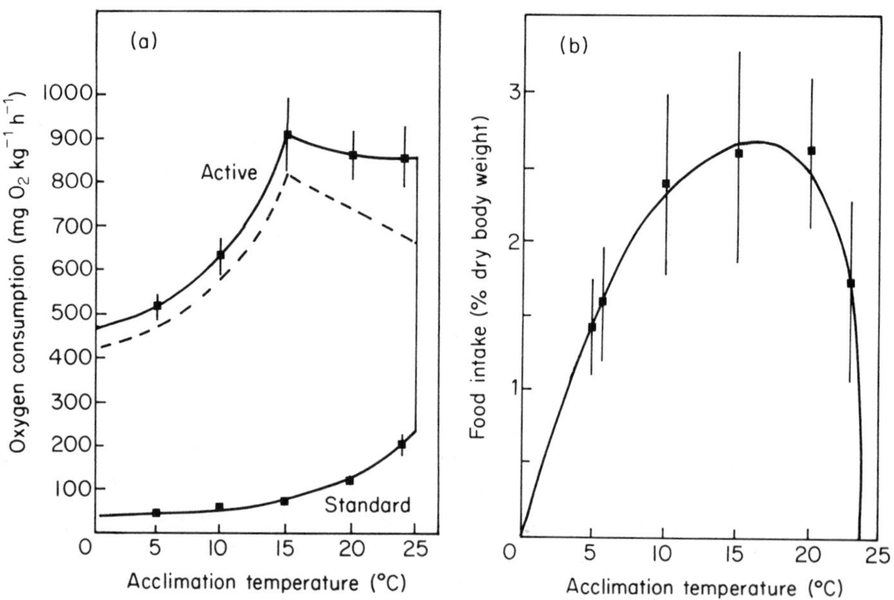

Figure 7.23 Effect of temperature on aspects of metabolism in the sockeye salmon. (a) Metabolic rate as a function of activity and temperature. Standard metabolism increases progressively with temperature but active metabolism (swimming speed 3–4 lengths s^{-1}) had a peak in the 15°C-acclimated group at 15°C. The broken line shows the estimated 'scope for activity' in these fish. (b) The curve showing the effect of temperature on food intake. Appetite is limited in the lower acclimation temperatures and is completely suppressed at 24°C, giving a broad optimum for appetite between the 10°C- and 20°C-acclimated groups. (After Brett, 1971.)

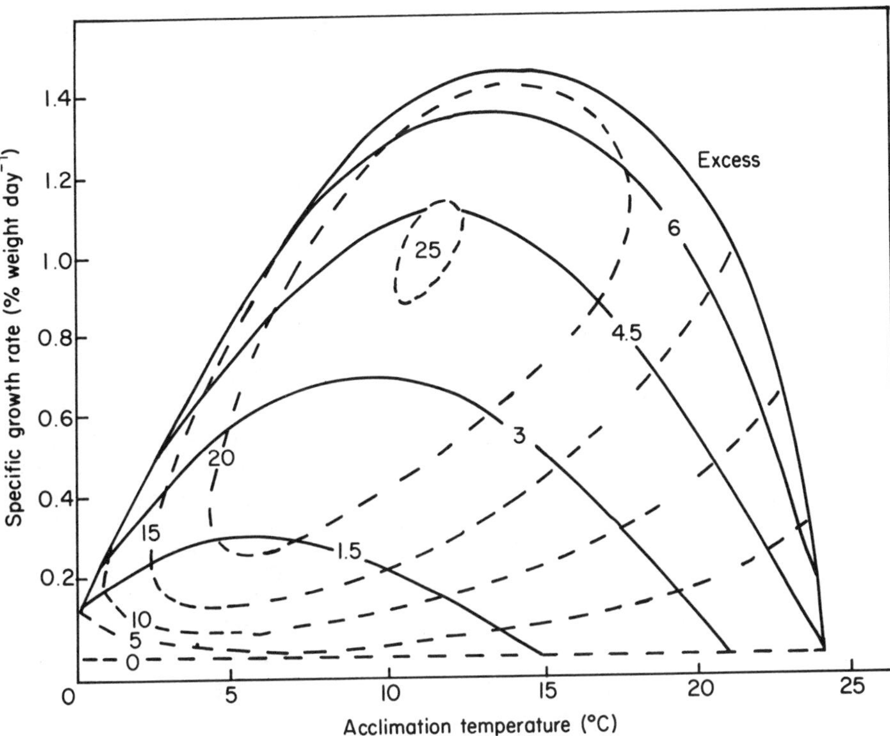

Figure 7.24 Gross conversion efficiency of young sockeye salmon as a function of temperature. Solid lines show growth curves at different rations (% body weight day^{-1}). The overlain broken lines are isopleths giving the percentage conversion efficiency of food into growth. A small area of 25% conversion efficiency is shown at about 11°C. A broad area of 20% conversion efficiency that runs diagonally from low growth at low temperatures to higher growth at about 15°C. Note, too, that as ration is limited the 'optimum' temperature for growth falls. (After Brett, 1971.)

limiting, then there was a progressive shift to lower acclimation groups for the temperature at which growth is maximal (Fig. 7.24). The dramatic increase in standard metabolism at higher acclimation temperatures results in correspondingly less energy being available for growth; hence the reduction in specific growth rates at higher acclimation temperatures when ration is limited. Appetite was shown to have a broad peak, but it was completely inhibited at 24°C (Fig. 7.23(b)). However, digestion time decreased progressively with increasing temperature acclimation, with a Q_{10} of approximately 2. Stomach emptying would then be some five to six times faster in fish acclimated to and living in warm compared with cold waters.

When these combined effects of temperature and ration are considered, it is possible to determine the conditions under which the fish convert food energy into flesh with the highest efficiency (percentage gross conversion efficiency). This is also shown in Fig. 7.24 and, as can be seen, a small area of 25% efficiency was determined to occur at about 11.5°C in fish on a ration of about 4% body weight day^{-1}. A much larger oval area encompasses the 20% efficiency band and, as can be seen, it passes diagonally across the graph from the low growth at lower temperatures upwards to the higher rates at the higher temperatures. It is obvious that temperature has as much effect as ration on the conversion efficiency of ingested food in this organism.

When several parameters – conversion efficiency, maximum meal size and digestion rate – are considered together, the depressing effect of low temperatures on growth can be seen. Above 20°C both conversion efficiency and appetite decline dramatically, which together result in a complete suppression of growth at 24°C. Brett was able to relate these physiological studies to the known thermal ecology of this species, which shows a diurnal vertical migration into deeper, colder waters below the thermocline in daytime. Feeding at the surface takes place at dawn and dusk. It appears that the fish are ration-limited and move into colder water. This may be a bioenergetic response, since at sub-optimum rations the optimum temperature for growth and conversion efficiency will be at temperatures somewhat lower than that of the surface waters (about 17°C) in which feeding takes place (Fig. 7.25).

Studies of that thoroughness do not exist for the effect of temperature on ectotherms other than fish, but have been confirmed for trout (Elliott, 1982) and *Cyprinodon* sp. (Kinne, 1960; Gerking and Lee, 1983). It would be reasonable to suggest that the model developed for the salmon by Brett (1971) will serve to describe the effect of temperature on growth in many other organisms. Although no detailed account of the interacting effects of temperature and nutrition on growth is available for them, it is still possible to lay down some guidelines. There are reported instances where the relationship between the amount of food required to attain a certain size or stage in development is the same at a variety of temperatures. In such cases the amount of food consumed per unit time was higher at higher temperatures, but the duration of the growth period was correspondingly shorter, so the total food consumed was independent of temperature. It might therefore be expected that rearing temperatures would have little effect on the final size.

In other cases the efficiency of conversion of food into growth and development falls with increasing temperature. This is because standard and active metabolism will be greatly stimulated by the higher temperatures, and consequently more food will be required to attain a particular growth rate or developmental stage at higher temperatures.

The reverse case is also found in some organisms, in that less food is used

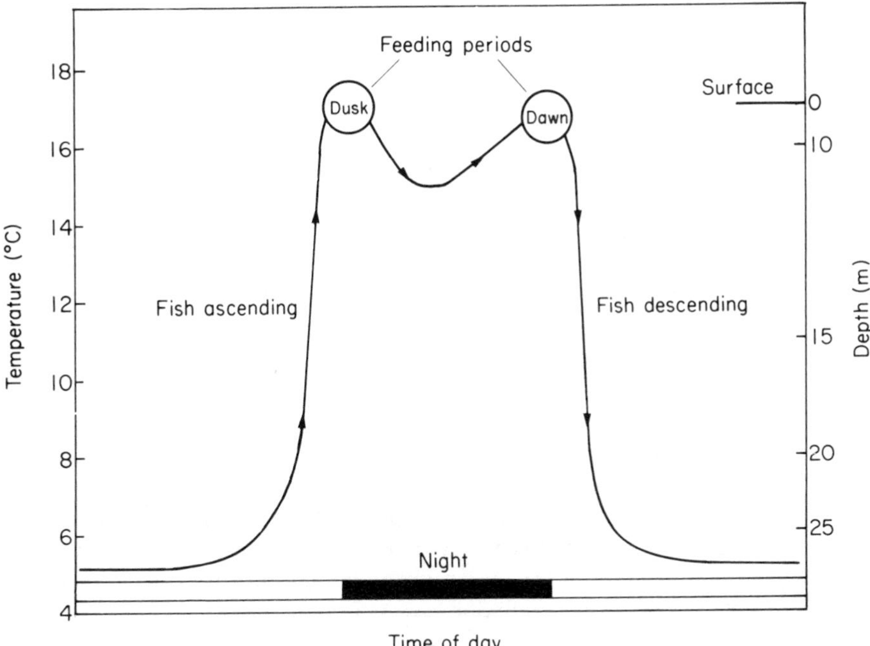

Figure 7.25 Diagram showing the daily movement patterns of sockeye salmon in summer. Feeding near the surface is restricted to two 1–2-h periods at dawn and dusk. The fish move into deeper (colder) water during the day. This may be of ecological importance, since if food is sub-optimal a move to lower temperatures will be necessary to optimize growth and conversion efficiency. (After Brett, 1971.)

to complete a developmental step at higher temperatures. In these instances larger individuals result from growth at low temperatures than at high temperatures. This is because development is accelerated by a rise in temperature faster than food intake, digestion and assimilation are.

This problem of the differential effect of temperature on the rate of growth and development and that of food conversion is particularly acute in developing eggs and embryos that have a fixed yolk supply. In salmon embryos, for example, the efficiency of food conversion is maximal at 10°C (79%), but at 7°C and 14°C it falls to 64%. Thus, the size at some rearing temperatures results from this differential effect of temperature on development and nutrition, when the exhaustion of food supplies is stimulated more rapidly than the rate of development. Gerking and Lee (1983) point to the potential practical applications of their studies on *Cyprinodon nevadensis*, where they found an 8°C differential in the lower limits for growth and egg production (Fig. 7.2). In commercial rearing of fish the metabolic cost of egg production may be avoided by the simple expedient of manipulating the temperature.

7.8 Populations

In earlier sections we have discussed the effect of temperature on a variety of organism functions, from gametogenesis to growth, that suggest that ambient temperature might have an important influence on fitness and adaptation to the environment. Consequently, temperature would be expected to have a significant influence on aspects of population biology.

Biologists have developed techniques to follow changes in the size of populations from considerations of reproduction and mortality. The derived parameter is called the innate capacity for increase (r_m), and is related to mean longevity, mean birth rate and rate of development; factors that are all strongly temperature dependent in ectotherms:

$$r_m = \log_e(R_0)/G \tag{7.9}$$

where R_0 is the multiplication rate per generation and G is the length of the generation; r_m is given as an instantaneous rate.

Pratt (1943) in a study on population growth in *Daphnia* showed a markedly different pattern depending on culture temperature. Figure 7.26 shows that at 25°C the population oscillated in numbers, whereas at 18°C a stable equilibrium population size was reached. Pratt showed that this occurred because at 25°C there was a delay in the effect that rising population density had on birth and death rates; thus, birth rates were quickly retarded by the rise in density but death rates rose later in the cycle. This lag effect causes repeated over- and under-shooting of numbers and no equilibrium density is achieved. How the interaction of factors, biotic and physical, cause an equilibrium density at 18°C but oscillations in numbers at 25°C are not understood. Lawton (1985) has pointed out that the

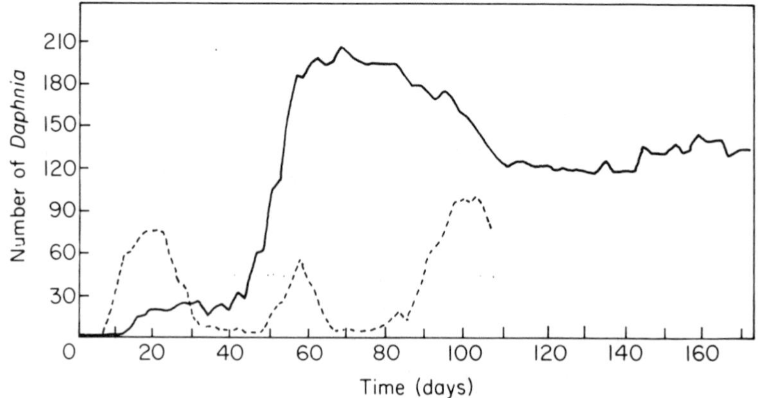

Figure 7.26 Population growth in *Daphnia magna* at two temperatures: 18°C (——) and 25°C (---). (After Pratt, 1943.)

interpretation of even simple experiments such as those carried out by Pratt is complex and probably does not satisfactorily explain the effect of temperature. In situations involving the interaction of physical factors, and the effect of temperature on predators or the food source, the picture becomes even more confused.

Birch (1953) has shown that the innate capacity for increase on the grain beetle *Calandra oryzae* is markedly affected by temperature and humidity (see Fig. 7.27). At a value of $r_m = 0$ the population does not change in size, so a contour map can be produced showing the conditions that permit a positive value for r_m. This is shown in Fig. 7.27 for both *Calandra* and *Rhizopertha*. Clearly sites should be cool and dry to minimize the damage caused by these pests. The ability of *Rhizopertha* to increase at higher temperatures and lower humidities than *Calandra* agrees well with their geographical distribution in Australia, where *Calandra* is temperate and *Rhizopertha* is tropical.

As an extension of this work Birch raised the two beetles in competition and showed the outcome to be strongly dependent upon experimental temperature. *Calandra* became dominant at 29°C, whereas *Rhizopertha* was the dominant species at 32°C. The values obtained for r_m for each species

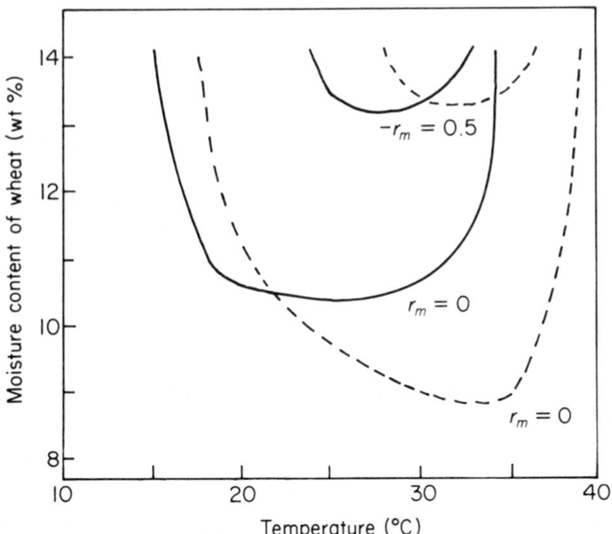

Figure 7.27 The contour lines for the effect of temperature and humidity on the innate capacity for increase (r_m) for two grain beetles, *Calandra oryzae* (—) and *Rhizopertha dominica* (----). At $r_m = 0$ the population does not change in size. The ability of *Rhizopertha* to increase at higher temperatures and lower humidities than *Calandra* can be seen, and is in accord with the distribution of the species. (After Birch, 1953.)

in competition were those predicted from the studies on the beetles in isolation, indicating that little interaction or interference occurred between the species.

Very little is known of the longer-term impact of climate on populations. George and Harris (1985) have recently reported a correlation between water temperature in June in Lake Windermere and the abundance of *Daphnia hyalina*, the most abundant species in the zooplankton. In the period 1940–80 both showed coincident 10-year cycles, but significantly the most abundant fish species, the perch, did not show this cycling of numbers. So these fish are neither affected by the cycles of numbers of *Daphnia*, nor do they influence the cycles. George and Harris also found that the 10-year cycles were observed in the northeastern Atlantic and suggest that they may reflect a much longer-scale pattern of climate change. This interesting observation raises many questions of the causal relationships between temperatures and population size in *Daphnia*, in particular about the mediating factors in the cycling phenomenon. It also raises the important question of whether climatic cycling might influence biological variation and so the evolutionary process.

There is very little information on adaptive changes in populations in relation to temperature. Lewontin and Birch (1966) showed that a fruit fly, *Dacus tryoni*, has extended its range in Australia from 1860 to the present (see Fig. 7.28).

This insect lays eggs on ripe fruit, and has spread southwards from the tropical rain forests of the north. The southward extension has been paralleled by an altitudinal extension in its range. Until about 1930 orchards above 500 m were little damaged, but at present orchards up to 800 m are affected by the pest. This range extension therefore seems to be limited by climate rather than host-plant availability.

Laboratory studies confirm this view for adult flies from tropical latitudes (17°S) have narrower heat *and* cold tolerances than flies from sub-tropical (27°S) or flies from more-southerly (38°S) latitudes. So the extension in geographical range is correlated with an increased tolerance to temperature. The source of this increase in tolerance seems to be an introgression of genes from a second species of *Dacus*, and this provides the genetic variation for selection. Lewontin and Birch raised mixed cultures of *D. tryoni* and *D. neohumeralis*, in the laboratory, at 20°C and 31.5°C. Hybridization between the two occurred, as can be found in nature, but did not produce populations better adapted than *D. tryoni* at all temperatures, for example hybrids produced fewer pupae per week. At 31.5°C, however, the hybrids had a clear advantage in the percentage of adults arising from eggs laid and in the fecundity and longevity of the adults. From these laboratory studies it is suggested that the introgression of new genetic material, as a result of hybridization, increases the tolerance of the species to extreme temperatures, and so permits a widening of its geographical range. This example is

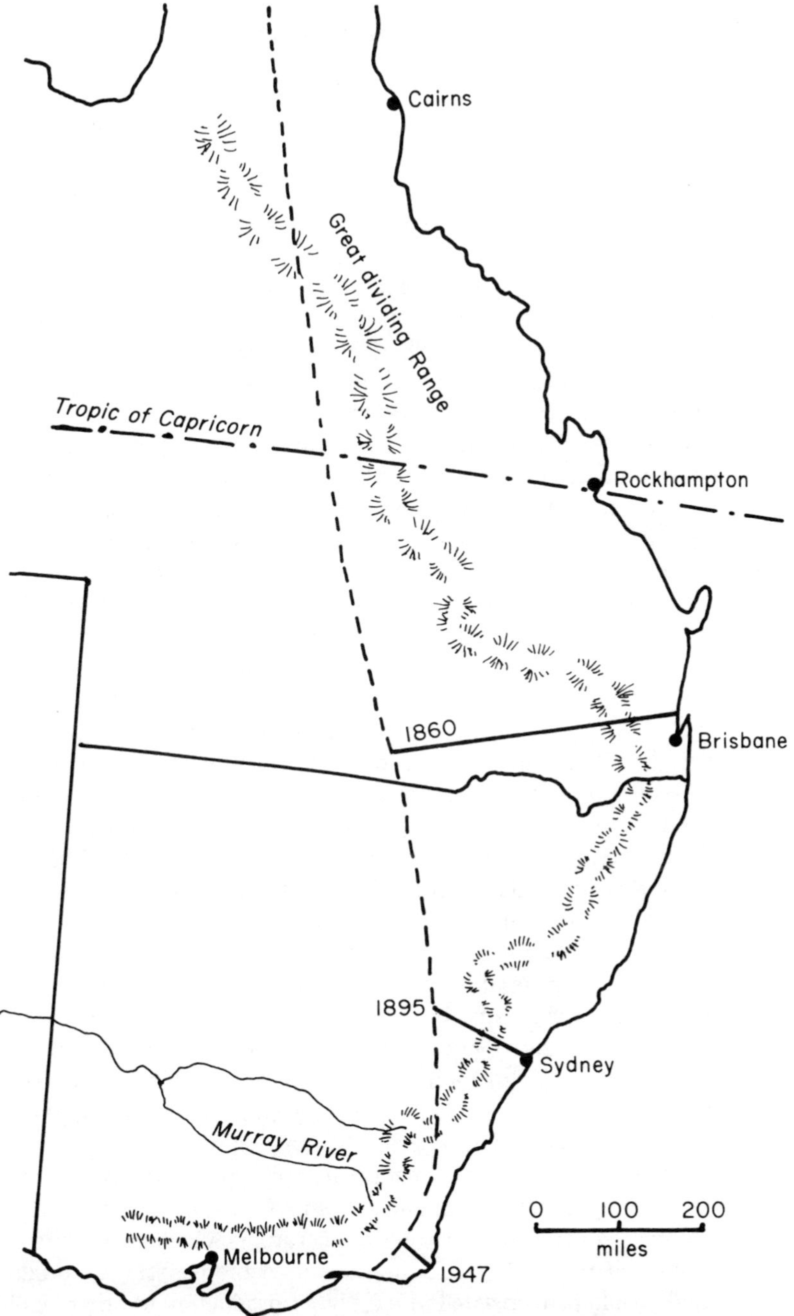

Figure 7.28 Distribution of *Dacus tryoni* in Eastern Australia. The species has spread southwards over about 100 years, west of the broken line. It has also extended its altitudinal range. (After Lewontin and Birch, 1966.)

unique in the demonstration that the distribution of a species, restricted by climatic factors, has been widened progressively within a century.

7.9 Conclusions

Kinne (1970) has remarked aptly that 'with regard to life on earth, temperature is – next to light – the most potent environmental component'. Temperature has two kinds of impact on ectotherms. First, as a direct factor affecting the physiological performance of individuals (Huey and Stevenson, 1979) which will, of course, have ecological implications for the species. Secondly, temperature can be a factor in evolution, acting as a selective force in speciation (Prosser, 1967).

Each species is uniquely fitted to the biotic and physical factors in its environment. This must mean that, for most ectotherms, they are adapted to withstand fluctuating temperatures, daily and seasonally. This is seen as a preadaption that allows compensation for, and behavioural response to, changes in environmental temperature.

In the few ectothermal species studied the most stenothermal processes are the complex 'changing reaction systems' (Laudien, 1973) of reproduction, development and growth. It must be stressed, however, that very few studies have used the fluctuating temperature regimes that are experienced in nature. The narrow thermal limits within which these processes occur must be a dominant factor in determining the pattern of life history and life cycle of ectotherms (Tauber and Tauber, 1982). In many species the timing of reproduction, development and growth has a seasonal basis related to climatic conditions and the availability of energy resources. The narrowness of the thermal limits for these processes must also be a powerful force in setting the biogeographical distribution of a species.

The increased use of natural waters for industrial cooling, particularly by the electricity-generating industry, presents a threat to the biological stability of those ecosystems. Thermal pollution may outstrip the capacity of the natural acclimational and adaptational responses of the organisms. Clearly, a population will show a negative response at water temperatures outside the normal range, as gametogenesis, fecundity and development will be impaired. The shift of ecological advantage may pass towards a predator or competitor species. So if the temperature change is marked and prolonged a species may become eliminated from the environment (Naylor, 1965). However, a small rise in water temperature, which represents an energy enhancement, may enrich the habitat. A higher primary production by algae would be translated into a higher production in other trophic levels (Gibbons, 1976). Subtle changes can also occur, such as a lengthening of the breeding season for a species. Barnett (1972) showed this to be the case in the whelk *Nassarius*, but this will only be of advantage to the species if the food source for its larvae also has the same seasonal extension. Aquatic ecosystems are complex and the factors involved in their stability and dynamic equilibria are not well understood. Thermal enhancement, in the form of thermal pollution, must threaten the integrity of an ecosystem. Persistent thermal pollution will change the species composition of a habitat, with at best native species being replaced by subtropical species, and at worst a restriction in species diversity (Naylor, 1965).

It is well documented in many species that the temperatures experienced during development affect body size, shape and colour, and also that at temperatures outside the normal range development is abnormal (Laudien, 1973). Studies on *Drosophila* suggest this is a result of the direct effect of temperature on the genome (Mitchell and Lipps, 1978). Temperature may then be a factor that causes phenotypic variation to arise, acting as a factor in canalization to produce adults of different fitness (Giesel *et al.*, 1982). There is also evidence that physiological performance, as well as the morphology of the organism is subject to canalization (Smith, 1957). The establishment of geographical races of a species in a cline may well have a strong thermal basis (Moore, 1949; Dehnel, 1955), for not only may thermally driven canalization produce different phenotypes but thermal selection of different genotypes might occur (Prosser, 1967). Whether temperature has an influence on the generation of mutations must also be considered. In laboratory cultures of *Drosophila* the frequency of mutations was greater at the higher rearing temperatures, but whether this can be extrapolated to field populations, experiencing normal temperature regimes is problematical (Prosser, 1967). However, random change in the genome, through mutation or hybridization, as in *Dacus* (Lewontin and Birch, 1966), may cause changes in the behaviour and thermal requirements of a species, and so allow a widening of the niche. However, the selection of genetic variants may lead to the formation of new varieties, races or sub-species, so environmental temperature may be an important factor in speciation.

References

Akhmerov, R.N. (1986) Quantitative difference in mitochondria of endothermic and ectothermic animals. *FEBS*, **198**, 251–5.

Alahiotis, S.N. (1983) Heat shock proteins. A new view of the temperature compensation. *Comp. Biochem. Physiol.*, **75B**, 379–87.

Alahiotis, S.N. and Stephanou, G. (1982) Temperature adaptation of *Drosophila* populations. The heat shock proteins system. *Comp. Biochem. Physiol.*, **73B**, 529–33.

Alderdice, D.F. (1972) Factor combinations: responses of marine poikilotherms to environmental factors acting in concert. In *Marine Ecology*, Vol. 1 (ed. O. Kinne), Wiley Interscience, London, pp. 1659–722.

Aleksiuk, M. (1971) Temperature-dependent shifts in the metabolism of a cool temperate reptile, *Thamnophis sirtalis parietalis. Comp. Biochem. Physiol.*, **39A**, 459–503.

Aleksiuk, M. (1976) Metabolic and behavioural adjustments to temperature change in the red-sided garter snake (*Thamnophis sirtalis parietalis*): An integrated approach. *J. Thermal Biol.*, **1**, 153–6.

Allen, T.E. and Bligh, J. (1969) A comparative study of the temporal patterns of cutaneous water vapour loss from some domesticated mammals with epitricial sweat glands. *Comp. Biochem. Physiol.*, **31**, 347–63.

Altman, P.L. and Dittmer, D.S. (1968) *Biological handbooks: Metabolism*, Bethesda: Federation of American Societies for Experimental Biology, 737 pp.

Anderson, G. 1978) Metabolic rate-temperature acclimation and resistance to high temperature of the soft-shell clam, *Mya arenaria*, as affected by shore level. *Comp. Biochem. Physiol.*, **61A**, 433–8.

Anderson, J.F. (1970) Metabolic rates of spiders. *Comp. Biochem. Physiol.*, **33**, 51–72.

Anderson, J.F. and Horsfall, W.R. (1965) Thermal stress and anomalous development of mosquitoes (Diptera: Culicidae). The effects on embryogeny of *Aedes stimulans. J. exp. Zool.*, **158**, 211–22.

Anderson, R.L., Minton, K.W., Hahn, G.C. and Hahn, G.M. (1981)

Temperature induced homeoviscous adaptations of Chinese Hamster ovary cells. *Biochim. Biophys. Acta*, **641**, 334–48.

Andrews, R.M. and Asato, T. (1977) Energy utilization of a tropical lizard. *Comp. Biochem. Physiol.*, **58A**, 57–62.

Armitage, K.B. and Lei, C-H. (1979) Temperature acclimation in the filtering rates and oxygen consumption of *Daphnia ambigua* Scourfield. *Comp. Biochem. Physiol.*, **62A**, 807–12.

Arrhenius, S. (1889) Über die Reaktiongeschwindigkeit bei der Inversion von Rohrzucker durch Sauren. *Z. Phys. Chem.*, **4**, 226–48.

Arrhenius, S. (1915) *Quantitative Laws in Biological Chemistry*. Bell, London.

Aschoff, J. (1981) Thermal conductance in mammals and birds: its dependence on body size and circadian phase. *Comp. Biochem. Physiol.*, **69A**, 611–20.

Aschoff, J. and Wever, R. (1958) Kern und Schale in warmehaushalt des Menschen. *Naturwissenschaften*, **45**, 477–85.

Avery, R.A. (1982) Field studies of body temperatures and thermoregulation. In *Biology of the Reptilia*, Vol. 12 (eds C. Gans and F.H. Pough), Academic Press, New York, 94–166.

Baggerman, B. (1957) An experimental study on the timing of breeding and migration in the three-spined stickleback (*Gasterosteus aculeatus* L.). *Archs neerl. Zool.*, **12**, 105–317.

Baker, M.A. (1979) A brain-cooling system in mammals. *Sci. Am.*, **240**, 114–23.

Baker, M.A. (1982) Brain cooling in endotherms in heat and exercise. *Ann. rev. Physiol.*, **44**, 85–96.

Bakken, G.S. (1976) An improved method for determining thermal conductance and equilibrium body temperature with cooling curve experiments. *J. Thermal Biol.*, **1**, 169–76.

Bakken, G.S. (1981) A two-dimensional operative-temperature model for thermal energy management by animals. *J. Thermal Biol.*, **6**, 23–30.

Baldwin, J. and P.W. Hochachka (1970) Functional significance of isoenzymes in thermal acclimation: acetylcholinesterase from trout brain. *Biochem. J.*, **116**, 883–7.

Barnes, H. (1963) Light, temperature and the breeding of *Balanus balanoides*. *J. mar. Biol. Ass. UK*, **43**, 717–27.

Barnes, H. and Barnes, M. (1969) Seasonal changes in the acutely determined oxygen consumption and effect of temperature for three common cirripedes, *Balanus balanoides* (L.), *B. balanus* (L.) and *Chthamalus stellatus* (Poli). *J. exp. mar. Biol. Ecol.*, **4**, 36–50.

Barnett, P.R.O. (1972) Effects of warm water effluents from power stations on marine life. *Proc. R. Soc. Lond.*, **B180**, 497–509.

Barnett, P.R.O. and Hardy, B.L.S. (1984) Thermal deformations. In *Marine Ecology*, Vol. 5 (ed. O. Kinne), Wiley Interscience, London, 1769–1964.

Bartholomew, G.A. (1964) The roles of physiology and behaviour in the maintenance of homeostasis in the desert environment. *Symp. Soc. Exp. Biol.*, **18**, 7–29.

Bartholomew, G.A. (1982) Physiological control of body temperature. In *Biology of the Reptilia*, Vol. 12 (eds C. Gans, F.H. Pough), Academic Press, New York, 536 pp.

Bartholomew, G.A. and Lasiewski, R.C. (1965) Heating and cooling rates, heart rate and simulated diving in the Galapagos marine iguana. *Comp. Biochem. Physiol.*, **16**, 573–82.

Bass, E.L. (1971) Temperature acclimation in the nervous system of the brown bullhead (*Ictalurus nebulosus*). *Comp. Biochem. Physiol.*, **40A**, 833–49.

Bass, E.L. (1977) Influences of temperature and salinity on oxygen consumption of tissues in the American oyster *Crassostrea virginica*. *Comp. Biochem. Physiol.*, **58B**, 125–30.

Baust, J.G. (1981) Biochemical correlates to cold hardening in insects. *Cryobiol.*, **18**, 186–98.

Baust, J.G. and Edwards, J.S. (1979) Mechanisms of freeze-tolerance in an antarctic midge, *Belgica antarctica*. *Physiol. Ent.*, **4**, 1–5.

Baust, J.G. and Lee, R.E. Jr. (1981) Divergent mechanisms of frost hardiness in two populations of the gall fly, *Eurosta solidaginis*. *J. Insect Physiol.*, **27**, 485–90.

Baust, J.G. and Rojas, R.R. (1985) Insect cold hardiness – facts and fancy. *J. Insect Physiol.*, **31**, 755–9.

Beamish, F.W.H. and Mookherjii, P.S. (1964) Respiration of fishes with special emphasis on standard oxygen consumption. I. Influence of weight and temperature on respiration of goldfish, *Carassius auratus* L. *Can J. Zool.*, **42**, 161–75.

Behrendt, K. (1963) Über die Eidiapause von *Aphis fabae*. Scop. (Homoptera Aphididae). *Zool. Jahrb.*, **70**, 309–98.

Bělehrádek, J. (1930) Temperature coefficients in biology. *Biol. Rev.*, **5**, 30–58.

Bělehrádek, J. (1935) *Temperature and living matter*, Gebrüder Borntraeger, Berlin.

Bělehrádek, J. (1957) Physiological aspects of heat and cold. *Ann. rev. Physiol.*, **19**, 59–82.

Bennett, A.F. and Ruben, J.A. (1979) Endothermy and activity in vertebrates. *Science*, **206**, 649–54.

Benzinger, T.H. (1969) Heat regulation: Homeostasis of central temperature regulation in man. *Physiol. Rev.*, **49**, 671–759.

Berthelot, M. (1862) Essai d'une theorie sur la formation des ethers. *Ann. Chim. Phys. 3e serie*, **66**, 110–28.

Birch, L.C. (1953) Experimental background to the study of the distribution and abundance of insects. I. The influence of temperature, moisture, and food on the innate capacity for increase of three grain beetles. *Ecology*, **34**, 698–711.

Blackmann, F.F. (1905) Optima and limiting factors. *Ann. Bot.*, **19**, 281–95.

Bligh, J. (1967) A thesis concerning the process of secretion and discharge of sweat. *Environ. Res.*, **1**, 28–45.

Bligh, J. (1973) *Temperature regulation in mammals and other vertebrates*, North Holland Pub. Comp., Amsterdam, 436 pp.

Bligh, J. (1984) Temperature regulation: A theoretical consideration incorporating Sherringtonian principles of central neurology. *J. Thermal Biol.*, **9**, 3–6.

Bligh, J. and Harthoorn, A.M. (1965) Continuous radiotelemetric records of deep body temperature of some unrestrained African mammals under near-natural conditions. *J. Physiol.*, **176**, 145–62.

Bligh, J. and Johnson, K.G. (1973) A glossary of terms for thermal physiology. *J. appl. Physiol.*, **35**, 941–61.

Block, W. and Young, S.R. (1978) Metabolic adaptations of antarctic terrestrial microarthropods. *Comp. Biochem. Physiol.*, **61A**, 363–8.

Bogert, C.M. and Del Campo, R.M. (1956) The Gila monster and its allies. *Bull. Am. Mus. Nat. Hist.*, **109**, 7–238.

Bond, T.F., Kelly, C.F., Morrison, S.R. and Pereira, N. (1967) Solar atmospheric and terrestrial radiation received by shaded and unshaded animals. *Trans. Amer. Soc. Agric. Eng.*, **10**, 622–5.

Bonner, W.N. (1979) Life in a cold climate. *Nature*, **282**, 234.

Boron, W.F. (1986) Intracellular pH regulation in epithelial cells. *Ann. Rev. Physiol.*, **38**, 377–88.

Bowler, K. (1963a) A study of the factors involved in acclimatisation to temperatures and death at high temperature in *Astacus pallipes*. I. Experiments on intact animals. *J. Cell. Comp. Physiol.*, **62**, 119–32.

Bowler, K. (1963) A study of the factors involved in acclimatisation to temperature and death at high temperatures in *Astacus pallipes*. II. Experiments at the tissue level. *J. Cell. Comp. Physiol.*, **62**, 133–46.

Bowler, K., Duncan, C.J., Gladwell, R.T. and Davison, T.F. (1973a) Cellular heat injury. *Comp. Biochem. Physiol.*, **45A**, 441–50.

Bowler, K., Gladwell, R.T. and Duncan, C.J. (1973b) Acclimatisation to temperature and death at high temperatures in the crayfish *Austropotamobius pallipes*. In *Freshwater Crayfish* (ed. S. Abrahamsson), Studentliteratur, Lund, Sweden, 122–31.

Bowler, K. and Kashmeery, A.M.S. (1979) Recovery from heat injury in the blowfly, *Calliphora arythrocephala*. *J. Thermal. Biol.*, **4**, 197–202.

Bowler, K., Laudien, H. and Laudien, I. (1983) Cellular heat injury. *J. Thermal Biol.*, **8**, 426–30.

Brattstrom, B.H. (1963) A preliminary review of the thermal requirements of amphibians. *Ecology*, **44**, 238–55.

Brattstrom, B.H. (1965) Body temperature of reptiles. *Am. Mid. Nat.*, **73**, 376–422.

Braune, H.J. (1971) Der Einfluss der Temperatur auf Eidiapause und Entwicklung von Weichwanzen (Heteroptera, Miridae). *Oecologica*, **8**, 233–66.

Brett, J.R. (1944) Some lethal temperature relations of Algonquin Park fishes. In *Univ. Toronto Studies*, Biol. Ser. No. 52, 1–49.

Brett, J.R. (1946) Rate of gain of heat-tolerance in goldfish (*Carassius auratus*). In *Univ. Toronto Studies*, Biol. Ser. No. 63, 1–28.

Brett, J.R. (1952) Temperature tolerance in young Pacific salmon *Oncorhynchus*. *J. Fisheries Res. Bd. Can.*, **9**, 265–323.

Brett, J.R. (1958) Implications and assessment of environmental stress. In *Investigations of Fish-Power Problems* (ed. P.A. Larkin), H.R. MacMillan Lectures in Fisheries, University of British Columbia, 69–83.

Brett, J.R. (1971) Energetic responses of salmon to temperature. A study of some thermal relations in the physiology and freshwater ecology of sockeye salmon (*Oncorhynchus nerka*). *Am. Zool.*, **11**, 99–113.

Brett, J.R. and Groves, T.D.D. (1979) In *Fish Physiology* (*Physiological energetics* Vol. 8) (eds. W.S. Hoar and D.J. Randle) Academic Press, London and New York, pp. 280–352.

Brett, J.R., Shelbourn, J.E. and Shoop, C.T. (1969) Growth rate and body composition of fingerling sockeye salmon, *Oncorhynchus nerka*, in relation to temperature and ration size. *J. Fisheries Res. Bd. Can.*, **26**, 2363–94.

Brock, T.D. (1985) Life at high temperatures. *Science*, **230**, 132–8.

Brown, A.V. and Fitzpatrick, L.C. (1981) Metabolic acclimation to temperature in the Ozark salamander *Plethodon dorsalis*, Angusticlavius. *Comp. Biochem. Physiol.*, **69A**, 499–503.

Brown, D.E., Johnson, F.H. and Marsland, D.A. (1942) The pressure, temperature relations of bacterial luminescence. *J. Cell. Comp. Physiol.*, **20**, 151–68.

Brown, M.E. (1946) The growth of brown trout (*Salmo trutta* Linn.). III. The effect of temperature on the growth of the two-year-old trout. *J. exp. Biol.*, **22**, 145–55.

Brück, K. and Schwennicke, H.P. (1971) Interaction of superficial and hypothalamic thermosensitive structures in the control of nonshivering thermogenesis. *Int. J. Biometeor.*, **15**, 156–61.

Brück, K. and Wünnenberg, B. (1965) Über die modi der Thermogenese beim neugeborenen Warmblüter. Untersuchungen am Meerschweinchen. *Pflugers. Arch. ges. Physiol.*, **282**, 362–75.

Buchanan, R.E. and Fulmer, E.I. (1930) *Physiology and biochemistry of bacteria*. II. *Effects of Environment upon Micro-organisms*. Baltimore Md: Williams and Wilkins, Baltimore.

Bulger, A.J. and Tremaine, S.C. (1985) Magnitude of seasonal effects on heat tolerance in *Fundulus heteroclitus. Physiol. Zool.*, **58**, 197–204.

Bull, J.J. (1980) Sex determination in reptiles. *Q. rev. Biol.*, **55**, 13–21.

Bullock, T.H. (1955) Compensation for temperature in the metabolism and activity of poikilotherms. *Biol. Rev.*, **30**, 311–42.

Burns, J.R. (1975) Seasonal changes in the respiration of pumpkinseed, *Lepomis gibbosus*, correlated with temperature, daylength and stage of reproductive development. *Physiol. Zool.*, **48**, 142–9.

Bursell, E. (1974) Environmental aspects in temperature. In *Physiology of the Insecta II* (ed. M. Rockstein), Academic Press, 2–43.

Bye, V.J. (1984) The role of environmental factors in the timing of reproductive cycles. In *Fish Reproduction: Strategies and Tactics* (eds G.W. Potts and R.S. Wotton), Academic Press, 187–206.

Cameron, J.N. (1978) Regulation of blood pH in teleost fish. *Resp. Physiol.*, **33**, 129–44.

Campbell, C.M. and Davies, P.S. (1975) Thermal acclimation in the teleost, *Blennius pholis (L). Comp. Biochem. Physiol.*, **52A**, 147–151.

Cannon, B. and Nedergaard, J. (1983) Biochemical aspects of acclimation to cold. *J. Thermal Biol.*, **8**, 85–90.

Carey, F.G. (1973) Fish with warm bodies. *Sci. Amer.*, **228**, 36–49.

Carey, F.G. and Teal, J.L. (1969) Regulation of body temperature by the bluefin tuna. *Comp. Biochem. Physiol.*, **28**, 205–13.

Carey, F.G., Teal, J.L., Kanwisher, J.W. and Lawson, K.D. (1971) Warm-bodied fish. *Am. Zool.*, **11**, 135–45.

Casey, T.M. and Hegel, J.R. (1981) Caterpillar setae: insulation for an ectotherm. *Science*, **214**, 1131–3.

Chappell, M.A. (1984) Temperature regulation and energetics of the solitary bee *Centris pallida* during foraging and intermale mate competition. *Physiol. Zool.*, **57**, 215–25.

Chatonnet, J. (1983) Some general characteristics of temperature regulation. *J. Thermal Biol.*, **8**, 33–6.

Clark, R.P. and Edholm, O.G. (1985) *Man and his thermal environment*, Edward Arnold, London.

Clarke, A. (1979) On living in cold water: K-strategies in antarctic benthos. *Mar. Biol.*, **55**, 111–9.

Clarke, A. (1980) A reappraisal of the concept of metabolic cold adaptation in polar marine invertebrates. *Biol. J. Linn. Soc.*, **14**, 77–92.

Clarke, J.M. and Maynard Smith, J. (1961) Two phases of ageing in *Drosophila subobscura. J. exp. Biol.*, **38**, 679–84.

Claussen, D.L. (1980) Thermal acclimation in the crayfish *Orconectes rusticus* and *O. virilis. Comp. Biochem. Physiol.*, **66A**, 377–84.

Cloudsley-Thompson, J.L. (1971) *The temperature and water relations of reptiles*, Merrow Press, Watford 159 pp.

Conover, D.O. (1984) Adaptive significance of temperature-dependent sex determination in a fish. *Am. Nat.*, **123**, 297–313.

Corbet, P.S. (1957) The life-history of the emperor dragonfly *Anax imperator* Leach (*Odonata: Aeshnidae*). *J. Animal Ecol.*, **26**, 1–29.

Cornwell, K. (1977) *The flow of heat*, Van Nostrand Reinhold, New York, 536 pp.

Corti, U.A. and Weber, M. (1948) Die matrix der fische. II. Intersuchungen über die vitalitat von fischern. *Schweiz. Z. Hydrol.*, **11**, 297–300.

Cossins, A.R. (1977) Adaptation of biological membranes to temperature. Effect of seasonal acclimation of goldfish upon the viscosity of synaptosomal membranes. *Biochim. Biophys. Acta*, **470**, 395–411.

Cossins, A.R. (1981) The adaptation of membrane dynamic structure to temperature. In *Fluorescent probes* (eds G. Beddard and M.A. West) Academic Press, New York and London, 39–80.

Cossins, A.R. (1983) The adaptations of membrane structure and function to changes in temperature. In *Cellular Acclimatization to Environmental Change* (eds A.R. Cossins and P. Sheterline) Cambridge University Press, UK, 3–32.

Cossins, A.R. and Bowler, K. (1976) Resistance adaptation of the fresh-water crayfish and thermal inactivation of membrane-bound enzymes. *J. Comp. Physiol.*, **111**, 15–24.

Cossins, A.R., Friedlander, M.J. and Prosser, C.L. (1977) Correlations between behavioural temperature adaptations of goldfish and the viscosity and fatty acid composition of their synaptic membranes. *J. Comp. Physiol.*, **120**, 109–21.

Cossins, A.R. and Prosser, C.L. (1978) Evolutionary adaptations of membranes to temperature. *Proc. Natl. Acad. Sci. USA*, **75**, 2040–3.

Cossins, A.R. and Sheterline, P.S. (1983) *Cellular acclimatisation to Environmental Change*, Cambridge University Press, UK.

Coutant, C.C. (1972) Biological aspects of thermal pollution. II. Scientific basis for water temperature standards at power plants. *C.R.C. Crit. Rev. Environ. Cont.*, **1**, 341–81.

Cowles, R.B. (1962) Semantics in biothermal studies. *Science*, **135**, 670.

Cowles, R.B. and Bogert, C.M. (1944) A preliminary study of the thermal requirements of desert reptiles. *Bull. Am. Mus. Nat. Hist.*, **83**, 261–96.

Crawshaw, L.I. (1984) Low-temperature dormancy in fish. *Am. J. Physiol.*, **246**, R479–R486.

Cress, A.E. and Gerner, E.W. (1980) Cholesterol levels inversely reflect the thermal sensitivity of mammalian cells in culture. *Nature*, **283**, 677–9.

Crisp, D.J. (1964) The effects of the severe winter of 1962–63 on marine life in Britain. *J. Animal Ecol.*, **33**, 165–210.

Crisp, D.J., Davenport, J. and Gabboth, P.A. (1977) Freezing tolerance in *Balanus balanoides*. *Comp. Biochem. Physiol.*, **57A**, 359–61.

Crozier, W.J. (1924) On the possibility of identifying chemical processes in living matter. *Proc. Natl. Acad. Sci. USA*, **10**, 461–4.

Crozier, W.J. (1925a) On the critical thermal increment for the locomotion of a diplopod. *J. Gen. Physiol.*, **7**, 123–36.

Crozier, W.J. (1925b) On biological oxidations as functions of temperature. *J. Gen. Physiol.*, **7**, 189–216.

Crozier, W.J. and Federighi, H. (1925a) Critical thermal increment for the movement of *Oscillatoria*. *J. Gen. Physiol.*, **7**, 137–50.

Crozier, W.J. and Federighi, H. (1925b) Temperature characteristic for heart rhythm of the silkworm. *J. Gen. Physiol.*, **7**, 565–70.

Crozier, W.J. and Stier, T.B. (1925a) Critical thermal increments for rhythmic respiratory movements of insects. *J. Gen. Physiol.*, **7**, 429–48.

Crozier, W.J. and Stier, T.B. (1925b) The temperature characteristic for pharyngeal breathing rhythm of the frog. *J. Gen. Physiol.*, **7**, 571–9.

Crozier, W.J. and Stier, T.B. (1925c) Critical increment for opercular breathing rhythm of the goldfish. *J. Gen. Physiol.*, **7**, 699–704.

Crozier, W.J. and Stier, T.B. (1925d) Temperature characteristic for heart beat frequency in *Limax*. *J. Gen. Physiol.*, **7**, 705–8.

Danilevskii, A.S. (1965) *Photoperiodism and seasonal development in insects*, Oliver and Boyd, Edinburgh, 283 pp.

Davidson, J. (1944) On the relationship between temperature and the rate of development of insects at constant temperatures. *J. Animal Ecol.*, **13**, 26–38.

Davies, P.M.C. and Bennett, E.L. (1981) Non-acclimatory latitude-dependent metabolic adaptation to temperature in juvenile Natricine snakes, *J. Comp. Physiol.*, **142**, 489–94.

Davies, P.S. and Tribe, M. (1969) Temperature dependence of metabolic rate in animals. *Nature*, **224**, 723–4.

Davison, T.F. (1971) The effect of temperature on oxidative phosphorylation in isolated flight muscle sarcosomes. *Comp. Biochem. Physiol.*, **38B**, 21–34.

Davison, T.F. and Bowler, K. (1971) Changes in the functional efficiency of flight muscle sarcosomes during heat death of adult *Calliphora erythrocephala*. *J. Cell Physiol.*, **78**, 37–48.

Dawson, W.R. and Bartholomew, G. (1956) Relation of oxygen consumption to body weight, temperature and temperature acclimation in lizards *Uta stansburiana and Sceloporus occidentalis*. *Physiol. Zool.*, **29**, 40–51.

Dawson, W.R. and Evans, F.C. (1960) Relation of growth and development to temperature regulation in nestling vesper sparrows. *Condor*, **62**, 329–40.

Dawson, W.R. and Hudson, J.W. (1970) Birds. In *Invertebrates and Non-mammalian Vertebrates* (*Comparative Physiology of Thermoregulation* Vol. 1) (ed. G.C. Whitton) Academic Press, New York, 223–310.

Dean, W.L. and Tanford, C. (1978) Properties of a delipidated, detergent-activated Ca^{2+}-ATPase. *Biochemistry*, **17**, 1683–90.

Dehnel, P.A. (1955) Rates of growth of gastropods as a function of latitude. *Physiol. Zool.*, 115–44.

Dehnel, P.A. and Segel, E. (1956) Acclimation of oxygen consumption to temperature in the American Cockroach *Periplaneta americana*. *Biol. Bull.*, **111**, 53–61.

De March, B.G.E. (1977) The effects of photoperiod and temperature on the induction and termination of reproductive resting stage in the freshwater amphipod *Hyalella azteca*. *Can. J. Zool.*, **55**, 1595–600.

DeVries, A.L. (1971) Freezing resistance in fishes. In *Fish Physiology* Vol. 6 (eds W.S. Hoar and D.J. Randall) Academic Press, London and New York, 157–90.

DeVries, A.L. (1974) Survival at freezing temperatures. In *Biochem. Biophys. Perspectives in Marine Biology* Vol. 1 (eds D.C. Malins and J.R. Sargent) Acadmic Press, London, 289–330.

DeVries, A.L. (1980) Biological antifreezes and survival in freezing environments. In *Animals and Environmental Fitness* (ed. R. Gilles) Pergamon, Elmswood, New York, 583–607.

DeVries, A.L. (1982) Biological antifreeze agents in coldwater fishes. *Comp. Biochem. Physiol.*, **73A**, 627–40.

DeVries, A.L. and Lin, Y. (1977) The role of glycoprotein antifreezes in the survival of antarctic fishes. In *Adaptations within Antarctic Ecosystems*, Smithsonian Institution, Gulf Publishing Company, Houston, Texas, 439–58.

De Zwann, A. and Wijsman, T.C.M. (1976) Anaerobic metabolism in bivalvia (Mollusca). Characteristics of anaerobic metabolism. *Comp. Biochem. Physiol.*, **54B**, 313–24.

Dierolf, B.M. and Macdonald, H.S. (1969) Effects of temperature acclimation on electrical properties of earthworm giant axons. *Z. vergl. Physiologie*, **62**, 284–90.

Dixon, M. and Webb, E.C. (1979) *The Enzymes*, 3rd Edition, Longman, London.

Dizon, A.E. and Brill, R.W. (1979) Thermoregulation in tunas. *Am. Zool.*, **19**, 249–65.

Doudoroff, P. (1942) The resistance and acclimatization of marine fishes to temperature changes. I. Experiments with *Girella nigricans* (Ayres). *Biol. Bull.*, **83**, 219–44.

Doudoroff, P. (1945) The resistance and acclimatization of marine fishes to temperature changes. II. Experiments with *Fundulus* and *Atherinopes*. *Biol. Bull.*, **88**, 194–206.

Duman, J.G. (1977) Variations in macromolecular antifreeze levels in larvae of the darkling beetle, *Meracantha contracta*. *J. exp. Biol.*, **201**, 85–92.

Duman, J.G. (1982) Insect antifreeze and ice-nucleating agents. *Cryobiology*, **19**, 613–27.

Duman, J.G., Morris, J.P. and Castellino, F.J. (1984) Purification and composition of an ice nucleating protein from queens of the hornet, *Vespula maculata*. *J. Comp. Physiol.*, **154**, 79–83.

Duman, J.G. and Patterson, J.L. (1978) The role of ice nucleators in the frost tolerance of overwintering queens of the bald faced hornet. *Comp. Biochem. Physiol.*, **59A**, 69–72.

Dunlap, D.G. (1972) Latitudinal effects on metabolic rates in the cricket frog, *Acris crepitans*: acutely measured rates in summer frogs. *Biol. Bull.*, **143**, 332–43.

Dunlap, D.G. (1980) Comparative effects of thermal acclimation and season on metabolic compensation to temperature in the hylid frogs, *Pseudacris triseriata* and *Acris crepitans*. *Comp. Biochem. Physiol.*, **66A**, 243–9.

Dusenbery, D.B., Anderson, G.L. and Anderson, E.A. (1978) Thermal acclimation more extensive for behavioural parameters than for oxygen consumption in the nematode *Caenorhabditis elegans*. *J. exp. Zool.*, **206**, 191–8.

Dutton, R.H. and Fitzpatrick, L.C. (1975) Metabolic compensation to seasonal temperatures in the rusty lizard *Sceloporus olivaceus*. *Comp. Biochem. Physiol.*, **51A**, 309–18.

Dwarakaneth, S.K. (1971) The influence of body size and temperature upon the oxygen consumption in the millipede, *Spirostreptus asthenes* (Pocock). *Comp. Biochem. Physiol.*, **38A**, 351–8.

Edney, E.B (1971) The body temperature of tenebrionid beetles in the Namib desert of Southern Africa. *J. exp. Biol.*, **55**, 253–72.

Edwards, G.S. and Irving, L. (1943) The influence of season and temperature upon the oxygen consumption of the beach flea, *Talorchestia megalopthalma*. *J. Cell. Comp. Physiol.*, **21**, 183–9.

Ege, R. and Krogh, A. (1914) On the relation between temperature and the respiratory exchange in fishes. *Int. Rev. ges. Hydrobiol.*, **7**, 48–55.

Elliott, J.M. (1982) The effects of temperature and ration size on the growth and energetics of salmonids in captivity. *Comp. Biochem. Physiol.*, **73B**, 81–91.

Else, P.L. and Hulbert, A.J. (1985) An allometric comparison of the mitochondria of mammalian and reptilian tissues: implications for the evolution of endothermy. *J. Comp. Physiol.*, **156**, 3–11.

Etzler, F.M. and Drost-Hansen, W. (1979) A role for water in biological rate processes. In *Cell-associated water* (eds W. Drost-Hansen and J.S. Clegg) Academic Press, New York, 125–64.

Evans, R.M., Purdie, F.C. and Hickman, C.P. (1962) The effect of temperature and photoperiod on the respiratory metabolism of rainbow trout. *Can. J. Zool.*, **40**, 107–18.

Feder, M.E. (1978) Environmental variability and thermal acclimation in neotropical and temperate zone salamanders. *Physiol. Zool.*, **51**, 7–16.

Fitzpatrick, L.C., Bristol, J.R. and Stokes, R.M. (1971) Thermal acclimation and metabolism in the Allegheny Mountain Salamander *Desmognathus ochrophaeus. Comp. Biochem. Physiol.*, **40A**, 681–8.

Fitzpatrick, L.C., Bristol, J.R. and Stokes, R.M. (1972) Thermal acclimation and metabolic rates in the Dusky Salamander *Desmognathus fuscus. Comp. Biochem. Physiol.*, **41A**, 89–96.

Fitzpatrick, L.C. and Brown, A.V. (1975) Metabolic compensation to temperature in the Salamander *Desmognathusochrophaeus* from a high elevation population. *Comp. Biochem. Physiol.*, **50A**, 733–7.

Foster, D.O. (1984) Quantitative contribution of brown adipose tissue thermogenesis to overall metabolism. *Can. J. Biochem. Cell Biol.*, **62**, 618–22.

Fowler, S.W., Small, L.F. and Keckes, S. (1971) Effects of temperature and size on molting of euphansiid crustaceans. *Marine Biol.*, **11**, 45–51.

Freed, J.M. (1971) Properties of muscle phosphofructokinase of cold- and warm-acclimated *Carassius auratus Comp. Biochem. Physiol.*, **39B**, 747–64.

Freeman, J.A. (1955) Oxygen consumption, brain metabolism and respiratory movements of goldfish during temperature acclimatisation with special reference to lowered temperature. *Biol. Bull.*, **99**, 416–24.

Friedlander, M.J., Kotchabhakdi, N. and Prosser, C.L. (1976) Effects of cold and heat on behaviour and cerebellar function in goldfish. *J. Comp. Physiol.*, **112**, 19–45.

Fries, G. (1964) Über die Einwirkung der Tugesperiodik und der Temperatur auf den Generaitions Wechsel, die Weibehergrösse und die Eier von *Daphnia magnus.* Straus. Z. *Morphal. Ökol Tiene*, **53**, 475–516.

Fry, F.E.J. (1957) The aquatic respiration of fish. In *The Physiology of Fishes*, Vol. 1 (ed. M.E. Brown) Academic Press, New York, 1–63.

Fry, F.E.J. (1967) Responses of vertebrate poikilotherms to temperature. In *Thermobiology* (ed. A.H. Rose). Academic Press, London, 375–409.

Fry, F.E.J. (1971) The effect of environmental factors on the physiology of fish. In *Fish Physiology* Vol. 6 (eds W.S. Hoar and D.J. Randall) Academic press, 1–98.

Fry, F.E.J. and Hart, J.S. (1948) Cruising speed of goldfish in relation to water temperature. *J. Fisheries Res. Bd. Can.*, **7**, 169–75.

Gagge, A.P. (1940) Standard operative temperature, a generalised temperature scale applicable to direct and partitional calorimetry. *Am. J. Physiol.*, **13**, 93–103.

Garside, E.T. (1970) Temperature – structural responses. In *Marine Ecology* Vol. 1 (ed. O. Kinne) J. Wiley & Sons Ltd, New York, 561–73.

Geiger, R. (1965) *The climate near the ground.* Harvard University Press, USA.

Geiser, F., Matwiejczyk, L. and Baudinette, R.V. (1986) From ectothermy to heterothermy: The energetics of the kowari, *Dasyuroides byrnei* (Marsupialia: Dasyuroidae). *Physiol. Zool.*, **59**, 220–9.

Gelineo, S. and Sokic, P. (1953) Sur l'apparition de l'homothermie chez le spermophila (*Citellus citellus*). *C.R. Soc. Biol. Paris*, **147**, 138–40.

George, D.G. and Harris, G.P. (1985) The effect of climate on long-term changes in the crustacean zooplankton biomass of Lake Windermere, UK. *Nature*, **316**, 536–9.

Gerking, S.D. and Lee, R.M. (1983) Thermal limits for growth and reproduction in the desert pupfish *Cyprinodon n. nevadensis. Physiol. Zool.*, **56**, 1–9.

Gibbons, J.W. (1976) Thermal alteration and the enhancement of species populations. In *Thermal Ecology II. EDRA Symposium Series* (Eds. G.W. Esch, and R.W. McFarlane).

Gibson, J.S., Ellory, J.C. and Cossins, A.R. (1985) Temperature acclimation of intestinal Na^+ transport in the carp (*Cyprinus carpio* L.). *J. exp. Biol.*, **114**, 355–64.

Giesel, J.T., Murphy, P.H. and Manlove, M. (1982) The influence of temperature on genetic interrelationships of life history traits in a population of *Drosophila melanogaster*: What tangled data sets we weave. *Am. Nat.*, **119**, 464–9.

Gladwell, R.T. (1975) Heat death in the crayfish *Austropotamobius pallipes*: Thermal inactivation of muscle membrane-bound ATPase in warm and cold adapted animals. *J. Thermal Biol.*, **1**, 95–100.

Gösswald, K. and Bier, K.H. (1955) Beeinflussung des geschlechtsverhalt-isses durch Temperatuveinwirkung bei *Formica rufa* L. *Naturwissen-schaften*, **42**, 133–4.

Grainger, J.N.R. (1956) Effects of changes of temperature on the respiration of certain Crustacea. *Nature*, **178**, 930–1.

Grainger, J.N.R. (1959) The effect of constant and varying temperatures on the developing eggs of *Rana temporaria* L. *Zoo. Anz.*, **163**, 267–77.

Grainger, J.N.R. (1975) Mechanisms of death at high temperatures in *Helix* and *Patella. J. Thermal Biol.*, **1**, 11–13.

Grainger, J.N.R. (1986) The effect of changing temperatures on the rate of development of poikilotherms. In *Temperature relations in animals and man, Biona report* 4 (ed. H. Laudien) Gustav Fischer, Stuttgart, 229 pp.

Griffiths, J.S. and Alderdice, D.F. (1972) Effects of acclimation and acute temperature experience on the swimming speed of juvenile coho salmon. *J. Fisheries Res. Bd. Can.*, **29**, 251–64.

Griffiths, M. (1978) *The Biology of the Monotremes*. Academic Press, New York, 367 pp.

Grigg, G.C. (1965) Studies on the Queensland Lungfish, *Neoceratodus forsteri* (Krefft) II. Thermal Acclimation *Aust. J. Zool.*, **13**, 407–11.

Grigg, G.C., Drane, C.R. and Courtice, G.P. (1979) Time constants of heating and cooling in the eastern water dragon, *Physignathus lesueruii* and some generalisations about heating and cooling in reptiles. *J. Thermal Biol.*, **4**, 95–103.

Hainsworth, F.R. and Strickler, E.M. (1970) Salivary cooling by rats in the heat. In *Physiological and Behavioural Thermoregulation* (eds J.D. Hardy, A.P. Gagge, and J.A.J. Stoliwijk) C.C. Thomas, Springfield, Illinois, 611–26.

Halcrow, K. and Boyd, C.M. (1967) The oxygen consumption and swimming activity of the amphipod *Gammarus oceanicus* at different temperatures. *Comp. Biochem. Physiol.*, **23**, 233–42.

Hamilton, A.G. (1956) Further studies on the relation of humidity and temperature to the development of two species of African locusts, *Locusta migratoria migratoroides* (R and F) and *Schistocerca gregaria* (Forsk). *Trans. roy. Entomol. Soc. London*, **101**, 1–58.

Hammell, H.T. (1955) Thermal properties of fur. *Am. J. Physiol.*, **182** 369–76.

Hand, S.C. and Somero, G.N. (1983) Phosphofructokinase of the hibernator, *Citellus beecheyi*: Temperature and pH regulation of activity via influences on the tetramer-dimer equilibrium. *Physiol. Zool.*, **56**, 380–8.

Hanegan, J.L. and Heath, J.E. (1970) Mechanisms for the control of body temperature in the moth *Hyalophora cecropia*. *J. exp. Biol.*, **53**, 349–62.

Hargens, A.R. and Shabica, S.V. (1973) Protection against lethal freezing temperatures by mucus in an Antarctic limpet. *Cryobiol.*, **10**, 331–7.

Harlow, H.J., Hillman, S.S. and Hoffman, M. (1976) The effect of temperature on digestive efficiency in the herbivorous lizard, *Dipsosaurus dorsalis*. *J. Comp. Physiol.*, **111**, 1–6.

Harri, M.N.E. and Talo, A. (1975) Effect of season and temperature acclimation on the heart rate-temperature relationship in the frog, *Rana temporaria*. *Comp. Biochem. Physiol.*, **50A**, 469–72.

Harrington, P.W. (1967) Environmentally controlled induction of primary male gonochorists from eggs of the self-fertilising hermaphroditic fish, *Rivulus marmoratus* Poely. *Biol. Bull. Woods Hole*, **132**, 174–99.

Hart, J.S. (1956) Seasonal changes in insulation of the fur. *Can. J. Zool.*, **34**, 53–7.

Haschemeyer, A.E.V. (1969a) Rates of polypeptide chain assembly in liver *in vivo*. Relation to the mechanism of temperature acclimation in *Opsanus tau*. *Proc. Natl. Acad. Sci. USA*, **62**, 128–35.

Haschemeyer. A.E.V. (1969b) Oxygen consumption of temperature–acclimated toadfish *Opsanus tau*. *Biol. Bull.*, **136**, 28–32.

Hazel, J.R. (1972) The effect of acclimation temperature upon succinic dehydrogenase activity from the epaxial muscle of the common goldfish (*Carassius auratus* L.) II. Lipid reactivation of the soluble enzyme. *Comp. Biochem. Physiol.*, **43B**, 863–82.

Hazel, J.R. (1979) Influence of thermal acclimation on membrane lipid composition of rainbow trout liver. *Am. J. Physiol.*, **236**, R91–R101.

Hazel, J.R. (1984) Effects of temperature on the structure and metabolism of cell membranes in fish. *Am. J. Physiol.*, **246**, R460–R470.

Hazel, J.R. and Prosser, C.L. (1970) Interpretation of inverse acclimation to temperature. *Z. vergl. Physiologie*, **67**, 217–28.

Hazel, J.R. and Prosser, C.L. (1974) Molecular mechanisms of temperature compensation. *Physiol. Rev.*, **54**, 620–77.

Heath, J.E. (1965) Temperature regulation and diurnal activity in horned lizards. *Univ. Calif. Pub. Zool.*, **64**, 97–136.

Heath, J.E. and Wilkin, P.J. (1970) Temperature responses of the desert cicada *Diceroprocta apache* (Homoptera, cicadidae). *Physiol. Zool.*, **43**, 145–54.

Heatwole, H. and Muir, R. (1979) Thermal microclimates in the pre-Saharan steppe of Tunisia. *J. Arid. Environments*, **2**, 119–36.

Heilbrünn, L.V. (1924) The colloid chemistry of protoplasm. *Am. J. Physiol.*, **69**, 190–9.

Heinrich, B. (1974) Thermoregulation in endothermic insects. *Science*, **185**, 747–56.

Heinrich, B. (1976) Heat exchange in relation to blood flow between thorax and abdomen in bumblebees. *J. exp. Biol.*, **64**, 561–85.

Heinrich, B. (1981) The mechanisms and energetics of honeybee swarm temperature regulation. *J. exp. Biol.*, **91**, 25–55.

Heinrich, B. (1982) *Insect Thermoregulation*, J. Wiley & Sons, New York.

Heinrich, B. and Bartholomew, G.A. (1979) Roles of endothermy and size in interspecific and intraspecific competition for elephant dung in an African dung beetle, *Scarabaeus laevistriatus*. *Physiol. Zool.*, **52**, 484–96.

Heinrich, B. and Casey, T.M. (1973) Metabolic rate and endothermy in sphinx moths. *J. Comp. Physiol.*, **82**, 195–206.

Heisler, H. (1979) Regulation of the acid-base status in fish. In *Environmental Physiology of Fishes* (ed. M.A. Ali) Plenum Press, New York and London, 123–62.

Heller, A.C. and Colliver, G.W. (1974) CNS regulation of body temperature during hibernation. *Am. J. Physiol.*, **227**, 583–9.

Heller, H.C., Colliver, G.W. and Beard, J. (1977) Thermoregulation during entrance to hibernation. *Pflugers Arch. ges. Physiol.*, **369**, 55–9.

Hellon, R.F., Cranston, W.I. and Townsend, Y. (1984) Central mediators of fever – new candidates. *J. Thermal Biol.*, **9**, 135–7.

Henshaw, R.E., Underwood, L.S. and Casey, T.M. (1972) Peripheral thermoregulation: foot temperature in two arctic canines. *Science*, **175**, 988–90.

Himms-Hagen, J. (1970) Regulation of metabolic processes in brown adipose tissue in relation to non-shivering thermogenesis. In *Advances in Enzyme Regulation* Vol. 8 (ed. G. Webber) Pergamon Press, Oxford, 131–51.

Hinton, H.E. (1960) Cryptobiosis in the larvae of *Polypedilum vanderplancii* Hint (Chironomidae). *J. Ins. Physiol.*, **5**, 286–300.

Hochachka, P.W. (1974) Regulation of heat production at the cellular level. *Fed. Proc.*, **33**, 2162–9.

Hochachka, P.W. and Hayes, F.R. (1962) The effect of temperature acclimation on pathways of glucose metabolism in the trout. *Can. J. Zool.*, **40**, 261–70.

Hochachka, P.W. and Somero, G.N. (1973) *Strategies of Biochemical Adaption*, Saunders, Philadelphia.

Hochachka, P.W. and Somero, G.N. (1984) *Biochemical Adaptation*, Princeton University Press, USA.

Hofer, R. (1979) The adaptation of digestive enzymes to temperature, season and diet in roach, *Rutilus rutilus* L. and rudd *Scardinius erythrophthalmus* L. 1. Amylase. *J. Fish Biol.*, **14**, 565–72.

Hoff, J.G. and Westman, J.R. (1966) The temperature tolerance of three species of marine fishes. *J. Marine Res.*, **24**, 131–40.

Hoffman, K.H. (1974) Wirkung von konstanten und tagesperiodisch alternierenden Temperaturen auf Lebensdauer, Nahrungsverwentung und Fertilitat adulter *Gryllys bimaculatus*. *Oecologia*, **17**, 39–54.

Hoffman, K.H. (1986) Hormones mediate temperature effects on reproduction in insects. In *Temperature relations in animals and man. Biona report 4* (ed. H. Laudien) Gustav Fischer Verlag, Stuttgart, 33–40.

Hogan, T.W. (1960) The effects of subzero temperatures on the embryonic diapause of *Acheta commodus* (Walk.) (Orthoptera). *Aust. J. Biol. Sci.*, **13**, 572–40.

Holeton, G.P. (1974) Metabolic cold adaptation of polar fish: fact or artefact? *Physiol. Zool.*, **47**, 137–52.

Hollingsworth, M.J. (1969) Temperature and length of life in *Drosophila*. *Exp. Gerontol.*, **4**, 49–55.

Holzman, N. and McManus, J.J. (1973) Effects of acclimation on metabolic rate and thermal tolerance in the carpenter frog, *Rana vergatipes*. *Comp. Biochem. Physiol.*, **45A**, 833–42.

Hope-Cawdrey, M.J. Gettinby, G. and Grainger, J.N.R. (1978) Mathematical models for predicting the prevalence of liver-fluke disease and its control from biological and meteorological data. In *Weather and parasite animal disease W.M.O. Technical Note 159* (ed. T.E. Gibson). World Meteorological Organisation, Geneva, 21–38.

Houlihan, D.F. and Allan, D. (1982) Oxygen consumption of some antarctic and British gastropods: an evaluation of cold adaptation. *Comp. Biochem. Physiol.*, **73A**, 383–7.

Hsieh, A.C.L. and Carlson, L.D. (1957) Role of adrenaline and noradrenaline in chemical regulation of heat production. *Am. J. Physiol.*, **190**, 243–6.

Huey, R.B. and Slatkin, M. (1976) Costs and benefits of lizard thermoregulation. *Q. rev. Biol.*, **51**, 363–84.

Huey, R.B. and Stevenson, R.D. (1979) Integrating thermal physiology and ecology of ectotherms: A discussion of approaches. *Am. Zool.*, **19**, 357–66.

Hutchins, L.W. (1947) The bases for temperature zonation in geographical distribution. *Ecol. Monogr.*, **17**, 325–35.

Hutchison, V.H. (1961) Critical thermal maxima in salamanders. *Physiol. Zool.*, **34**, 92–125.

Iggo, A. (1969) Cutaneous thermoreceptors in primates and subprimates. *J. Physiol.*, **200**, 403–30.

Irving, L. (1956) Physiological insulation of swine as bareskinned mammals. *J. appl. Physiol.*, **9**, 414–20.

Irving, L. (1964) Terrestrial animals in cold: birds and mammals. In *Handbook of Physiology: Adaptation to the Environment* Vol. 4 (eds B.D. Dill, E.F. Adolf and C.G. Wilber), Amer. Physiol. Soc. Washington, 361–77.

Irving, L. and Krog, J. (1955) Temperature of skin in arctic as a regulator of heat. *J. appl. Physiol.*, **7**, 355–64.

Jacobson, E.R. and Whitford, W.G. (1970) The effect of acclimation on physiological responses to temperature in the snakes *Thamnophis proximus and Natrix rhombiera. Comp. Biochem. Physiol.*, **35**, 439–49.

Jähnig. F. and Bramhall, J. (1982) The origin of a break in Arrhenius plots of membrane processes. *Biochim. Biophys. Acta*, **690**, 310–3.

Janisch, E. (1925) Über die temperaturabhangigkeit biologischer vorgange und ihre kurvenmässige analyse. *Pflugers Arch.*, **209**, 414–36.

Janisch, E. (1927) Das Exponentialgesetz als Grindlage einer vergleichenden Biologie. *Abh. theorie Org. Entwicklung*, **2**, 1–371.

Jansky, L. (1962) Maximal metabolism and organ thermogenesis in mammals. In *Comparative Physiology of Temperature Regulation* (eds J.P. Hannon and E. Viereck) Arctic Aeromedical Lab, Ft. Wainwight, Alaska, 133–74.

Jensen, D.W. (1972) The effect of temperature on transmission at the neuromuscular junction of the sartorius muscle of *Rana pipiens. Comp. Biochem. Physiol.*, **41A**, 685–95.

Johnson, F.H., Eyring, H. and Stover, B.J. (1974) *The theory of rate processes in biology and medicine*, Wiley Interscience, New York and London.

Johnston, I.A. (1979) Calcium regulatory proteins and temperature acclimation of actomyosin from a eurythermal teleost *Carassius auratus* L. *J. Comp. Physiol.*, **129**, 163–7.

Johnston, I.A. (1982) Capillarisation, oxygen diffusion distances and mitochondrial content of carp muscles following acclimation to summer and winter temperatures. *Cell Tissue Res.*, **222**, 325–37.

Johnston, I.A. and Altringham, J.D. (1985) Evolutionary adaptation of muscle power output to environmental temperature: force-velocity characteristics of skinned fibres isolated from antarctic, temperate and tropical marine fish. *Pflugers Arch.*, **405**, 136–40.

Johnston, I.A. and Lucking, M. (1978) Temperature-induced variation in the distribution of different types of muscle fibre in the goldfish *Carassius auratus*. *J. Comp. Physiol.*, **124**, 111–16.

Johnston, I.A. and Maitland, B. (1980) Temperature acclimation in crucian carp, *Carassius carassius* L. morphometric analysis of muscle fibre ultrastructure. *J. Fish Biol.*, **17**, 113–25.

Johnston, I.A., Sidell, D.B. and Driedzic, W.R. (1985) Force-velocity characteristics and metabolism of carp muscle fibres following temperature acclimation. *J. exp. Biol.*, **119**, 239–99.

Johnston, I.A. and Walesby, N.J. (1977) Molecular mechanisms of temperature adaptation in fish myofibrillar adenosine triphosphatases. *J. Comp. Physiol.*, **119**, 195–206.

Jones, P.L. and Sidell, B.D. (1982) Metabolic responses of striped bass (*Morone saxatilis*) to temperature acclimation. II. Alterations in metabolic carbon sources and distributions of fibre types in locomotory muscle. *J. exp. Zool.*, **219**, 163–71.

Kanungo, M.S. and Prosser, C.L. (1959a) Physiology and biochemical adaptation of goldfish to cold and warm temperatures. I. Standard and active oxygen consumptions of cold and warm-acclimated goldfish at various temperatures. *J. Cell. Comp. Physiol.*, **54**, 259–63.

Kanungo, M.S. and Prosser, C.L. (1959b) Physiology and biochemical adaptation of goldfish to cold and warm temperatures. II. Oxygen consumption of liver homogenates: oxygen consumption and oxidative phosphorylation of liver mitochondria. *J. Cell. Comp. Physiol.*, **54**, 265–73.

Kanwisher, J. (1959) Histology and metabolism of frozen intertidal animals. *Biol. Bull.*, **16**, 258–64.

Kaufman, O. (1932) Einige Bemerkungen uber den Einfluss von Temperaturschwankungen auf die Entwicklungsdauer und Streuung bei Insekten und seine graphische Darstellung durch Kettelinie und Hyperbel. *Z. Morph. U. Okol. Tiere.*, **25**, 353–61.

Kavanau, J.L. (1950) Enzyme kinetics and the rate of biological processes. *J. Gen. Physiol.*, **34**, 193–209.

Keen, R. and Parker, D.L. (1979) Determining expected duration of development under conditions of alternating temperatures. *J. theor. Biol.*, **81**, 599–607.

Kenny, R. (1969) Temperature tolerance of the polychaete worms *Diopatra cuprea* and *Clymenella torquata*. *Mar. Biol.*, **4**, 219–23.

Kent, J.D. and Prosser, C.L. (1980) Effects of incubation and acclimation temperatures on incorporation of U-[^{14}C] glycine into mitochondrial protein of liver cells and slices from green sunfish, *Lepomis cyanellus*, *Physiol. Zool.*, **53**, 293–304.

Kinne, O. (1960) Growth, food intake, and food conversion in a euryplastic fish exposed to different temperatures and salinities. *Physiol. Zool.*, **33**, 288–317.

Kinne, O. (1961) Growth molting frequency, heart beat, number of eggs and incubation time in *Gammarus zaddachi* exposed to different environments. *Crustaceana*, **2**, 26–36.

Kinne, O. (1964) The effects of temperature and salinity on marine and brackish water animals. 2. Salinity and temperature-salinity combinations. *Oceanogr. mar. Biol. A. Rev.*, **2**, 281–339.

Kinne, O. (1970) Temperature. In *Marine Ecology* Vol. 1, J. Wiley and Sons, New York, 407–514.

Kinne, O. and Kinne, E.M. (1962) Rates of development in embryos of a cyprinodont fish exposed to different temperature-salinity-oxygen concentrations. *Can. J. Zool.*, **40**, 231–53.

Kirberger, C. (1953) Untersuchungen über die temperature abhang-igkeit von Lebensprozessen bei verschiedenen Wirbellosen. *Z. Vergeich. Physiol.*, **35**, 175–98.

Kistiakowsky, G.B. and Lumry, R.L. (1949) Anomalous temperature effects in the hydrolysis of urea by urease. *J. Am. Chem. Soc.*, **71**, 2006–13.

Kluger, M.J. (1978) The evolution and adaptive value of fever. *Am. Sci.*, **66**, 38–43.

Koban, M. (1986) Can cultured hepatocytes show temperature acclimation? *Am. J. Physiol.*, **250**, R211–R220.

Kohshima, S. (1984) A novel cold-tolerant insect found in a Himalayan glacier. *Nature*, **310**, 225–7.

Konishi, J. and Hickman, Jr., C.P. (1964) Temperature acclimation in the central nervous system of rainbow trout *Salmo gairdneri*. *Comp. Biochem Physiol.*, **13**, 433–42.

Krogh, A. (1914a) On the influence of the temperature on the rate of embryonic development. *Zeit. allg. Physiol.*, **16**, 163.

Krogh, A. (1914b) The quantitative relation between temperature and standard metabolism in animals. *Int. Z. Phys. Chem. Biol.*, **1**, 491–508.

Kruger, G. (1969) Uber die temperaturadaptation des Bitterlings *Rhodeus amarus* Bloch *Z. wiss. Zool.*, **167**, 87–104.

Kruuv, J., Glofcheski, D., Cheng, K.-H. *et al.* (1983) Factors influencing survival and growth of mammalian cells exposed to hypothermia. I. Effects of temperature and membrane perturbers. *J. Cell Physiol.*, **115**, 179–85.

Kumamoto, J., Raison, J.K. and Lyons, J.M. (1971) Temperature 'breaks' in

Arrhenius plots: A thermodynamic consequence of a phase change. *J. theor. Biol.*, **31**, 47–51.

Kuramoto, M. (1976) Embryonic temperature adaptation in development rate of frogs. *Physiol. Zool.*, **48**, 360–6.

Lagerspetz, K.Y.H. (1974) Temperature acclimation and the nervous system. *Biol. Rev.*, **49**, 477–514.

Lagerspetz, K.Y.H. and Skytta, M. (1979) Temperature compensation of sodium transport and ATPase activity in frog skin. *Acta Physiol. Scand.*, **106**, 151–8.

Lagerspetz, K.Y.H. and Talo, A. (1967) Temperature acclimation of the functional parameters of the giant nerve fibres in *Lumbricus terrestris* L. 1. Conduction velocity and the duration of the rising and falling phase of action potential. *J. exp. Biol.*, **47**, 471–80.

Lamb, M.J. (1977) *Biology of Ageing*. Blackie, Glasgow and London, 181 pp.

Lasiewski, R.C. and Snyder, G.K. (1969) Responses to high temperature in nestling double-crested and pelagic cormorants. *Auk*, **86**, 529–40.

Laudien, H. (1973) Changing reaction systems. In *Temperature and Life* (eds H. Precht, J. Christophersen, H. Hensel and W. Larcher) Springer-Verlag, Berlin, 779 pp.

Lawton, J.H. (1985) Water fleas as population model. *Nature*, **316**, 577–8.

Lee, J.A.C. and Cossins, A.R. (1987) Adaptation of intestinal morphology in the temperature acclimated carp *Cyprinus carpio*. *Cell and Tissue Research*, in Press.

Lee, R.E., Jr. and Baust, J.G. (1982a) Respiratory metabolism of the antarctic tick, *Ixodes uriae*. *Comp. Biochem. Physiol.*, **72A**, 167–71.

Lee, R.E., Jr. and Baust, J.G. (1982b) Absence of metabolic cold adaptation and compensatory acclimation in the Antarctic fly, *Belgica antarctica*. *J. Insect Physiol.*, **28**, 725–9.

Lees, A.D. (1955) *The Physiology of diapause in Arthropods*, Cambridge University Press, UK, 151 pp.

Lees, A.D. (1959) The role of photoperiod and temperature in the determination of parthenogenetic and sexual forms in the aphid *Megoura viciae* Buckton. III. Further properties of the maternal switching mechanism in apterous aphids. *J. Insect Physiol.*, **9**, 153–64.

Lewis, J.R. (1964) *The Ecology of Rocky Shores*. Hodder and Stoughton, London.

Lewontin, R.C. and Birch, L.C. (1966) Hybridization as a source of variation for adaptation to new environments. *Evolution*, **20**, 315–36.

Li, G.C. and Werb, Z. (1982) Correlation between synthesis of heat shock proteins and development of thermotolerance in Chinese hamster fibroblasts. *Proc. Natl. Acad. Sci. USA*, **79**, 3218–22.

Licht, P. (1965) Effects of temperature on heart rates of lizards during rest and activity. *Physiol. Zool.*, **38**, 129–37.

Lin, Y. (1979) Environmental regulation of gene expression. *J. Biol. Chem.*, **254**, 1422–6.

Lin, Y. (1983) Regulation of the seasonal biosynthesis of antifreeze peptides in cold-adapted fish. In *Cellular Acclimatisation for Environmental Change* (eds A.R. Cossins and P. Sheterline) Cambridge University Press, UK, 217–26.

Loughna, P.T. and Goldspink, G. (1985) Muscle protein synthesis rates during temperature acclimation in a eurythermal *Cyprinus carpio* and a stenothermal *Salmo gairdneri (species of teleost). J. exp. Biol.*, **118**, 267–76.

Louw, G.N. and Seely, M.K. (1982) *Ecology of Desert Organisms*. Longman, Harlow, UK, 194 pp.

Lovelock, J.E. (1953a) The haemolysis of human red blood cells by freezing and thawing. *Biochim. Biophys. Acta*, **10**, 414–26.

Lovelock, J.E. (1953b) Mechanism of the protective action of glycerol against haemolysis by freezing and thawing. *Biochim. Biophys. Acta*, **11**, 28–36.

Low, P.S., Bada, J.L. and Somero, G.N. (1973) Temperature adaptation of enzymes: roles of the free energy, the enthalpy and the entropy of activation. *Proc. Natl. Acad. Sci. USA*, **70**, 430–2.

Low, P.S. and Somero, G.N. (1974) Temperature adaptation of enzymes: A proposed molecular basis for the different catalytic efficiencies of enzymes from ectotherms and endotherms. *Comp. Biochem. Physiol.*, **49B**, 307–12.

Lowe, C.H. and Heath, W.G. (1969) Behavioural and physiological responses to temperature in the desert pupfish *Cyprinodon macularis*, *Physiol. Zool.*, **42**, 53–9.

Lyman, C.P. (1948) The oxygen consumption and temperature regulation of hibernating hamsters. *J. exp. Zool.*, **109**, 55–78.

Lyons, J.M. (1973) Chilling injury in plants. *Ann. Rev. Plant. Physiol.*, **24**, 445–66.

Lyons, J.M. and Raison, J.K. (1970) A temperature-induced transition in mitochondrial oxidation: contrasts between cold and warm-blooded animals. *Comp. Biochem. Physiol.*, **37**, 405–11.

Macdonald, J.A. (1981) Temperature compensation in the peripheral nervous system: antarctic vs temperate poikilotherms. *J. Comp. Physiol.*, **142**, 411–18.

Macdonald, J.A. and Balnave, R.J. (1984) Miniature end plate currents from teleost extraocular muscle. *J. Comp. Physiol.*, **155A**, 649–59.

McFarlane, W.V. (1964) Terrestrial animals in dry heat: ungulates. In *Adaptation to the Environment Handbook of Physiology*, Vol. 4 (eds B.D. Dill, E.F. Adolf and C.G. Wilber), Amer. Physiol. Soc. Washington D.C., pp. 509–39.

McLaren, I.A. (1965) Temperature and frog eggs. A reconsideration of metabolic control. *J. gen. Physiol.*, **48**, 1071–9.

McLeese, D.W. (1956) Effects of temperature, salinity and oxygen on the survival of the American lobster. *J. Fisheries Res. Bd. Can.*, **13**, 247–72.

McLeese, D.W. and Wilder, D.G. (1958) The activity and catchability of the lobster *Homarus americanus* in relation to temperature. *J. Fisheries Res. Bd. Can.*, **15**, 1345–54.

McMahon, R.F. (1979) Response to temperature and hypoxia in the oxygen consumption of the introduced asiatic freshwater clam, *Corbicula fluminea* (Muller). *Comp. Biochem. Physiol.*, **63A**, 380–3.

McMahon, R.F. and Russell-Hunter, W.D. (1977) Temperature relations of aerial and aquatic respiration in six littoral snails in relation to their vertical zonation. *Biol. Bull.*, **152**, 182–98.

McMahon, T. (1973) Size and shape in biology. *Science*, **179**, 1201–4.

Maetz, J. and Evans, D.H. (1972) Effects of temperature on bronchial sodium-exchange and extrusion mechanisms in the seawater-adapted flounder *Platichthys flesus* L. *J. exp. Biol.*, **56**, 565–85.

Magnuson, J.J., Crowder, L.B. and Medvick, P.A. (1979) Temperature as an ecological resource. *Am. Zool.*, **19**, 331–43.

Malan, A. (1982) Respiration and acid-base state in hibernation. In *Hibernation and Torpor in mammals and birds* (eds C.P. Lyman, J.S. Willis, A. Malan and L.C.H. Wang) Academic Press, New York.

Malan, A., Rodeau, J.L. and Daull, F. (1985) Intracellular pH in hibernation and respiratory acidosis in the European hamster. *J. Comp. Physiol.*, **156B**, 251–8.

Maness, J.D. and Hutchison, V.H. (1980) Acute adjustment of thermal tolerance in vertebrate ectotherms following exposure to critical thermal maxima. *J. Thermal Biol.*, **5**, 225–34.

Mangum, C.P., Oakes, M.J. and Shick, J.M. (1972) Rate-temperature responses in scyphozoan medusae and polyps. *Mar. Biol.*, **15**, 295–303.

Markel, R.P. (1974) Aspects of the physiology of temperature acclimation in the limpet *Acmaea limatuia* Carpenter (1864). An integrated field and laboratory study. *Physiol. Zool.*, **47**, 99–109.

Marzush, K (1952) Untersuchungen über die Temperature ab hängig keit von hebens prozessen bei lnsekten unter besonderer Berück sichtigung winterschlafender kartoffeläger. Z. *vergl. Physiol.*, **34**, 75–92.

Maxwell, J.G.H. (1977) Aspects of the biology and ecology of selected antarctic invertebrates. Ph.D. thesis, University of Aberdeen, 131 pp.

May, M.L. (1976) Thermoregulation and adaptation to temperature in dragonflies (*Odonato: anisoptera*). *Ecol. Monogr.*, **46**, 1–32.

May, M.L. (1979) Insect thermoregulation. *Ann. rev. Entomol.*, **24**, 313–49.

Mayhew, W.W. (1965) Hibernation in the horned lizard, *Phyrnosoma m'calli. Comp. Biochem. Physiol.*, **16**, 103–19.

Mazur, P. (1984) Freezing of living cells: mechanisms and implications. *Am. J. Physiol.*, **247**, C125–C142.

Meats, A. and Khoo, K.C. (1976) The dynamics of ovarian maturation and oocyte resorption in Queensland fruit fly, *Dacus tryoni*, in daily-rhythmic and constant temperature regimes. *Physiol. Entomol.*, **1**, 213–21.

Meryman, H.T. (1974) Freezing injury and its prevention in living cells. *Ann. Rev. Biophys.*, **3**, 341–63.

Meuvis, A.L. and Heuts, M.J. (1957) Temperature dependence of breathing rate in Carp. *Biol. Bull.*, **112**, 97–107.

Milkman, R. (1962) Temperature effects on day old *Drosophila* pupae. *J. gen. Physiol.*, **45**, 777–99.

Miller, L.K. (1969) Freezing tolerance in an adult insect. *Science*, **166**, 105–6.

Miller, K.B. (1978) Physical and chemical changes associated with seasonal alterations in freezing tolerance in the adult northern tenebriorid *Upis caramboides*. *J. Insect Physiol.*, **24**, 791–6.

Minton, K.W., Stevenson, M.A. and Kendig, J. *et al.* (1980) Pressure inhibits thermal killing of Chinese hamster ovary fibroblasts. *Nature*, **285**, 482–3.

Mitchell, H.K. and Lipps, L.S. (1978) Heat shock and phenocopy induction in *Drosophila*. *Cell*, **15**, 907–18.

Moberly, W.R. (1963) Hibernation in the desert iguana *Dipsosaurus dorsalis*. *Physiol. Zool.*, **36**, 152–60.

Moerland, T.S. and Sidell, B.D. (1981) Characterisation of metabolic carbon flow in hepatocytes isolated from thermally acclimated killifish *Fundulus heteroclitus*. *Physiol. Zool.*, **54**, 379–89.

Moeur, J.E. and Eriksen, C.H. (1972) Metabolic responses to temperature of a desert spider, *Lycosa* (*Paradosa*) *carolinensis* (Lycosidae). *Physiol. Zool.*, **45**, 290–301.

Monin, A.S., Kamenovitch, V.M. and Kort, V.M. (1974) *Variability of the oceans*. John Wiley, New York.

Monroe, S. and Pelham, H. (1985) What turns on heat shock genes. *Nature*, **317**, 477.

Monteith, J.L. (1973) *Principles of Environmental Physics*. Edward Arnold, London, 241 pp.

Montgomery, J.C and Macdonald, J.A. (1984) Performance of motor systems in Antarctic fish. *J. Comp. Physiol.*, **154A**, 241–8.

Montgomery, J.C., McVean, A.R. and McCarthy, D. (1983) The effects of lowered temperature on spontaneous eye movements in a teleost fish. *Comp. Biochem. Physiol.*, **75A**, 363–8.

Moon, T.W. and Hochachka, P.W. (1971) Temperature and enzyme activity in poikilotherms: isocitrate dehydrogenases in rainbow trout liver. *Biochem. J.*, **123**, 695–705.

Moore, J.A. (1949) Geographic variation of adaptive characters in *Rana pipiens*. *Evolution*, **3**, 1–21.

Morris, R.W. (1962) Body size and temperature sensitivity in the cichlid fish, *Aequidens portalgnesis* (Hensel) *Am. Nat.*, **96**, 35–50.

Morris, R.W. (1965) Thermal acclimation of metabolism of the yellow Bullhead, *Ictalurus natalis* (Le Sueur). *Physiol. Zool.*, **38**, 219–27.

Morrison, W.W. and Milkman, R. (1978) Modification of heat resistance in *Drosophila* by selection. *Nature*, **273**, 49–50.

Morrissey, R.E. and Baust, J.G. (1976) The ontogeny of cold tolerance in the gall fly *Eurosta solidaginis*. *J. Insect Physiol.*, **22**, 431–7.

Mrosovsky, N. (1980) Thermal biology of sea turtles. *Am. Zool.*, **20**, 531–47.

Murphy, D.J. (1979) A comparative study of the freezing tolerance of the marine snails *Littorina littorea* (L.) and *Nassarius obsoletus* (Say). *Physiol. Zool.*, **52**, 219–30.

Murphy, D.J. (1983) Freezing resistance in intertidal invertebrates. *Ann. Rev. Physiol.*, **45**, 289–99.

Murphy, D.J. and Johnson, L.C. (1980) Physical and temporal factors influencing the freezing tolerance of the marine snail *Littorina littorea* (L.). *Biol. Bull.*, **158**, 220–32.

Murphy, D.J. and Pierce, S.K. (1975) The physiological basis for changes in the freezing tolerance of intertidal molluscs. 1. Response to subfreezing temperatures and the influence of salinity and temperate acclimation. *J. Exp. Zool.*, **193**, 313–22.

Murrish, D.E. and Vance, V. (1968) Physiological responses to temperature acclimation in the lizard *Uta mearosi*. *Comp. Biochem. Physiol.*, **27**, 329–37.

Nagy, K.A. (1983) Ecological energetics. In *Lizard Ecology: Studies of a model organism* (eds R.B. Huey and E.R. Painka), Harvard Univ. Press, Cambridge, USA, 24–54.

Naylor, E. (1965) Biological effects of a heated effluent in docks at Swansea, S. Wales. *Proc. Zool. Soc. Lond.*, **144**, 253–68.

Newell, R.C. (1966) The effect of temperature on the metabolism of poikilotherms. *Nature*, **212**, 426–28.

Newell, R.C. and Kofoed, L.H. (1977) Adjustments of the components of energy balance in the gastropod *Crepidula fornicata* in response to the thermal acclimation. *Marine Biol.*, **44**, 275 –86.

Newell, R.C. and Northcroft, H.R. (1967) A reinterpretation of the effect of temperature on the metabolism of certain marine invertebrates. *J. Zool. Lond.*, **151**, 277–98.

Newell, R.C. and Pye, V.I. (1970) Influence of thermal acclimation on the relation between oxygen consumption and temperature in *Littorina littorea* L. and *Mytilus edulis* L. *Comp. Biochem. Physiol.*, **34**, 385–97.

Newell, R.C. and Pye, V.I. (1971) Quantitative aspects of the relationship between metabolism and temperature in the winkle *Littorina littorea* (L.). *Comp. Biochem. Physiol.*, **38B**, 635—50.

Nicholls, D.G. and Locke, R.M. (1984) Thermogenic mechanisms in brown fat. *Physiol. Rev.*, **64**, 1–64.

Nicholls, D.G. and Rial, E. (1984) Brown fat mitochondria. *Trends in Biochem. Science*, **9**, 489–91.

Nielsen, J.B.K., Plant, P.W. and Haschemeyer, A.V. (1977) Control of protein synthesis in temperature acclimation. II. Correlation of elongation factor-1 activity with elongation rate *in vivo. Physiol. Zool.*, **50**, 22–30.

Norris, K.S. (1963) The functions of temperature in the ecology of the perciod fish *Girella nigricans* (Ayres). *Ecol. Monogr.*, **33**, 23–62.

Nuccitelli, R. and Deamer, D.W. (1982) *Intracellular pH: its measurement regulation and utilisation in cellular functions.* Alan R. Liss, New York.

Oliphant, L.N. (1983) First observations of brown fat in birds. *Condor*, **85**, 350–4.

Owen, T.G. and Wiggs, A.J. (1971) Thermal compensation in the stomach of the brook trout *Salvelinus fontinalis* (Mitchill). *Comp. Biochem. Physiol.*, **40B**, 465–73.

Packard, G.C. (1972) Inverse compensation for temperature in oxygen consumption of the hylid frog, *Pseudacris triseriata. Physiol. Zool.*, **45**, 270–5.

Paladino, F.V. and Spotila, J.R. (1978) The effect of arsenic on the thermal tolerance of newly hatched muskellunge fry (*Esox masquinongy*) *J. Thermal Biol.*, **3**, 223–7.

Paladino, F.V., Spotila, J.R., Schubauer, J.P. and Kowalski, K.T. (1980) The critical thermal maximum: a techinque used to elucidate physiological stress and adaptation in fish. *Revue Can. Biol.*, **39**, 115–22.

Patterson, J.W. and Davies, P.M.C. (1978a) Thermal acclimation in temperate lizards. *Nature*, **275**, 646–7.

Patterson, J.W. and Davies, P.M.C. (1978b) Energy expenditure and metabolic adaptation during winter dormancy in the lizard, *Lacerta vivipara* Jacquin. *J. Thermal Biol.*, **3**, 183–6.

Paul, J. (1986) Body temperature and the specific heat of water. *Nature*, **323**, 300–000.

Pearl, R. (1928) *The rate of living*. Knopf, New York, 185 pp.

Penny, R.K. and Goldspink, G. (1980) Temperature adaptation of sarcoplasmic reticulum of fish muscle. *J. Thermal Biol.*, **5**, 63–8.

Percy, J.A. (1974) Thermal adaptation in the boreo-arctic echinoid *Stronglylocentrotus droebachiensis* (O.F. Muller 1776) iv. Acclimation in the laboratory. *Physiol. Zool.*, **47**, 163–71.

Peterson, R.H. and Anderson, J.M. (1969) Influence of temperature change on spontaneous locomotor activity and oxygen consumption of Atlantic Salmon, *Salmo salar*, acclimated to temperatures. *J. Fisheries Res. Bd. Can.*, **26**, 93–109.

Piatt, J. (1971) Effect of temperature differentials upon reconstitution of embryonic primordia in *Ambystoma*. *J. Embryol. exp. Morph.*, **25**, 339–45.

Pitkow, R.B. (1960) Cold death in the guppy. *Biol. Bull.*, **119**, 231–45.

Porter, W.P., Mitchell, J.W., Beckman, W.A. and DeWitt, C.B. (1973) Behavioural implications of mechanistic ecology. *Oecologia*, **13**, 1–54.

Pratt, D.M. (1943) Analysis of population development in *Daphnia* at different temperatures. *Biol. Bull. Woods Hole*, **85**, 116–40.

Precht, H. (1951) Der Einfluss der temperatur gu die atming and aufeinige fermente beim aal *Anguilla vulgaris*. L. *Biol. Zentr.*, **70**, 71–85.

Precht, H. (1958) Concepts of the temperature adaptation of unchanging reaction systems of cold-blooded animals. In *Physiological Adaptation* (ed. C.L. Prosser) Amer. Assn. Adv. Science, Wash. D.C. 50–78.

Precht, H., Christophersen, J., Hensel, H. and Larcher, W. (1973) *Temperature and life*, Springer Verlag, Berlin.

Prosser, C.L. (1967) Molecular mechanisms of temperature adaptation in relation to speciation. In *Molecular mechanisms of temperature adaptation* (ed. C.L. Prosser) Amer. Ass. Adv. Science, Washington, 351–76.

Prosser, C.L. (1973) *Comparative animal physiology*, Saunders, Philadelphia.

Prosser, C.L. and Nelson, D.O. (1981) The role of nervous systems in temperature adaptation of poikilotherms. *Ann. Rev. Physiol.*, **43**, 281–300.

Pütter, A. (1914) Temperaturkoeffizienten. *Zeits. allg. Physiol.*, **16**, 574–627.

Quinn, P.J. (1981) The fluidity of cell membranes and its regulation. *Prog. Biophys. Mol. Biol.*, **38**, 1–104.

Ragland, I.M., Wit, L.C. and Sellers, J.C. (1981) Temperature acclimation in the lizards *cnemidophorus sexlineatus* and *Anolis carolinensis*. *Comp. Biochem. Physiol.*, **70A**, 33–6.

Raison, J.K. (1973) The influence of temperature-induced phase changes on the kinetics of respiratory and other membrane-associated enzyme systems. *Bioenergetics*, **4**, 285–309.

Raison, J.K. and Lyons, J.M. (1971) Hibernation: alteration of mitochondrial membranes as a requisite for metabolism at low temperature. *Proc. Natl. Acad. Sci. USA*, **68**, 2092–4.

Rao, K.P. and Bullock, T.H. (1954) Q_{10} as a function of size and habitat temperature in poikilotherms. *Am. Nat.*, **88**, 33–44.

Raymond, J.A. and DeVries, A.L. (1977) Adsorption inhibition as a mechanism of freezing resistance in polar fishes. *Proc. Natl. Acad. Sci. USA*, **74**, 2589–93.

Read, K.R.H. (1967) Thermostability of proteins in poikilotherms. In *Molecular Mechanisms of Temperature Adaptation* (ed. C.L. Prosser), Am. Assoc. Adv. Science, Washington.

Reeves, R.B. (1977) The interaction of body temperature and acid-base balance in ectothermic vertebrates. *Ann. Rev. Physiol.*, **39**, 559–86.

Reynolds, W.W. and Casterlin, M.E. (1979) Behavioral thermoregulation and the 'final preferendum'. *Am. Zool.*, **19**, 211–24.

Rising, T.L. and Armitage, K.B. (1969) Acclimation to temperature by the terrestrial gastropods *Limax maximus* and *Philomycus carolinianus*, Oxygen consumption and temperature preference. *Comp. Biochem. Physiol.*, **30**, 1091–114.

Robert, M. and Gray, I. (1972) Enzymatic mechanisms during temperature acclimation of the Blue Crab, *Callinectes sapidus*. II. Kinetic and thermodynamic studies of glucose-6-phosphate dehydrogenase and 6-phosphogluconate dehydrogenase. *Comp. Biochem. Physiol.*, **42B**, 389–402.

Roberts, J.L. (1957) Thermal acclimation of metabolism in the Crab *Pachygrapsus crassipes Randall*. II. Mechanism and the influence of season and latitude. *Physiol. Zool.*, **30**, 242–55.

Roberts. J.L. (1964) Metabolic responses of freshwater sunfish to seasonal photoperiods and temperatures. *Helgolander Wiss. Meeresuntersuch*, **9**, 459–73.

Roberts, J.L. (1967) Metabolic compensations for temperature in sunfish. In *Molecular Mechanism of Temperature Adaptation* (ed. C.L. Rosser) Amer. Assoc. Adv. Sci. Symp. 84, Washington D.C. pp. 245–62.

Roberts, J.L. (1979) Seasonal modulation of thermal acclimation and behavioral thermoregulation in aquatic animals. In *Marine Pollution: Functional Responses* (eds W.B. Vernberg, A. Calabrese, F.P. Thurberg and F.J. Vernberg) Academic Press, 365–88.

Rome, L.C., Loughna, P.W. and Goldspink, G. (1984) Muscle fibre activity in carp as a function of swim speed and muscle temperature. *Am. J. Physiol.*, **247**, R272-R279.

Roots, B.I. and Prosser, C.L. (1962) Temperature acclimation and the nervous system in fish. *J. exp. Biol.*, **39**, 617–29.

Rozin, P.N. and Mayer, J. (1961) Thermal reinforcement and thermoregulatory behavior in the goldfish *Carassius auratus*. *Science*, **134**, 942–3.

Rule, G.S., Frim, J. and Thompson, J.E. *et al.* (1978) The effect of membrane lipid perturbers on survival of mammalian cells to cold. *Cryobiology*, **15**, 408–14.

Salt, R.W. (1959) The role of glycerol in the cold-hardening of *Bracon cephi*. *Can. J. Zool.*, **37**, 59–69.

Saunders, D.S. (1971) The temperature-compensated photoperiodic clock 'programming' development and pupal diapause in the flesh-fly, *Sarcophaga argyrostoma*. *J. Insect Physiol.*, **17**, 801–12.

Schmeing-Engberding, F. (1953) Die Vorsungs temperaturen einiger Kochenfische und ihre physiologie Bedeutung. *Z. Fischerei (N.F.)*, **2**, 125–55.

Schmidt, J., Laudien, H. and Bowler, K. (1984) Acute adjustments to high temperature in FHM-cells from *Pimephales promelas* (*Pisces, Cyprinidae*). *Comp. Biochem. Physiol.*, **78A**, 823–8.

Schmidt, W.D. (1982) Survival of frogs at low temperature. *Science*, **215**, 697–8.

Schmidt-Nielsen, K. (1983) *Animal Physiology: Adaptation and Environment*, Cambridge University Press, Cambridge, 619 pp.

Schmidt-Nielsen, K. (1984) *Scaling – why is animal size important*, Cambridge University Press, Cambridge, 241 pp.

Schmidt-Neilsen, K. and Danson, W.R. (1964) Terrestrial animals in dry heat: desert reptiles. In *Handbook of Physiology. Section 4: Adaption to the Environment* (ed. D.B. Dill) Physiol. Soc., Washington D.C. pp. 467–80.

Schmidt-Neilsen, K., Hainsworth, F.R. and Murrish, D.E. (1970) Countercurrent heat exchangers in respiratory passages: Effect on water and heat balance. *Resp. Physiol.*, **9**, 263–74.

Schmidt-Nielsen, K., Schmidt-Nielsen, B., Jarnum, S.A. and Houpt, T.R. (1957) Body temperature of the camel and its relation to water economy. *Am. J. Physiol.*, **188**, 103–12.

Schmidt-Nielsen, K., Taylor, C.R. and Shkolnik, A. (1972) Desert snails: problems of survival. *Symp. Zool. Soc. Lond.*, **31**, 1–13.

Scholander, P.F., Flagg, W., Walters, V. and Irving, L. (1953) Climatic adaptation in arctic and tropical poikilotherms. *Physiol. Zool.*, **26**, 67–92.

Scholander, P.F., Hock, R. and Walters, V. *et al.* (1950b) Heat regulation in some arctic and tropical mammals and birds. *Biol. Bull.*, **99**, 237–58.

Scholander, P.F., Walters, V., Hock, R. and Irving, L. (1950a) Body insulation of some arctic and tropical mammals and birds. *Biol. Bull.*, **99**, 225–36.

Schv̂enke, M. and Wodtke, E. (1983) Cold-induced increase of Δ^9- and Δ^6-desaturase activities in endoplasmic reticulum. *Biochim. Biophys. Acta*, **734**, 70–75.

Scott, D.B.C. (1979) Environmental timing and the control of reproduction in teleost fish. *Symp. Zool. Soc. Lond.*, **44**, 105–32.

Segal, E. (1961) Acclimation in molluscs. *Am. Zool.*, **1**, 235–44.

Segal, E. (1963) A temperature dependent developmental abnormality in the slug *Limax flavus* L. 1. Appearance and incidence. *J. exp. Zool.*, **153**, 159–70.

Shaklee, J., Christiansen, J.A. and Sidell, B.D. *et al.* (1977) Molecular aspects of temperature acclimation in fish: Contributions of changes in enzyme activities and isozyme patterns to metabolic reorganisation in the green sunfish. *J. exp. Zool.*, **201**, 1–20.

Shrode, J.B. and Gerking, S.D. (1977) Effects of constant and fluctuating temperatures on reproductive performance of a desert pupfish *Cyprinodon n. nevadensis. Physiol. Zool.*, **50**, 1–10.

Sidell, B.D. (1977) Turnover of cytochrome c in skeletal muscle of green sunfish *Lepomis cyanellus*, R. during thermal acclimation *J. exp. Zool.*, **199**, 233–50.

Sidell, B.D. (1980) Responses of goldfish *Carassius auratus* L. muscle to acclimation temperature: Alterations in biochemistry and proportions of different fibre types. *Physiol. Zool.*, **53**, 98–107.

Sidell, B.D. (1983) Cellular acclimatisation to environmental change by quantitative alterations in enzymes and organelles. In *Cellular acclimatisation to environmental change* (eds A.R. Cossins and P. Sheterline) Cambridge University Press, UK, 103–20.

Sidell, B.D. and Johnston, I.A. (1985) Thermal sensitivity of contractile function in chain pickerel, *Esox niger*. *Can. J. Zool.*, **63**, 811–16.

Silvius, J.R. and McElhaney, R.N. (1980) Membrane lipid physical state and modulation of the Na^+, Mg^{2+}-ATPase activity in *Acholeplasma laidlawii* B. *Proc. Natl. Acad. Sci. USA*, **77**, 1255–9.

Silvius, J.R. and McElhaney, R.N. (1981) Non-linear Arrhenius plots and the analysis of reaction and motional rates in biological membranes. *J. theor. Biol.*, **88**, 135–52.

Simon, E. (1981) Effects of CNS temperature on the generation and transmission of temperature signals in homeotherms. A common concept for mammalian and avian thermoregulation. Pflugers. Arch. ges. Physiol., **392**, 79–88.

Simonson, J.R. (1975) *Engineering heat transfer*, Macmillan Press, London.

Simpson, S. and Galbraith, J.J. (1905) An investigation into the diurnal variations of the body temperature of nocturnal and other birds, and a few mammals. *J. Physiol.*, **33**, 225–38.

Sinensky, M., Pinkerton, F., Sutherland, E. and Simon, F.R. (1979) Rate limitation of ($Na^+ + K^+$)-stimulated ATPase by membrane acyl chain ordering. *Proc. Natl. Acad. Sci. USA*, **76**, 4893–7.

Singer, S.J. and Nicholson, G.L. (1972) The fluid-mosaic model of the structure of cell membranes. *Science*, **175**, 720–31.

Smit, H. (1967) Influence of temperature on the rate of gastric juice secretion by the brown bullhead, *Ictalurus nebulosus*. *Comp. Biochem. Physiol.*, **21**, 125–32.

Smith, B.K. and Dawson, T.J. (1984) Changes in the thermal balance of a marsupial (*Dasyuroides byrnei*) during cold and warm acclimation. *J. Thermal Biol.*, **9**, 199–204.

Smith, J.M. (1957) Temperature tolerance and acclimatization in *Drosophila subobscura*, *J. exp. Biol.*, **34**, 85–96.

Smith, M.W. (1966) Influence of temperature acclimatisation on sodium-glucose interactions in the goldfish intestine. *J. Physiol.*, **182**, 574–90.

Smith, M.W. (1976) Temperature adaptation in fish. *Biochem. Soc. Symp.*, **41**, 43–60.

Smith, M.W. and Ellory, J.C. (1971) Temperature-induced changes in sodium transport and Na^+/K^+-adenosine triphosphatase activity in the intestine of goldfish *Carassius auratus* L. *Comp. Biochem. Physiol.*, **39A**, 209–18.

Smith, W.K. and Miller, P.C. (1973) The thermal ecology of two south Florida fiddler crabs: *Uca rapax*, Smith and *U. pugilator*, Bosc. *Physiol. Zool.*, **26**, 186–207.

Snyder, C.D. (1908) A comparative study of the temperature coefficients of the velocities of various physiological actions. *Am. J. Physiol.*, **22**, 309–34.

Snyder, C.D. (1911) On the meaning of variations in the magnitude of temperature coefficients of physiological processes. *Am. J. Physiol.*, **28**, 167–75.

Somero, G.N. (1981) pH-temperature interactions on proteins: principle of optimal pH and buffer system design. *Marine Biol. Letters*, **2**, 163–78.

Somero, G. and DeVries, A.L. (1967) Temperature tolerance of some Antarctic fishes. *Science*, **156**, 257–8.

Sømme, L. and Conradi-Larsen, E.M. (1977) Cold-hardiness of collembolans and oribatid mites from windswept mountain ridges. *Oikos*, **29**, 118–26.

Southwick, E.E. (1983) The honeybee cluster as a homeothermic superorganism. *Comp. Biochem. Physiol.*, **75**, 641–5.

Spoor, W.A. (1946) A quantitative study of the relationship between the activity and oxygen consumption of the goldfish and its application to the measurement of respiratory metabolism of fishes. *Biol. Bull.*, **91**, 312–25.

Spray, D.C. (1975) Effect of reduced acclimation temperature on responses of frog cold receptors. *Comp. Biochem. Physiol.*, **50A**, 391–5.

Standora, E.A., Spotila, J.R. and Foley, R.E. (1982) Regional endothermy in the sea turtle, *Chelonia mydas*. *J. Thermal Biol.*, **7**, 159–65.

Stebbins, R.C. and Barwick, R.E. (1968) Radiotelemetric study of thermoregulation in a lace monitor. *Copeia*, **3**, 541–7.

Stevens, E.D. and Dizon, A.E. (1982) Energetics of locomotion in warm-bodied fish. *Ann. rev. Physiol.*, **44**, 121–31.

Stevens, E.D. and Neill, W.H. (1978) Body temperature regulation of tunas, especially skipjack. In *Fish Physiology* Vol. VII (eds W.S. Hoar and D.J. Randall) Academic Press, 315–359.

Stitt, J.T. (1983) Hypothalamic generation of effector signals. *J. Thermal Biol.*, **8**, 113–17.

Stone, B.B. and Sidell, B.D. (1981) Metabolic responses of striped bass *Morone saxatilis* to temperature acclimation. I. Alterations in carbon sources for hepatic energy metabolism *J. exp. Zool.*, **218**, 371–9.

Storey, J.M. and Storey, K.B. (1985) Freezing and cellular metabolism in the gall fly larva, *Eurosta solidaginis*. *J. Comp. Physiol.*, **155**, 333–7.

Storey, K.B. and Storey, J.M. (1986) Freeze tolerance and intolerance as strategies of winter survival in terrestrially-hibernating amphibians. *Comp. Biochem. Physiol.*, **83A**, 613–9.

Strumwasser, F. (1960) Mammalian hibernation. XV. Some physiological principles governing hibernation in *Citellus beecheyi*. *Bull. Mus. Comp. Zool.*, **124**, 285–320.

Suhrmann, R. (1955) Weitere versuche über die temperatur-adaptation der karauschen *Carassius vulgaris* NILS. *Biol. Zentr.*, **74**, 432–48.

Taigen, T.L. (1983) Activity metabolism of anuran amphibians: implications for the origin of endothermy. *Am. Nat.*, **121**, 94–109.

Talo, A. and Lagerspetz, K.Y.H. (1967) Temperature acclimation of the functional parameters of the giant nerve fibres in *Lumbricus terrestris* L. II. The refractory period. *J. exp. Biol.*, **47**, 481–4.

Tanguay, G.M. (1983) Genetic regulation during heat shock and function of heat shock proteins: a review. *Can. J. Biochem. Cell. Biol.*, **61**, 387–94.

Tauber, C.A. and Tauber, M.J. (1982) Evolution of seasonal adaptations and life history traits in *Chrysopa*: Response to diverse selection pressures. In *Evolution and Genetics of Life Histories* (eds. H. Dingle and J.P. Hegmann) Springer-Verlag, New York, 51–72.

Taylor, C.R. (1970) Dehydration and heat: effects on temperature regulation of East African ungulates. *Am. J. Physiol.*, **219**, 1136–9.

Tissieres, A., Mitchell, H.K. and Tracy, U.M. (1974) Protein synthesis in salivary glands of *D. melanogaster*. Relation to chromosome puffs. *J. mol. Biol.*, **84**, 389–98.

Touzeau, J. (1966) Influence des temperatures sur l'emergence des papillons printaniers due carpocapse des pommes (*Laspeyresia pomonella* L.) dans le sud-ouest de la France. *Rev. Zool. Agric. Appl.*, **65**, 41–9.

Treagust, D.F., Folk, G.E., Randall, W. and Folk, M.A. (1979) The circadian rhythm of body temperature of unrestrained opossums, *Didelphis virginiana*. *J. Thermal Biol.*, **4**, 251–5.

Tribe, M.A. and Bowler, K. (1968) Temperature dependence of standard metabolic rate in a poikilotherm. *Comp. Biochem. Physiol.*, **25**, 427–36.

Trout, W.E. and Kaplan, W.D. (1970) A relation between longevity, metabolic rate, and activity in shaker mutants of *Drosophila melanogaster*. *Expt. Gerontol.*, **5**, 83–92.

Tsukuda, H. (1960) The mortality and the occurrence of vertebral column abnormalities in the guppies reared at high temperature. *Physiol. Ecol.*, **9**, 79–83.

Turner, J.S. and Schroter, R.C. (1985) Why are small homeotherms born naked? Insulation and the critical radius concept. *J. Thermal Biol.*, **10**, 233–8.

Tyler, A.V. (1966) Some lethal temperature relations of two minnows of the genus *Chrosomus*. *Can J. Zool.*, **44**, 349–64.

Tyler, S. and Sidell, B.D. (1984) Changes in mitochondrial distribution and diffusion distances in muscle of goldfish upon acclimation to worm and cold temperatures. *J. exp. Zool.*, **232**, 1–9.

Tyndale-Biscoe, H. (1973) *Life of Marsupials*. Edward Arnold, London, 254 pp.

Ulbricht, R.J. (1973) Effect of temperature acclimation on the metabolic rate of sea urchins. *Marine Biol.*, **19**, 273–7.

Umminger, B.L. (1978) The role of hormones in the acclimation of fish to low temperatures. *Naturwissenshaften*, **65**, 144–50.

Ushakov, B.P. (1964) Thermostability of cells and proteins in poikilotherms. *Physiol. Rev.*, **44**, 518–60.

Uvarov, B.P. (1931) Insects and climate. *Trans. Ent. Soc. Lond.*, **79**, 1–232.

Van't Hoff, J.H. (1896) *Studies in chemical dynamics*. Muller, Amsterdam.

Vendrick, A.J.H. (1959) The regulation of body temperature in man. *Ned. Tijdschr. Geneesk.*, **103**, 240–4.

Vernberg, F.J. (1952) The oxygen consumption of two species of Salamanders of different seasons of the year. *Physiol. Zool.*, **25**, 243–9.

Vernberg, F.J. (1969) Acclimation of intertidal crabs. *Am. Zool.*, **9**, 333–41.

Vernberg, F.J. and Costlow, J.D., Jr. (1966) Studies on the physiological variation between tropical and temperate zone fiddler crabs of genus *Uca*. IV. Oxygen consumption of larvae and young crabs reared in the laboratory. *Physiol. Zool.*, **39**, 36–52.

Vernberg, F.J. and Vernberg, W.B. (1964) Metabolic adaptation of animals from different latitudes. *Helgo. wiss. Meers*, **9**, 476–87.

Vernberg, W.B., DeCoursey, P.J. and O'Hara, J. (1974) Multiple environmental factor effects on physiology and behaviour of the fiddler crabs, *Uca pugilator*. In *Pollution and the Physiological Ecology of Estuarine and Coastal Water Organisms* (ed. F.J. Vernberg) Academic Press, New York, 381–426.

Walsberg, G.E. and Weathers, W.W. (1986) A simple technique for estimating operative environmental temperature. *J. Thermal Biol.*, **11**, 67–72.

Walsh, P.J., Foster, G.D. and Moon, T.W. (1983) The effects of temperature on metabolism of the American eel *Anguilla rostrata* (Le Seur): Compensation in the summer and torpor in the winter. *Physiol. Zool.*, **56**, 532–540.

Wang, L.C. (1979) Time patterns and metabolic rates of natural torpor in the Richardson's ground squirrel. *Can. J. Zool.*, **57**, 149–55.

Watson, J.D. (1975) *Molecular Biology of the Gene*, 3rd edn, Benjamin, Menlo Park California.

Weiser, W. (1973) Temperature relations of ectotherms – a speculative review. In *Effects of temperature on ectothermic organisms* (ed. W. Weiser) Springer-Verlag, Berlin, 1–23.

Wekstein, D.R. and Zolman, J.F. (1967) Homeothermic development of the young chick. *Proc. Soc. Exp. Biol. Med.*, **125**, 294–7.

Wheeler, P.E. (1978) Elaborate CNS cooling structures in large dinosaurs. *Nature*, **275**, 441–3.

Widdows, J. and Bayne, B.L. (1971) Temperature acclimation of *Mytilus edulis* with reference to its energy budget. *J. Mar. Biol. Assoc. UK*, **51**, 827–43.

Willis, J.S. (1982) The mystery of periodic arousal. In *Hibernation and torpor in mammals and birds*, Chap. 6 (eds C.P. Lyman, J.S. Willis, A. Malan and L.C. Wang) Academic Press, New York, 92–103.

Wilson, F.R. (1973) Quantitative changes of enzymes of the goldfish *Carassius auratus* L. in response to temperature acclimation: an immunological approach. Ph.D. thesis, University of Illinois.

Wilson, J.M. and McMurdo, A.C. (1981) Chilling injury in plants. In *Effects of Low Temperatures on Biological Membranes* (eds G.J. Morris and A. Clarke) Academic Press, London, 145–72.

Wodtke, E. (1981) Temperature adaptation of biological membranes. Compensation of the molar activity of cytochrome c oxidase in the mitochondrial energy-transducing membrane during thermal acclimation of the carp *Cyprinus carpio* L. *Biochim. Biophys. Acta*, **640**, 710–20.

Wohlschlag, D.E. (1960) Metabolism of an antarctic fish and the phenomenon of cold adaptation. *Ecology*, **41**, 287–92.

Wohlschlag, D.E. (1964) Respiratory metabolism and ecological characteristics of some fishes in McMurdo Sound, Antarctica. In *Biology of the Antarctic Seas* (*Research Series*, Vol. 1) (ed. H.W. Wells) American Geographical Union, Washington D.C.

Wolf, L.L. and Hainsworth, E.R. (1972) Environmental influence on regulated body temperature in torpid humming birds. *Comp. Biochem. Physiol.*, **41**, 167–73.

Wood, D.H. (1978) Temperature adaptation in the freshwater snail, *Helisoma trivolis* (Say), in an artificially heated reservoir in the southeastern United States. *J. Thermal Biol.*, **3**, 187–94.

Wood, S.C., Lykkeboe, G. and Johansen, D. *et al.* (1978) Temperature acclimation in the pancake tortoise *Malacochersus tornieri*: Metabolic rate, blood, pH, oxygen affinity and red cell organic phosphates. *Comp. Biochem. Physiol.*, **59A**, 155–60.

Yatvin, M.B. (1977) Influence of membrane lipid composition and procaine and hypothermic death of cells. *Int. J. Radiat. Biol.*, **32**, 512–21.

Zachariassen, K.E. (1985) Physiology of cold tolerance in insects. *Physiol.* Rev., **65**, 799–832.

Zecevic, D. and Levitan, H. (1980) Temperature acclimation: effects on membrane physiology of an identified snail neuron. *Am. J. Physiol.*, **239**, C47–C57.

Zwölfer, W. (1933) Studien zur ökologie, insbesondere zur Bevölkungslehne, der, Nonne, *Lymantria monarcha. Z. Angew., Ent.*, **20**, 1–50.

Index

Thermal inactivation,
  of enzymes, 237
  of myofibrillar ATPase, 197